Management of Insect Pests with Semiochemicals

CONCEPTS AND PRACTICE

Management of Insect Pests with Semiochemicals

CONCEPTS AND PRACTICE

Edited by
Everett R. Mitchell
US Department of Agriculture
Gainesville, Florida

PLENUM PRESS • NEW YORK AND LONDON

Library of Congress Cataloging in Publication Data

Main entry under title:

Management of insect pests with semiochemicals.

Proceedings of an international symposium presented jointly by the Insect Attractants, Behavior, and Basic Biology Research Laboratory, USDA, and the Dept. of Entomology and Nematology, University of Florida, and held Mar. 23-28, 1980, in Gainesville, Fla.

Includes indexes.

1. Insect control–Biological control–Congresses. 2. Insects baits and repellents–Congresses. I. Mitchell, Everett R. II. Insect Attractants, Behavior, and Basic Biology Research Laboratory (U.S.) III. University of Florida. Dept. of Entomology and Nematology. IV. Title: Semiochemicals.

SB933.3.M36 628.9'657 81-570

ISBN 0-306-40630-6 AACR2

Proceedings of an international colloquium on Management of Insect Pests with Semiochemicals, held March 23–28, 1980, in Gainesville, Florida.

A Division of Plenum Publishing Corporation
227 West 17th Street, New York, N.Y. 10011

Printed in the United States of America

FOREWORD

Perhaps the best expression of our intent in organizing this gathering is found in the definition of the word colloquy and its derivations. A gathering allowing *familiar and informal conversation* among colleagues with similar interests was our objective. Our motives were, of course, complex. Our main intent was not, however, to add to the list of books competing for the time of the scientific community at-large. However, while informality was our objective, a lasting document exists in the form of this publication of the presentations forming the skeleton on which we built less formal but meatier communications. We hope you can reconstruct on these bones a perception of the state of the art in the subject at hand.

The members of this assemblage are specialists in one or more subdisciplines. Their formal communications are found in texts and journals appropriate to their broader disciplines. Often their friends alone are privy to their less formal thoughts, intuitions, hopes, and especially fears and failures. We hoped by organizing this colloquium to develop familiar and informal conversation among those most interested and active in applying semiochemicals in pest control. That community, like others also shared by Gainesville entomologists, has little or no formal organization or means for assemblage. We proposed on this and future occasions to offer the opportunity to this and similar groups to gather, though we do not presume too much to lead but rather to facilitate conversation.

After the fact of the first colloquium, I believe the participants realized most of the benefits we envisioned. The likelihood that subsequent colloquia will be successful has been greatly increased by the success of this, the first. I am grateful to Dr. E. R. Mitchell and his organizing committee for the high standards they have set.

D. L. Chambers

PREFACE

Semiochemicals -- pheromones, kairomones, oviposition deterrents, and other behavior-modifying chemicals -- offer enormous potential for controlling many of the world's most devastating insect pests when used in imaginative, truly integrated pest management schemes. Considerable effort is being directed worldwide toward developing semiochemicals to detect and monitor insect pests and to time conventional insecticide sprays. In addition they are used to annihilate or suppress insect pest populations using a variety of innovative approaches including mass trapping, mating disruption, and various combinations of attractant baits, trap crops, and toxicants.

Many of the world's leading experts in the use of insect semiochemicals, representing the views of governmental, industrial, and academic spheres, convened in Gainesville, Florida, March 23-28, 1980 to participate in a colloquium entitled "Management of Insect Pests with Semiochemicals." The colloquium was presented jointly by the Insect Attractants, Behavior, and Basic Biology Research Laboratory, USDA and the Department of Entomology and Nematology, University of Florida. This volume is the proceedings of that colloquium.

By bringing together the different experts concerned with the development of insect semiochemicals for formal presentations, general discussions, and informal exchanges, it was hoped that methods for evaluating the effects of semiochemicals would be improved and recommendations would be forthcoming for achieving recognition that these chemicals are safe, effective, and environmentally acceptable. We also aimed to stimulate the search for new materials, to promote research on the behavioral and ecological effects of semiochemicals used in insect control programs, and to encourage the development of efficient formulations for dispensing pheromones and other behavior-modifying chemicals.

The program was divided into five sessions: biomonitoring; mass trapping; mating disruption; formulations, toxicology, and registration; and the use of semiochemicals for manipulating phytophagous and entomophagous insects. Papers were presented on the use of insect attractants for monitoring pests in field and orchard crops, forests, and stored products; traps to suppress insect pest populations in common ground and recreational areas of cities, in forests, in unseasoned lumber at sawmills, and in field crops; the formulation, toxicology, and registration of semiochemicals for insect control; oviposition-deterring pheromones to control fruit flies; and semiochemicals for manipulating insect parasites and predators.

Of particular interest was a report that damage to cotton caused by the pink bollworm was reduced to subeconomic levels in fields treated with pheromone, although mating in the pheromone-treated areas was not totally eliminated. Similarly, economic control of the pink bollworm was achieved in mass-trapping experiments, although traps were less than 100% efficient. Several other successful applications of semiochemicals for control of lepidopterous and coleopterous pests were also reported. Successes such as these are very encouraging and demonstrate the potential of semiochemicals in agricultural and silvicultural insect pest control, particularly when integrated with more conventional control methods involving insecticides, parasites and predators, pathogens, resistant varieties, and good cultural practices.

The primary obstacles to effective application and use of semiochemicals were identified as the high cost of active ingredients, formulations technology, and the development of clearly defined marketing strategies. The latter requires that applied biologists, consultants, extension personnel, and others responsible for developing and implementing insect control practices develop an awareness of the potential value of semiochemicals in their programs.

The colloquium was timely, and the participants were optimistic that semiochemicals can and should play a significant role in managing insect pest populations in a wide variety of situations around the world. It is impossible to capture in this volume the enthusiasm shown by those in attendance, but I hope that the reader will find the papers published here as informative as did the delegates at the colloquium.

Arranging such an international gathering requires considerable planning and organization. I am indebted to the Program Committee -- Drs. J. R. McLaughlin, J. H. Tumlinson, K. W. Vick, C. S. Barfield, J. F. Butler and J. L. Nation -- for their tireless support and boundless energies in organizing the program and selecting the participants. Among the many individuals involved in local arrangements before, during, and after the conference were the following: Ms. Marianne J. Donato, serving as administrative assistant for the colloquium, contributed greatly to its overall success, and to this volume, through her organizational and creative skills; Mrs. Betty

F. Brooke, Mrs. Diane G. Crispell, and Mrs. Elaine S. Miller, manuscripts; Mrs. Elaine S. Turner, facilitator of manuscripts; Mr. Felix L. Lee, recordings; and Mr. Robert R. Rutter, projection and transportation. I am especially grateful for the editorial assistance provided by Drs. T. R. Ashley, M. D. Huettel, and J. R. McLaughlin of this laboratory, and Mrs. Jane D. Wall, National Technical Editor - Entomology, USDA, Beltsville, MD. Appreciation is also extended to Dr. Dave Carlson for design of the program cover.

The colloquium was partially supported by contributions provided by Conrel, Inc.; Hercon, Inc.; IFAS, University of Florida; U. S. Department of Agriculture; and Zoecon, Inc. The support of the many state, federal, and foreign governmental organizations which financed the attendees at our conference is also gratefully acknowledged. Special thanks are extended to Dr. D. L. Chambers, Director, Insect Attractants, Behavior, and Basic Biology Research Laboratory, USDA and Dr. G. C. Smart, Acting Chairman, Department of Entomology and Nematology, University of Florida for their support and encouragement throughout the planning and conduct of the conference and the preparation of these proceedings.

And finally, sincere appreciation is extended to all of the participants who deserve the major credit for the success of the colloquium, both for their formal contributions and their participation in the informal meetings and discussions.

Everett R. Mitchell

CONTENTS

SESSION I. BIOMONITORING

SESSION II. MASS TRAPPING

SESSION III. MATING DISRUPTION

PRACTICAL APPLICATION OF PHEROMONES IN REGULATORY PEST MANAGEMENT PROGRAMS

John W. Kennedy

John W. Kennedy Consultants, Inc.
608 Washington Boulevard
Laurel, Maryland 20810 USA

INTRODUCTION

Plant Protection and Quarantine Program (PPQP) activities are designed to prevent the entry, establishment, and spread of foreign pests into the United States and to suppress periodic outbreaks of certain native pests. Activities include surveys, regulatory inspection, eradication (elimination) or regulatory pest management. The objective of the regulatory agencies is to obtain maximum pest control or elimination using chemicals or other methods which will have the minimum effect on the environment. New control tools, including pesticides which have a unique mode of action, cultural techniques, sterile insect release, and other methods are evaluated constantly for use in these programs. Biological control techniques are substituted for chemicals when the objective of the program can be met. Semiochemicals are becoming one of the more important tools used for survey, control, and evaluation in PPQP activities.

Foreign Quarantine Program

There is an increasing reliance upon pheromone and food-baited traps to detect foreign pests at ports of arrival. Surveillance activities are usually concentrated on specific pests, and provide base information to managers who develop strategies to prevent or reduce their entry into the United States on carriers such as aircraft or ships. This increased emphasis on surveillance is being made because it is more efficient to prevent the entry and establishment of foreign pests than it is to eliminate or regulate them once in the United States. Activities regarding surveillance are increasing with the availability of pheromones for important quarantine pests such as the khapra beetle, Trogoderma granarium (Everts). Extensive trapping programs for fruit flies are continuing at ports of entry where there is a high risk of entry and subsequent establishment. When establishment of economically important pests have occurred on a small scale, the elimination of such infestations has been successful.

Domestic Quarantine Programs

Pest management systems for insects are based on a sound surveillance system using pheromone and food-baited traps or other less sophisticated methods such as egg or larval counts. Regulatory programs have become more efficient and effective because of the economy of the trapping systems in detecting the presence of an insect, delimiting (defining the boundaries of the infestation) by placing a grid of traps around a positive detection trap, and monitoring population levels using relative numbers of captured insects in each trap set in an array. These surveillance activities along with information on parasite or predator activity, presence of disease, and weather data aid regulatory managers in determining if the pest population is increasing or decreasing. Pest management objectives can be established and programs planned based upon such surveillance activities.

Intervention systems, including direct control using pesticides, biologicals, parasites and predators, pathogens, traps, sterile insects, mating disruption, or other strategies are considered based on surveillance results. Whenever possible, pheromones or biologicals are used in environmentally sensitive areas to mitigate the pest population or eliminate it entirely. In less environmentally sensitive areas, chemical pesticides are used because of the availability, ease of handling, higher efficacy, and lower cost of materials and applications.

An evaluation system is designed which is compatible with objectives, and it is usually based upon determining insect populations and other factors after the control activity is initiated. Monitoring using attractants in traps usually is the primary tool used to measure the effect of the control activities. One drawback in using pheromones in a mating disruption control program is that evaluation of the program cannot take place until the following year. Pheromones used in mating disruption permeate the atmosphere with sex lure making traps less effective in capturing a sample of the population.

CASE STUDIES

Gypsy Moth

The gypsy moth, Lymantria dispar (L.), was introduced into

the United States at Medford, MA, in 1869. Though this insect has been in this country for more than 100 years, it still occupies a relatively small portion of its potential ecological range. Only 10% of the susceptible forest area is presently infested, leaving nearly 90% to be attacked with potential defoliation and possible mortality. The present average annual economic loss to gypsy moth is estimated at $268 million with a conservative projected annual loss of $2,247 million if the susceptible areas become infested. A quarantine was imposed in October 1912, and has continued with revisions. Activities regarding the regulatory and historical aspects on the gypsy moth quarantine are reviewed by Kennedy (1979).

Synthesis of the gypsy moth sex pheromone, disparlure, by with the sex attractant for the male gypsy moth, necessary to more precisely determine the presence and the relative number of insects in an economically feasible manner. Presently the pheromone is dispensed in delta traps using a Hercon® wick with a slow-release mechanism. This trapping system has been the main tool in determining the presence of new infestations throughout the United States.

During the 1979 season, approximately 95,000 sex pheromone traps were set in 38 states to detect the presence of the gypsy moth. In addition, approximately 1,473 traps were used in a pilot project to monitor gypsy moth populations in preparation for a management program along the leading edge of the spreading gypsy moth populations in the central part of Pennsylvania and northern Maryland. Tight arrays served to define the boundaries of any possible infestation to determine the intervention (control) activities for the next season. The information gained on detection in the large-scale dispersement of traps throughout the United States, monitoring the movement of the gypsy moth along the leading edge, and the delimiting of possible isolated infestations are the main surveillance activities used in the gypsy moth programs.

An extensive array of gypsy moth traps was placed in Appleton, WI, to further define the boundaries of an infestation which has been present for at least the last 4 years. The trap catches in this area have been decreasing yearly, indicating that the population is declining. Reasons for the population reduction are not apparent, although environmental factors may be playing a large role. Because of the large number of traps in this area, and the possible marginal ecological effects on the gypsy moths, there are indications that the intensive trapping may be lowering population levels.

Disruption of mating by gypsy moths was theorized if concentrations of the synthetic pheromone in the atmosphere approached the amount of pheromone naturally present in populations. Beroza (1960) suggested that concentrations equal to or greater than the amount occurring in natural populations may cause the male moths to become confused regarding the location of the females.

Several methods of providing a slow release mechanism have been tried, including microcapsular liquid latex materials, Hercon and Conrel® systems. In the 1979 seasons, racemic disparlure pheromone in Hercon flakes were applied by helicopter to approximately 170 ha of a populated residential area in Oconomowoc, WI. Prior to the pheromone treatment, a larval treatment of virus (NPV) was applied by air to 129.5 ha to the same area to determine if eradication was possible using such a combined pheromone-virus attack. Results were encouraging, but evaluation will only be possible in the 1980 trapping season. In Pennsylvania, approximately 4,047 ha were treated with microencapsulated pheromone. Previous to the adult pheromone confusion treatment, 2,023 of these ha were treated with _Bacillus thuringiensus_ (Bt). A control area was present to indicate if there was an effect of the combined pheromone-bacterial treatment in the 1979 season and surveys will continue to monitor gypsy moth populations over a few years. An evaluation system to measure the effect of the intervention activities will be difficult, but a relative measure of the success of the confusion technique combined with biologicals will be attempted.

Pheromone traps have encouraged the development of a relatively sophisticated regulatory pest management system for the gypsy moth. Extensive trap tests conducted under the direction of the Animal and Plant Health Inspection Service, Plant Protection and Quarantine Programs (APHIS, PPQP), Environmental Evaluation Staff and others have provided new information regarding the relationships involved in the mechanics of trapping. Variables including pheromone type (racemic vs. (+) lure), trap size, slow release mechanism (Hercon vs. cotton wicks), distance between traps, and other factors were examined. Data compiled by Embody (1980) indicate that properly designed trapping systems have potential for use in pest management strategies, including population control or reduction, even at moderate gypsy moth levels.

Japanese Beetle

The Japanese beetle, _Popillia japonica_ Newman, was first found in the United States near Riverton, NJ, in 1916. In 1918, the U. S. Department of Agriculture and New Jersey began a program to eliminate this pest, but the infestation was so well established that it could not be eradicated by the control measures then known and with the funds available.

Extremely high populations occurred in contiguous areas for many years, with approximately $10 million lost annually to this pest in over 300 different ornamental and food crops. The beetle is still a major pest of turf in towns, golf courses, and pastures, with grubs feeding on grass roots. It occurs in parts of coastal and adjacent states from Maine to Alabama, with isolated infestations outside this area westward to the Mississippi River. An historical review of the quarantine, regulatory, and pest management activities are presented in the Environmental Impact Statement -- Japanese Beetle (1978, 1979).

The goal of the current regulatory program for the Japanese beetle is to prevent the movement of this pest from the 24 eastern states to 7 currently uninfested western states. The main thrust is preventing establishment of the Japanese beetle in western states by an intensive management program at major eastern airports to prevent beetles from hitchhiking on aircraft. Surveillance is conducted at the airports to determine population numbers.

The food-bait components, combinations, and proportions have changed many times (Fleming 1976). McGovern et al. (1973) patented the phenylethyl propionate (PEP)-eugenol in a 7:3 mixture, used by the U. S. Department of Agriculture since 1970. Klein et al. (1973) found that traps baited with virgin females and food lures captured more beetles than those with either lure alone. Tumlinson et al. (1977) produced the first synthetic sex lure, providing the regulatory program with an improved surveillance tool with capabilities in an intervention program.

The use of large numbers of Japanese beetle traps to suppress populations was conducted at Dulles International Airrport in 1979. Tests by Embody (unpublished data) using traps in arrays indicated that the beetle population could be reduced using traps alone. Approximately 637 traps baited with the sex pheromone food attractant combination were set out at Dulles Airport which had been previously declared regulated in the 1977 and 1978 seasons due to Japanese beetle populations. The trap arrays caught hundreds of pounds of beetles with only a few beetles reaching the innermost traps in the arrays. Only a few beetles were recovered on aircraft arriving in western states from Dulles, indicating population suppression to a degree with traps alone, although no definite conclusions were possible.

Other tests conducted included the placement of traps in highly attractive host plant concentrations to further enhance the attractiveness to the Japanese beetle. The "trap-crops" concept was applied, with the strips of host material treated with insecticide at weekly intervals to kill the large accumulation of beetles. Using this principle, the sex lure and food bait attractant combination in traps could be used in noncrop segments for the same purpose. Large numbers of beetles attracted to traps either do not enter the traps or they escape. After mating, females deposit eggs in turf surrounding the traps. Persistent soil insecticide could be applied to the turf surrounding the traps to eliminate larval populations concentrating in these areas.

By monitoring Japanese beetle populations with traps, only those areas with high populations need to be treated. If Japanese beetle populations are relatively uniform in distribution, a "trap crop" combination can be used and only the highly attractive crops baited with food and sex lure would require treatment. Tests this season indicated that the use of large numbers of traps in relatively high beetle populations may be responsible for lower population levels. Large concentrations of traps may not only lower populations, but they can also provide an evaluation of intervention if soil insecticides were applied during the fall or spring season.

The Pink Bollworm

The pink bollworm, *Pectinophora gossypiella* (Saunders), first appeared in the United States near Hearne, Robertson County, TX, in 1917. It is generally established in all cotton growing areas of the world except parts of the United States. This insect causes losses in other countries ranging from 15-25% of the annual yield. Extensive annual surveys are conducted to detect possible spread of the insect from Arizona and New Mexico into adjacent states and to monitor the movement of the adults into the San Joaquin Valley of California. Cultural practices are maintained in southern Texas, where the pest represents the greatest threat because of the long growing season.

The synthetic sex attractant was made available by Hummel et al. (1973) and Bierl and Moffit (1974) which opened the way for a sensitive surveillance system. Hummel et al. (1973) also suggested the confusion technique as a control.

The sterile moth release program along with cultural control and pheromone traps has been attributed to the continual low populations of the pink bollworm in the San Joaquin Valley of California. In 1979, a total of 20,591 pheromone traps were employed over much of the southern San Joaquin Valley as a surveillance tool to detect initial location of the yearly influx of pink bollworm, and to delimit the boundaries of possible concentrations of catches. The relative population levels of the native moths caught in that particular area determine future aerial drops of sterile moths. With surveillance information collected by surveys, the sterile moths can be released more judiciously into high populations of native moths providing a more cost-effective program using dispersal of a relatively costly intervention tool. The traps also provide information for an evaluation system which determines the effect of the release of the sterile moths throughout the season. Other activities include trial programs such as the elimination of the moth using sterile release techniques in the U. S. Virgin Islands.

During the 1979 season, both Hercon and Conrel slow-release systems were used in experiments to suppress mating by the pink bollworm using the confusion technique. Results from these tests show promise as a control method, and large scale pilot programs are planned for the 1980 season.

Fruit Flies

The fruit flies are some of the most economically important insects known to man, and monumental efforts have been made to decrease damage in the infested areas. The U. S. Department of Agriculture has successfully launched many eradication programs to prevent permanent establishment in the United States and to retard the movement of these pests into the country.

Synthetic chemical attractants lure primarily male fruit flies. Although the mode of action has yet to be fully explained, little or no competing attractants are known in the environment. Methyl eugenol is used for the oriental fruit fly, *Dacus dorsalis* Herdel, cuelure for the melon fly, *D. cucurbitae* Coquillet, and medlure for the Mediterranean fruit fly, *Ceratitis capitata* (Wiedemann) (Beroza, 1972). These lures are used extensively in quarantine surveillance programs to detect, delimit, and monitor populations. In addition, methyl eugenol has been used as an intervention tool in oriental fruit fly programs. The attractant is used with the pesticide Naled® on a slow release material, and acts as a bait-and-kill point source. Results were reported of a successful male annihilation experiment by Steiner et al. (1965) and Steiner (1969) on Rota near Guam. Other eradication attempts have been successful in California using this technique.

The use of traps to detect the presence and monitor population levels of the Mediterrean fruit fly continues, as the fruit fly moves northward in Mexico. Large numbers of traps were deployed and carefully maintained to provide information on where chemical treatments might be best applied to provide the maximum results. Limited funding requires the most judicious use of pesticides, men, and materials in an almost overwhelming situation. In addition to the monitoring activities along the leading edge of the infestation in southern Mexico and Guatemala, large numbers of detection traps were placed along the border of the United States and in Florida and California where infestations in the U. S. are highly probable.

Should an eradication program be mounted against all 3 species of fruit flies in the Hawaiian Islands, great reliance will be placed on synthetic chemical attractants for surveillance, intervention, and evaluation activities. The use of various techniques involving bait-and-kill using a combination attractant and killing agent are necessary. Again, the use of a relatively environmentally acceptable procedure can provide the desired quarantine objectives with the use of relatively limited amounts of chemical pesticides.

Hercon flakes show great promise as a slow release mechanism which can provide long lasting effectiveness compared to present day practices.

Boll Weevil

The boll weevil, Anthonomus grandis Boheman, entered the United States from Mexico in the late 1800s and has caused economic losses in the billions of dollars. The amount of chemical pesticides used for the boll weevil is astronomical, and indicates a disregard for pest management activities. Brazzel and Newsome (1959) discovered that the boll weevil enters diapause late in the growing season. Subsequently, Taft et al. (1972) showed that the numbers of diapausing weevils entering overwintering sites late in the season could be drastically reduced by the timely application of insecticide sprays.

Synthesis by Tumlinson et al. (1969, 1971) of the boll weevil pheromone, Grandlure®, has provided a highly improved surveillance system with intervention capabilities. Monitoring of the boll weevil population immediately adjacent to and on the Texas High Plains provides managers with needed information. There are indications that tight arrays of traps may also provide a measure of suppression in low level boll weevil populations in a program integrated with other intervention activities (Knipling and McGuire, 1966).

Pheromones were used in traps or in the "trap crop" procedure, where cigarette filters impregnated with the sex pheromone are dropped at measured intervals in the highly attractive rows of cotton purposely planted early or which show a high attractability to the boll weevil (Boyd, 1976). Row crop bait and kill techniques have been shown to play a role in boll weevil management programs and eradication trials that have been conducted. Managers are placing emphasis on techniques incorporating pheromones to provide the control needed. The insecticide and pheromone may be combined into one Hercon wick or other system for longer effectiveness.

Indications are that large numbers of highly efficient boll weevil traps on the side of fields may lower influx of populations. Traps also are an aid in determining the relative abundance of overwintered boll weevils in selected areas.

Concepts and Possible Applications

Increased emphasis should be placed on the bait-and-kill systems for many insect pests. Pheromones and food baits should be combined with killing agents such as Baygon® or Vapona® to provide efficient point source attract-and-kill stations. Applied with a sticker by ground or air, these tiny point source granules, flakes, or fibers can be applied to selective sites to eliminate specific insects. This can be incorporated into several programs including those which only now incorporate the confusion technique. Examples are the gypsy moth, pink bollworm, and the boll weevil.

Use of pheromones in Hercon, Conrel, or other granular strips or fibers should be increased in highly congested locations to either confuse or bait-and-kill insects.

The pheromone or food bait combined with a pesticide in a slow-release system should also be assessed for use against fire ants. A relatively high toxicity material may be found to be released at low rates in a slow-release system when combined with a food or pheromone base.

An increased emphasis on the research and development of trapping systems is needed. There are indications that more efficient and attractive traps may provide a control mechanism, or even elimination of insect populations at low levels. Additional attention should be directed to provide more efficient, effective and economical traps and other system components for such purposes.

Traps or other mechanisms should be designed to spread disease or virus within an insect population by capturing, infecting the insect with a disease or virus, and then releasing them into the environment. Theoretically, this can cause an epidemic effect since these insects would then mate with others, spreading the disease or the virus.

The use of pheromones or food baits may be used to attract low-level populations of insects into areas where mechanical destruction would occur. One possibility would be the attraction of the insects over a body of water whereby the trap itself or other mechanism would automatically eject the insects into the water.

Increased emphasis on more economical formulations of pheromones and biologicals are needed. We must also improve systems for dispersal of the currently used slow-release systems.

Additional efforts should be directed at finding other approaches to using the pheromones that have been isolated and synthesized. Experience has indicated that pheromones can be efficiently used to obtain objectives of regulatory pest programs with reduced dependence on chemical pesticides.

REFERENCES

Anonymous, 1978, Environmental impact statement, Japanese beetle program, USDA-APHIS-ADM 78-1-F.

Anonymous, 1979, Environmental impact statement, Japanese beetle program, USDA-APHIS-ADM 79-1-F.

Beroza, M., 1960, Insect attractants are taking hold, J. Agri. Chem., 15(7): 37.

Beroza, M., 1972, Attractants and repellents for insect pest control, in: "Pest Control Strategies for the Future," National Academy of Science, Washington, D. C.

Beroza, M. and Knipling, E. F., 1972, Gypsy moth control with sex attractant pheromone, Science, 177: 19(4043).

Beroza, M., Stevens, L. J., Bierl, B. A., Philips, F. M., and Tardif, J. G. R., 1973, Pre- and post-season field tests using disparlure, the sex pheromone of the gypsy moth, to prevent mating, Environ. Entomol., 2(6): 1051.

Beroza, M., Hood, C. S., Trefrey, D., Leonard, D. E., Knipling, E. F., and Klassen, W., 1975, Field trials with disparlure in Massachusetts to suppress mating of gypsy moth, Environ. Entomol., 4(5): 705.

Bierl, B. A., and Moffit, H. R., 1974, Sex pheromones: (E,E)-8, 10-dodecadien-1-ol in the codling moth, Science, 183(4120): 89.

Bierl, B. A., Beroza, M., and Collier, C. W., 1970, Potent sex attractant of the gypsy moth, Porthetria dispar (L.): Its isolation, identification, and synthesis, Science, 170(3953): 87.

Boyd, F. J., 1976, Boll weevil population levels during the inseason and reproduction-diapause control phases of the pilot boll weevil eradication experiment, in: Proc. Boll Weevil Suppression, Management, and Elimination Technology, Feb. 13-15, 1974, Memphis, TN, USDA S-71, Washington, D.C. pp. 75-81.

Brazzel, J. R. and Newsome, L. D., 1959, Diapause in Anthonomus grandis, Boh. J. Econ. Entomol., 52(4): 603.

Cameron, E. A., 1973, Disparlure: A potential tool by gypsy moth population manipulation, Bull. Entomol. Soc. Am., 19(1): 15.

Cameron, E. A., Schwalbe, C. P., Beroza, M., and Knipling, E. F., 1974, Disruption of gypsy moth mating with microencapsulated disparlure, Science, 183(4128): 972.

Embody, D. R., 1980, Arrays of sex pheromone trap and the important variables affecting catches of gypsy moth, Lymantria dispar (Linnaeus) Lepidoptera: Lymantridae, USDA APHIS, 81-32, 64 pp.

Fleming, W. E., 1976, Integrating control of the Japanese beetle -- an historical review, USDA Tech. Bull. 1545, 53 pp.

Hummel, H. E., Gaston, L. K., Shorey, H. H., Kaae, R. S., Byrne, R. S., and Silverstein, R. M., 1973, Clarification of the chemical status of the pink bollworm sex pheromone, Science, 181(4102): 873.

Kennedy, J. W., 1979, Gypsy moth and browntail moth quarantine and regulations, Proc. 1979 USDA Public Hearings, Chicago, IL.

Klein, M. G., Ladd, T. L., Jr., and Lawrence, K. O., 1973, Simultaneous exposure of phenylethyl propionate-eugenol (7:3) and virgin Japanese beetles as a lure, J. Econ. Entomol., 66: 373.

Knipling, E. F. and McGuire, J. U., Jr., 1966, Population models to test theoretical effects of sex attractants used for insect control, USDA Agri. Info. Bull. 308, 20 pp.

McGovern, T. P., Beroza, M., and Ladd, T. L., Jr., 1973, Phenylethyl propionate and eugenol, a potent attractant for the Japanese beetle (Popillia japonica Newman), U. S. Patent 3,761,584.

Richerson, J. V., Cameron, E. A., and Brown, E. A., 1976, Sexual activity of the gypsy moth, Am. Midland Naturalist, 95(2): 299.

Steiner, L. F., 1969, Control and eradication of fruit flies on citrus, Proc. 1st Int. Citrus Symp. 2: 881.

Steiner, L. F., Harris, E. J., Mitchell, W. C., Fujimoto, M. S., and Christensen, L. D., 1965, Melon fly eradication by over-flooding with sterile flies, J. Econ. Entomol. 58(3): 519.

Stevens, L. J. and Beroza, M., 1972, Mating inhibition field tests using disparlure, the synthetic gypsy moth sex pheromone, J. Econ. Entomol., 65(4): 1090.

Taft, H. M., Hopkins, A. R., and Roach, S. H., 1972, Suppression of emerging early-season boll weevils using integrated control, J. Econ. Entomol., 65(6): 1663.

Tumlinson, J. H., Hardee, D. D., Gueldner, R. C., Thompson, A. C., Hedin, P. A., and Minyard, J. P., 1969, Sex pheromones produced by male boll weevil: Isolation, identification and synthesis, Science, 166(3908): 1010.

Tumlinson, J. H., Gueldner, R. C., Hardee, D. D., Thompson, A. C., Hedin, P. A., and Minyard, J. P., 1971, Identification and synthesis of the four compounds comprising the boll weevil sex attractant, J. Org. Chem. 36(18): 2616.

Tumlinson, J. H., Klein, M. G., Doolittle, R. E., Ladd, T. L., and Proveaux, A. T., 1977, Identification of the female Japanese beetle sex pheromone: Inhibition of male response by an enantiomer, Science, 197: 789.

MONITORING THE BOLL WEEVIL IN TECHNICAL AND COMMERCIAL OPERATIONS

D. D. Hardee

Pest Management Specialists, Inc.
Post Office Box 364
Starkville, MS 39759 USA

INTRODUCTION

Since the introduction of the boll weevil, Anthonomus grandis Boheman, into the United States about 1892, it has been the most costly insect to American agriculture; in fact, 17 years ago it was called the "$10 billion insect" (Dunn, 1964). Losses in cotton production because of the boll weevil average about $200 to $300 million annually. To prevent even greater losses, growers spend about $70 million/year for its control (Knipling, 1964). Rainwater (1962) estimated that about 1/3 of all insecticides used in agriculture are used to control the boll weevil. After the onslaught of Heliothis spp. [especially the tobacco budworm, H. virescens (Fab.)] in the last 20 years, these estimates are probably high. However, increased environmental pollution and destruction of beneficial insects by boll weevil insecticides (which, in turn, release pests like the tobacco budworm) make its presence highly unwelcome.

Considerable research was done in the last few years to try to control the boll weevil and, in fact, eliminate it from its entire distribution range in the United States. A highly successful method has been the use of male-produced pheromone, grandlure, alongside traps and trap crops. This paper discusses using traps and grandlure in monitoring programs by research and extension personnel as well as commercial applications.

DEVELOPMENT OF GRANDLURE

Keller et al. (1964) and Cross and Mitchell (1966) discovered that the male boll weevil produces a wind-borne pheromone for the female; this, in turn, made it possible to begin monitoring this insect. In addition, Cross and Hardee (1968) initially demonstrated, Bradley et al. (1968) confirmed, and Hardee et al. (1969) explicitly showed that the male pheromone also acts as an aggregating pheromone for both sexes in early- and late-season. After identifying and synthesizing grandlure's 4 components (Tumlinson et al., 1969), grandlure use in traps was improved by overcoming the logistics of using caged males in traps. Grandlure was subsequently refined by developing longer-lasting and slow-releasing formulations (Hardee et al., 1975a).

Cross and Hardee (1968) developed the first trap, "wing trap," for monitoring studies in Mississippi; subsequently, trap modifications were used to study the boll weevil's range and intensity in the High and Rolling Plains of Texas (Hardee et al., 1970). This study showed that boll weevils will disperse up to 72.4 km while searching for cotton or other boll weevils (Davich et al., 1970), and, in turn, enabled the researcher to obtain positive correlation between overwintering weevils captured and weevils observed in the field (Roach et al., 1971). Later trap improvements (Leggett and Cross, 1971; Mitchell and Hardee, 1974; Hardee et al., 1975b) along with the aforesaid grandlure improvements enabled progress to continue toward refining the technique for monitoring the boll weevil.

Improved traps and grandlure formulations provided the following information through boll weevil monitoring studies:

1. In Mississippi, date of diapause entry determines date of emergence the next spring; i.e., weevils entering diapause earliest in the fall are the latest to emerge the next spring (Mitchell et al., 1973).

2. In west Texas, weevil populations in fields during October and November were the most important in starting the new generation the next year (Wade and Rummel, 1978).

3. The boll weevil's long-range movement is primarily windborne; furthermore, weevils cannot detect a pheromone further than 152-182 m and move randomly until they come within this distance of the pheromone (Mitchell and Hardee, 1976).

4. Traps and pheromones can detect a potential for early field populations on individual farms and can now be substituted for woods trash examinations for this purpose (Hopkins et al., 1977).

5. Of the 4 methods used for monitoring small populations of boll weevils, use of in-field traps baited with grandlure was the superior method (Merkl et al., 1978).

6. In-field traps baited with grandlure were more than 8X as effective as manual examinations of punctured squares when detecting very low-level boll weevil infestations (Hardee, 1976).

COMMERCIALIZATION OF GRANDLURE

Traps and grandlure are now recommended for commercial use the following ways:

1. Trap row placement surrounding cotton fields to detect overwintered boll weevils for initiation and sequence of insecticide spraying of trap rows (Hardee et al., 1975b);

2. Placement in trap crops planted early and surrounding cotton fields to detect overwintered boll weevils for initiation and sequence of insecticide spraying (Bodegas et al., 1977);

3. Area-wide program use in Mississippi and south Texas to see if and when to initiate pinhead-square-stage insecticide applications for boll weevils (R. R. Frisbie, J. L. Hamer, personal communication).

4. In addition to being used to suppress weevils in the Trial Boll Weevil Eradication Program in North Carolina and Virginia, traps and grandlure are used to detect low-level and isolated populations that might escape other treatments, as well as to determine immigration into the test area from untreated areas (M. P. Ganyard, personal communication).

Traps and grandlure have greatly helped us to attain current knowledge on boll weevil behavior. They will continue to increase their importance in monitoring and control programs.

REFERENCES

Bodegas, P. Rene, Flores, R., and Coss Flores, M., 1977, La Utilizacion de cultivos trampa para el combate del picudo del algodonero en el Soconusco, Chiapas, Mexico, _Centro de Investigaciones Ecologicas del Sureste Inform. Bull._, 7: 1.

Bradley, J. R., Jr., Clower, D. F., and Graves, J. B., 1968, Field studies of sex attraction in the boll weevil, _J. Econ. Entomol._, 61: 1457.

Cross, W. H., and Hardee, D. D., 1968, Traps for survey of overwintered boll weevil populations, _USDA Coop. Econ. Insect Rept._, 18: 430.

Cross, W. H., and Mitchell, H. C., 1966, Mating behavior of the female boll weevil, _J. Econ. Entomol._, 59: 1503.

Davich, T. B., Hardee, D. D., and Alcala, J. M., 1970, Long-range dispersal of boll weevils determined with wing traps baited with males, _J. Econ. Entomol._, 63: 1706.

Dunn, H. A., 1964, Cotton boll weevil, *Anthonomus grandis* Boh.,: Abstracts of research publication, 1843-1960, *USDA Coop. States Research Service Misc. Publ.* 985, p. 1.

Hardee, D. D., 1976, Development of boll weevil trapping technology, Proceedings of Conference on Boll Weevil Suppression, Management and Elimination Technology, *USDA Tech. Bull.* ARS-71, 34 pp.

Hardee, D. D., Cross, W. H., and Mitchell, E. B., 1969, Male boll weevils are more attractive than cotton plants to boll weevils, *J. Econ. Entomol.*, 62: 165.

Hardee, D. D., Cross, W. H., Huddleston, P. M., and Davich, T. B., 1970, Survey and control of the boll weevil in west Texas with traps baited with males, *J. Econ. Entomol.*, 63: 1041.

Hardee, D. D., McKibben, G. H., and Huddleston, P. M., 1975a, Grandlure for boll weevils: Controlled release with a plastic laminated dispenser, *J. Econ. Entomol.*, 68: 477.

Hardee, D. D., Moody, R., Lowe, J., and Pitts, A., 1975b, Grandlure, in-field traps, and insecticides in population management of the boll weevil, *J. Econ. Entomol.*, 68: 502.

Hopkins, A. R., Taft, H. M., and Roach, S. H., 1977, Boll weevils: Leggett traps as a substitute for woods trash examinations as an indicator of potential field populations, *J. Econ. Entomol.*, 70: 445.

Keller, J. C., Mitchell, E. B., McKibben, G., and Davich, T. B., 1964, A sex attractant for female boll weevils from males, *J. Econ. Entomol.*, 57: 609.

Knipling, E. F., 1964, The potential role of the sterility method for insect population control with special references to combining this method with conventional methods, *USDA Tech. Bull.*, *ARS*-33-98, 26 pp.

Leggett, J. E., and Cross, W. H., 1971, A new trap for capturing boll weevils, *USDA Coop. Econ. Insect Rept.*, 21: 773.

Merkl, M. E., Cross, W. H., and Johnson, W. L., 1978, Boll weevils: Detection and monitoring of small populations with in-field traps, *J. Econ. Entomol.*, 71: 29.

Mitchell, E. B., and Hardee, D. D., 1974, In-field traps: A new concept in survey and suppression of low populations of boll weevils, *J. Econ. Entomol.*, 67: 506.

Mitchell, E. B., and Hardee, D. D., 1976, Boll weevils: Attractancy to pheromone in relation to distance and wind direction, *J. Ga. Entomol. Soc.*, 11: 113.

Mitchell, E. B., Huddleston, P. M., Wilson, N. M., and Hardee, D. D., 1973, Boll weevils: Relationship between time of entry into diapause and time of emergence from overwintering, *J. Econ. Entomol.*, 66: 1230.

Rainwater, C. F., 1962, Where we stand on boll weevil control and research, Proceedings, Boll Weevil Research Symposium, State College, Mississippi, March 21, 1962, 11 pp.

Tumlinson, J. H., Hardee, D. D., Gueldner, R. C., Thompson, A. C., Hedin, P. A., and Minyard, J. P., 1969, Sex pheromones produced by male boll weevils: Isolation, identification, and synthesis, Science, 166: 1010.

Roach, S. H., Ray, L., Taft, H. M., and Hopkins, A. R., 1971, Wing traps baited with male boll weevils for determining spring emergence of overwintered weevils and subsequent infestations in cotton, J. Econ. Entomol., 64: 107.

Wade, L. J., and Rummel, D. R., 1978, Boll weevil immigration into winter habitat and subsequent spring and summer emergence, J. Econ. Entomol., 71: 173.

RECENT DEVELOPMENTS IN THE USE OF PHEROMONES TO MONITOR *PLODIA INTERPUNCTELLA* AND *EPHESTIA CAUTELLA*

K. W. Vick

J. A. Coffelt

R. W. Mankin

Insect Attractants, Behavior, and Basic
Biology Research Laboratory
AR-SEA, USDA
P. O. Box 14565
Gainesville, FL 32604 USA

E. L. Soderstrom

Stored-Product Insects Laboratory
AR-SEA, USDA
5578 Air Terminal Drive
Fresno, CA 93727 USA

INTRODUCTION

Stored commodities are particularly susceptible to insect infestation, in part, because food is rarely limited in storage facilities, and temperature and climatic conditions tend to be ideal for insect development. Furthermore, the handling involved in milling, processing, and transporting greatly increases the possibility of introducing insect pests into the commodity. These factors plus the very stringent requirements for absence of insects in processed commodities make sensitive insect monitoring tools necessary. Numerous entomologists have suggested that traps baited with sex pheromones may be ideally suited to this use.

A sex pheromone of the almond moth, *Ephestia cautella* (Walker), and the Indian meal moth, *Plodia interpunctella* (Hübner), was identified as (*Z*,*E*)-9,12-tetradecadien-1-ol acetate (Z9,E12-14:Ac) by Kuwahara et al. (1971b) and Brady et al. (1971). This chemical has since been identified as a sex pheromone of 3 other important stored-product moths; the Mediterranean flour moth, Anagasta kuehniella (Zeller) (Kuwahara et al., 1971a); raisin moth, Cadra figulilella (Gregson) (Brady and Daley, 1972); and tobacco moth, E. elutella (Hübner) (Brady and Nordlund, 1971). The fact that Z9,E12-14:Ac attracts 5 of the most important moth species to traps greatly simplifies the use of pheromone-baited traps for insect detection. This report concerns the use of Z9,E12-14:Ac for monitoring the moth population in 2 warehouses, the discovery of a new sex pheromone component for the Indian meal moth, and some calculations concerning dispersal of the pheromone in warehouses.

WAREHOUSE MONITORING

Although the vast majority of papers reporting results of insect monitoring with sex pheromones have been concerned with field crop, orchard, or forest insects, a few papers have reported data on warehouse studies. For example, Vick et al. (1979) found that the pheromone of the Indian meal moth and the Angoumois grain moth, Sitotroga cerealella (Olivier), could be dispersed from the same trap. They also investigated the effects of different rates of release of the pheromone and different trap designs on insect catch. Reichmuth et al. (1976, 1978, 1980) investigated populations of E. elutella in warehouses containing traps baited with sex pheromone. Hoppe and Levinson (1979) monitored populations of E. elutella, E. cautella, and P. interpunctella in a chocolate factory. All these scientists emphasized the use of pheromone-baited traps in timing applications of insecticides and in subsequent monitoring of effectiveness.

The monitoring study reported here was conducted in both commercial peanut warehouses and in military food warehouses. The traps were baited with Z9,E12-14:Ac obtained from Farchan Division of Storey Chemical Co., Willoughby, OH. It was purified so as to contain less than 2% of the Z,Z-isomer by elution through a 1.5 x 50-cm glass column containing 20% $AgNo_3$ on 60/200 mesh silica gel (Hi-Flosil-Ag®). Overall purity was about 98% as determined by GLC analysis on a 1.8 m x 2 mm ID column packed with 5% Carbowax 20M® on 100/120 mesh Chromosorb®. Ten mg of pheromone was dispersed from 250-ml capacity polyethylene caps (size #3, BEEM® polyethylene embedding capsule C) hung midway between the top and bottom of a Pherocon 1C® trap. New pheromone bait was placed in the trap every 2 weeks.

In the peanut warehouse, 4 traps were placed ca. 0.6 to 1.0 m above the peanut surface and ca. equal distances apart. Traps were checked 2 or 3 times/week depending upon the number of moths captured. This test ran for 1 year, from October 1977 to September 1978. During this period, the only lepidopteran insects trapped were E. cautella and P. interpunctella; about 20X more of the former than the latter were present in each collection. Fig. 1 shows the numbers of insects caught/trap per week. Catch increased steadily for several months after harvest and then declined during the cold months of February and March. The warehouse was fumigated with aluminum phosphine in February, but population levels, as reflected by trap catch, had returned to a high level by the middle of April. After the warehouse was emptied in June, the population declined, but insects were trapped at a fairly constant rate throughout the summer.

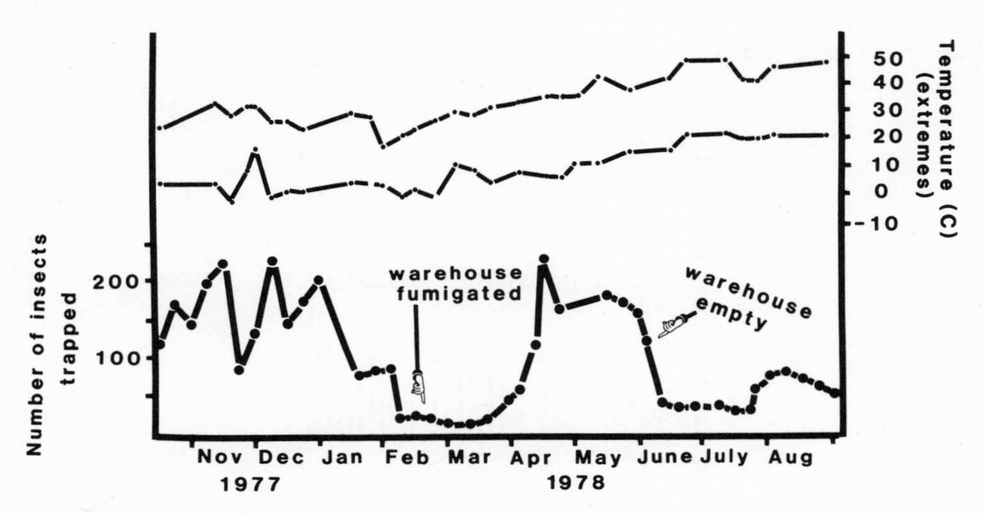

Fig. 1 Monthly pheromone trap catches of E. cautella in a peanut warehouse.

The results therefore showed the need for better insect monitoring tools for use in warehouses. Even in situations such as this where high insect populations can be tolerated, some gauge is needed to determine when a population has reached the economic threshold. For instance, in the peanut warehouse, the fumigant was applied inappropriately at a time when the population of adults was declining. Consequently, the fumigation was at best only partially successful.

The military food warehouse contained mostly processed, packaged food and was a modern facility that had been designed as a warehouse and had good lighting, wide aisles, etc. The warehouse-person in charge professed to know of no insect populations in the building. Six traps were placed in the warehouse, and one was placed outside on the loading dock. Traps in the warehouse were positioned near commodities that were more likely to be infested; i.e., dry cereals, dog food, cake mix, spaghetti, and flour, and were checked twice/week. Fig. 2 shows the weekly catches on the 6 traps for 17 weeks of the test. (The trap on the loading dock caught only a few insects.) Again, only E. cautella and P. interpunctella were caught in the warehouse. Insect densities as measured by trap catch, were not uniform throughout the warehouses. Some traps seemed to be situated in particularly infested areas. For instance, trap 5 caught more insects than any of the other traps in 13 of the 17 weeks. Trap catches were particularly heavy in traps 5 and 6 for weeks 14 and 15. Catches in traps 1 and 4 were consistently low, particularly during weeks when others were catching many more.

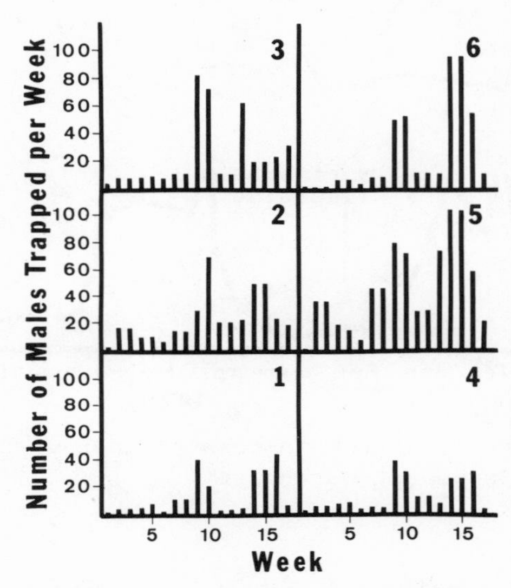

Fig. 2 Weekly pheromone trap catches of E. cautella and P. interpunctella at 6 different positions in a food storage warehouse.

This food warehouse like others, presented a difficult problem. Insect population levels must always be kept as low as possible; and detection of hidden infestations is especially important when the food is to be distributed over large geographical areas. Nevertheless, survey traps clearly showed an insect problem that had gone undetected. Additionally, the pheromone-baited traps seemed to pinpoint areas of high insect infestation. Knowing even the approximate locations of such areas is of great help in applying control measures.

IMPROVED PHEROMONE FOR P. INTERPUNCTELLA

A 2nd compound, (Z,E)-9,12-tetradecadien-1-ol (Z9,E12-14:OH), that is produced and released by P. interpunctella females was identified by Sower et al. (1974), Sower and Fish (1975) and Coffelt et al. (1978). Sower et al. (1974) could not ascribe a conspecific function for this alcohol in their laboratory tests, but they felt that the compound might also contribute to the sexual communication of P. interpunctella. Therefore, we attempted to determine the efficacy of Z9,E12-14 as a trap bait relative to female-baited traps, to discover if the alcohol is attractive to male moths and to evaluate the binary mixture as a trap bait.

Two tests were conducted in fall 1979 in a building that contained several hundred pounds of cull figs, which harbored a population of P. interpunctella. In both cases, we used Pherocon 1C traps containing 3 virgin females in a 7- x 5- x 4-cm screen cage or similar traps containing the test compounds. The test compounds were volatilized from 1 cm^2 fiberglass-coated PVC window screen. Traps were rotated to a different position each day, and one complete rotation (4 days) constituted one replication.

Table 1. Numbers of male P. interpunctella captured in traps baited with females, Z9,E12-14:Ac, Z9,E12-14:OH, or Z9,E12-14:Ac + Z9,E12-14:OH.

Treatment	Dose	No. Males Captured[1]
Z9,E12-14:Ac	4 mg	107 a
Z9,E12-14:OH	6 mg	10 b
Z9,E12-14:Ac + Z9,E12-14:OH	4 mg + 6 mg	1417 c
3 Females	-	492 d

[1]Numbers followed by different letters are not significantly different as determined by Duncan's multiple range test.

In the 1st test, we compared male captures in traps baited with females, Z9,E12-14:Ac, Z9,E12-14:OH, or a 40:60 mixture of the acetate and alcohol (Table 1). The acetate-baited traps captured significantly more moths than did the alcohol-baited traps. Similarly, traps containing the binary mixture captured ca. 14X more than those baited with the acetate alone and 3X as many as did female-baited traps.

In the 2nd test, we examined the relationships between dosage applied (binary mixture) and male trap capture. Traps baited with 3 females were included as the standard. Pheromone used in these tests was dispensed from window screen evaporators that were loaded with 10, 1, or 0.1 mg of the 60:40 (alcohol:acetate) mixture. Trap captures were highly correlated ($r^2 = 0.999$) with applied dosage (Fig. 3).

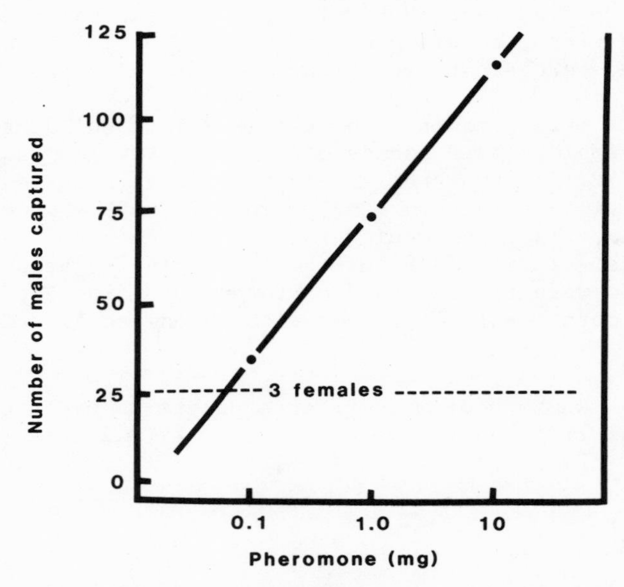

Fig. 3 Number of male P. interpunctella trapped in pheromone traps as a function of pheromone dose.

The results provided new information regarding the sex pheromone system of P. interpunctella and clearly showed that a mixture of Z9,E12-14:Ac and Z9,E12-14:OH was a more effective trap bait for this species than the acetate alone. The next step is to conduct detailed behavioral analyses and define the behavioral mechanisms involved in the sex pheromone response of male P. interpunctella. Studies already are in progress to establish the most effective release rates and substrates for these rather labile compounds.

PHEROMONE DISPERSAL IN A WAREHOUSE

In attempting to pinpoint sources of infestation, it is helpful to know the distance from which a trap attracts insects. Consequently, Mankin et al. (1980a) developed a model of pheromone dispersal in open and confined spaces that deals with the relationship among the major physical parameters of a warehouse environment. The pheromone concentration at some distance, r, from a source that has emitted pheromone for time, t, depends upon the emission rate, the positions of boundary surfaces, the pheromone diffusion coefficient, and the deposition velocity of pheromone to the boundary

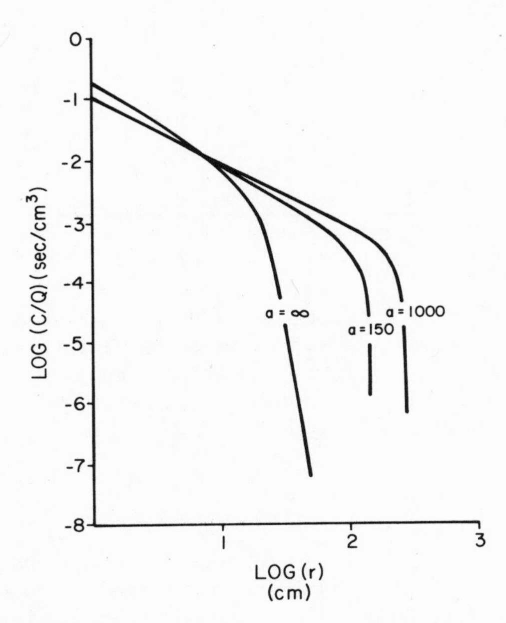

Fig. 4 Spatial distribution of the relative concentration of pheromone, C_r= Concentration/Release rate, at t = 60 sec after the start of release. The distance from pheromone source is designated by r, the position of boundary by a in units of cm.

surfaces. (The latter term is defined as the rate of deposition to a unit surface area/unit concentration.) Here we first consider some general aspects of these relationships and then present an example of how the model can be used to solve practical problems.

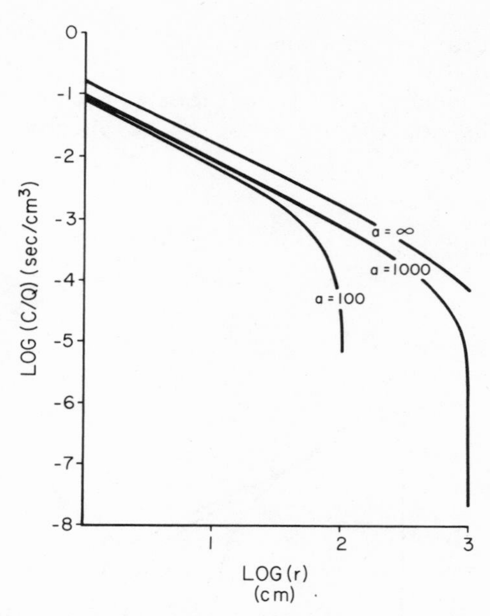

Fig. 5 Spatial distribution of the relative concentration of pheromone, C_r = Concentration/Release rate, at t = 2.5d after the start of release. The distance from pheromone source is designated by r, the position of boundary by a in units of cm.

Fig. 4 shows the effect of distance and boundary position on the ratio of pheromone concentration to its emission rate, C/Q, for the case where t = 60 sec, the diffusion coefficient, D, is 1 cm^2/sec, and the deposition velocity, V_d, is 1 cm/sec. The latter 2 values were chosen because they are fairly typical of warehouse conditions. Fig. 5 shows the identical case except that t has been extended to a duration more typical of a trapping situation, 8.6×10^5 sec. It is assumed that there is only one boundary surface, a sphere of radius a. Note that the smaller the boundary radius, the smaller the change in C/Q as the duration of emission increases. For example, the curve for a = ∞ changes considerably as the duration increases from 60 sec to 8.6×10^5 sec, but the curve for a = 150 cm at 60 sec differs little from the curve for a = 100 cm at 8.6×10^5 sec. Also, the difference between the magnitudes of C/Q in bounded and unbounded spaces decreases as the emission duration increases. Indeed, so long as $r/a < 0.9$, the effect of the boundary is negligible at t = 8.6×10^5 sec, so we can assume that after long emission, the effect of a boundary on the pheromone concentration is negligible except at points very near the boundary.

To demonstrate a practical application of the model, we determined the active space of a trap used to capture male P. interpunctella in a warehouse and compared this with the active space of the pheromone-producing female. Previous studies indicate that the behavioral threshold of the male for its sex pheromone, Z9,E12-14:Ac, is about 1×10^{-8} ng/cm^3 (Mankin et al., 1980b). Typical pheromone traps for these insects emit 0.01 - 0.76 ng/sec (Vick et al., 1979). Over this range of emission rates, essentially the entire volume inside a boundary of either 100 or 1000 cm radius is an active space according to Fig. 5. The ratios $1 \times 10^{-8}/0.01 = 1 \times 10^{-6}$ and $1 \times 10^{-8}/0.76 = 1.3 \times 10^{-8}$ intersect the 2 curves for a = 100 and a = 1000 very near the respective boundaries. By contrast, a female Plodia, which emits about 8×10^{-4} ng/sec (Sower and Fish, 1975), has an active space of about 250 cm radius in a sphere of 1000 cm radius after it has called for 60 sec (see Fig. 4). In a sphere of 150 cm radius the active space covers the entire volume, but in an unbounded environment the active space has a radius of about 40 cm.

The model does have some defects, and should therefore be used with caution. For example, the traps in the food warehouse baited with 10 mg Z9,E12-14:Ac, emitting 0.76 ng/sec (Vick et al., 1979), theoretically have an active space greater than 10 m in radius. From this, one would not expect to find the large differences among the trap catches in Fig. 2. The effects of trap competition or habituation to the pheromone may have been a factor here.

Nevertheless, the model can be applied in a general way to predict optimal spacing and release rates for traps in closed environments, and the 2 insect surveys herein serve as an additional guide in planning future warehouse monitoring programs. We are optimistic that new developments, such as the improved pheromone blend for P. interpunctella, will increase the efficacy and the use of pheromone traps as monitors of insect pests in warehouses.

REFERENCES

Brady, U. E. and Daley, R. C., 1972, Identification of a sex pheromone from the female raisin moth, Cadra figulilella, Ann. Entomol. Soc. Am., 65: 1356.

Brady, U. E. and Nordlund, D. A., 1971, Cis-9,trans-12-tetradecadien-1-yl acetate in the female tobacco moth, Ephestia elutella (Hübner), and evidence for an additional component of the sex pheromone, Life Sci., 10: 797.

Brady, U. E., Tumlinson, J. H., Brownlee, R. G., and Silverstein, R. M., 1971, Sex stimulant and attractant in the Indian meal moth and in the almond moth, Science, 171: 802.

Coffelt, J. A., Sower, L. L., and Vick, K. W., 1978, Quantitative analysis of identified compounds in pheromone gland rinses of *Plodia interpunctella* and *Ephestia cautella* at different times of day, *Environ. Entomol.*, 7: 502.

Hoppe, T. and Levinson, H. Z., 1979, Befallserkennung und Population-suberwachung vorratsschadlicher Motten (Phycitinae) in liner Schokoladenfalrik mit Hilfe pheromonbekoderten Klubefallen, *Anz. Schaedlingskde. Planz. Umwbeltschutz.*, 52: 177.

Kuwahara, Y., Hara, H., Ishii, S., and Fukami, H., 1971a, Sex pheromone of the Mediterranean flour moth, *Agric. Biol. Chem.*, 35: 447.

Kuwahara, Y. C., Kitamura, C., Takahoshi, S., Hara, H., Ishu, S., and Fukami, H., 1971b, Sex pheromone of the almond moth and the Indian meal moth: *cis*-9,*trans*-12-tetradecadienyl acetate, *Science*, 171: 801.

Mankin, R. W., Vick, K. W., Mayer, M. S., Coffelt, J. A., and Callahan, P. S., 1980a, Dispersal of vapors in open and confined spaces with applications to sex pheromone trapping in a warehouse, *J. Chem. Ecol.*, (In press).

Mankin, R. W., Vick, K. W., Mayer, M. S., and Coffelt, J. A., 1980b, Anemotactic response threshold of the Indian meal moth, *Plodia interpunctella* (Hübner) (Lepidoptera, Pyralidae), to its sex pheromone, *J. Chem. Ecol.*, (In press).

Reichmuth, C., Wohlgemuth, R., Levinson, A. R., Levinson, H. Z., 1976, Untersuchungen uber den Einsatz von pheromonbekoderten Klebefallen zur Bekampfung von Motten im Vorratsschutz, *Z. Angew. Entomol.*, 82: 95.

Reichmuth, C., Schmidt, H. U., and Levinson, H. Z., 1978, Die Fangigkeit pheromonbekoderter Klebefallen fur Specichermotten (*Ephestia elutella* Hbn) in unterschildlich dicht befallenen Getreidelagern, *Z. Angew. Entomol.*, 86: 205.

Reichmuth, C., Schmidt, H. U., Levinson, A. R., and Levinson, H. Z., 1980, Das jahreszeitliche Auftreten von Specichermotten (*Ephestia elutella* Hbn) in Berliner Getreideschuttbodenlagern sowie der, Zeitentsprechende Einsatz von Bekompfungsmassmahmen, *Z. Angew. Entomol.*, 89: 104.

Sower, L. L., Vick, K. W., and Tumlinson, J. H., 1974, (*Z*,*E*)-9,12-Tetradecadien-1-ol: A chemical released by female *Plodia interpunctella* that inhibits the sex pheromone response of male *Cadra cautella*, *Environ. Entomol.*, 3: 120.

Sower, L. L. and Fish, J. C., 1975, Rate of release of the sex pheromone of the female Indian meal moth, *Environ. Entomol.*, 4: 168.

Vick, K. W., Kvenberg, J., Coffelt, J. A., and Steward, C., 1979, Investigation of sex pheromone traps for simultaneous detection of Indian meal moths and Angoumois grain moths, *J. Econ. Entomol.*, 72: 245.

BIOMONITORING FOR STORED-PRODUCT INSECTS

W. E. Burkholder

Stored Product and Household
Insects Laboratory
AR-SEA, USDA
Department of Entomology
University of Wisconsin
Madison, WI 53706 USA

INTRODUCTION

Monitoring of storage insects by traps is an essential but hardly a new procedure. Food attractants, light traps and physical traps have been used for this purpose for years. However, food attractants have the disadvantage in that they usually do not attract over a very long distance, and food traps necessarily compete with the natural food and may provide nutriment for a new infestation; light traps are dependent on electricity, the light wavelength, space and other factors, and physical traps such as perforated-cylinder grain probes and pitfall traps do not have appreciable luring capability. Pheromones, however, have added a powerful new dimension to this important procedure.

Therefore, efforts have been made during recent years to use pheromones in traps as a means of producing a highly effective monitoring system for stored-product insects. The first step has been to discover the pheromones in a pest species. Next has come isolation and identification coupled with preliminary field efficacy tests. We will soon have this type of data for many of our important stored-product pests. Tables 1 and 2 summarize the current array of such pheromones available for monitoring.

The problem now facing us is how can we use these pheromones most effectively. Certainly we need to continue studies of the biology and behavior of the insects in an effort to devise innovative monitoring systems. We also need to continue our studies of each pheromone in order to determine the most effective rate of release and the behavior in an actual storage facility. Most of all we need to develop innovative pheromone monitoring systems that can be coordinated with control methods. These needs are not simple. Although food storage facilities tend to be regarded as uniform and stable environments that lend themselves to effective trapping, such a concept is unrealistic. There is considerable variability among and within facilities that must be considered in the design of any pheromone monitoring system.

Table 1. Major Pheromone Components of Some Stored-Product Coleoptera (Anobiidae, Bostrichidae, Bruchidae and Dermestidae).

Species and Sex which Produce Pheromones	Pheromones
Stegobium paniceum ♀	2,3-Dihydro-2,3,5-trimethyl-6 (1-methyl-2-oxobutyl)-4*H*-pyran-4-one (Kuwahara et al., 1975, 1978)
Lasioderma serricorne ♀	4,6-Dimethyl-7-hydroxy-nonan-3-one (Coffelt & Burkholder, 1972; Chuman et al., 1979a,b).
Rhyzopertha dominica ♂	1-Methylbutyl (*E*)-2-methyl-2-pentenoate and 1-methylbutyl(*E*)-2,4-dimethyl-2-pentenoate (Khorramshahi & Burkholder, 1980; Williams et al., 1980).
Acanthoscelides obtectus ♂	(*E*)-(-)-Methyl-2,4,5-tetradecatrienoate (Hope et al., 1967; Horler, 1970).
Attagenus megatoma ♀	(*E*,*Z*)-3,5-Tetradecadienoic acid (Burkholder and Dicke, 1966; Silverstein et al., 1967).
Attagenus elongatulus ♀	(*Z*,*Z*)-3,5-Tetradecadienoic acid (Barak and Burkholder, 1977a,b; Fukui et al., 1977).
Anthrenus flavipes ♀	(*Z*)-3-Decenoic acid (Burkholder et al., 1974; Fukui et al., 1974).
Trogoderma inclusum and *T. variabile* ♀	(*Z*)-14-Methyl-8-hexadecen-1-ol and (*Z*)-14-methyl-8-hexadecenal (Burkholder and Dicke, 1966; Rodin et al., 1969; Cross et al., 1976).

Species and Sex which Produce Pheromones	Pheromones
Trogoderma glabrum ♀	(*E*)-14-Methyl-8-hexadecen-1-ol and (*E*)-14-methyl-8-hexadecenal (Burkholder and Dicke, 1966; Yarger et al., 1975; Cross et al., 1976).
Trogoderma granarium ♀	92:8 (*Z*:*E*)-14-Methyl-8-hexadecenal (Levinson and Barilan, 1967; Cross et al., 1976).

Table 2. Major Pheromone Components of Some Stored-Product Lepidoptera (Gelechiidae and Pyralidae).

Species and Sex which Produce Pheromones	Pheromones
Sitotroga cerealella ♀	(*Z*,*E*)-7,11-Hexadecadien-1-ol acetate (Keys and Mills, 1968; Vick et al., 1974).
Ephestia elutella ♀	(*Z*,*E*)-9,12-Tetradecadien-1-ol acetate (Brady, 1973; Brady and Nordlund, 1971; Brady and Daley, 1972; Brady et al., 1971a,b; Kuwahara and Casida, 1973; Kuwahara et al., 1971a,b).
Plodia interpunctella ♀	
Ephestia cautella ♀	
Anagasta kuehniella ♀	
Cadra figulilella ♀	
Amyelois transitella ♀	(*Z*,*Z*)-11,13-Hexadecadienal (Coffelt et al., 1979a,b).

COLEOPTERA

Dermestid Beetles

An extensive pheromone-monitoring program for dermestid beetles, especially *Trogoderma* spp., has been underway for several years in the United States. The pheromone is routinely used by the Plant Protection and Quarantine Program, USDA, APHIS at the major U.S. port facilities where large numbers, mostly T. variabile Ballion, are caught consistently. It also has been used in the interior of California to survey khapra beetles, T. granarium Everts, and Levinson and Levinson (1979) evaluated it in Morocco against this beetle. In addition, Trogoderma pheromone traps have been evaluated also by the U. S. Army in warehouses in Tennessee and California. In these studies, the location of the traps received particular attention. Traps near the exterior walls accounted for most of the captured Trogoderma. The pheromone-baited traps were significantly better at catching Trogoderma than the controls. Although trap counts were highest in floor traps, there was a tendency for more insects to appear in aerial traps near the walls and doors of a warehouse, and significant differences in the 2 kinds of traps were not obtained. Dichlorvos was the killing agent in these traps.

In earlier studies (Barak and Burkholder, 1976), the alcohol component of the Trogoderma pheromone was evaluated with megatomoic acid, the sex pheromone of Attagenus megatoma (F.), the black carpet beetle. Traps treated with the pheromones, either singly or in combination, were placed in 3 warehouse and grain elevator locations in Milwaukee, WI. The traps treated with pheromones caught significantly more target insects than control traps. In particular, a hidden population of T. variabile in one warehouse was discovered. Also, the seasonal emergence of A. megatoma was observed and charted over a 2-year period.

Studies of Trogoderma trapping with pheromone are now concerned with proper trap location, density, pheromone concentration and release rate, and trap design. Also, wheat germ oil has been tested as an attractant for Trogoderma larva in some of the floor-placed traps. Several of the attractive components in wheat germ oil have been identified in the Madison Laboratory, and they will now be used to increase our effectiveness in larval monitoring of Trogoderma. In addition, the physical design of the trap has been improved and the most effective enantiomeric composition of the pheromone has been determined (Silverstein et al., 1980), as have the theoretical maximum communication distances for the pheromone (Shapas and Burkholder, 1978).

Finally, it appeared that the use of pathogens, in addition to insecticides, as a killing agent in the pheromone trap may provide some insurance should the insecticides fail. A population of T. glabrum (Herbst) was successfully suppressed by using the pheromone to lure this dermestid to a protozoan pathogen. The pathogen was then disseminated effectively by the beetles as they moved out into the storage area (Shapas et al., 1977).

Lesser Grain Borers

Male lesser grain borers, *Rhyzopertha dominica* (F.), produce an aggregating pheromone that attracts both sexes (Khorramshahi and Burkholder, 1980). In fact, this pheromone is responsible for the characteristic "sweetish" odor of grain infested with the borer. Monitoring studies for the species were conducted in a variety of warehouses in Texas; corrugated traps placed on the floor were compared with aerial traps. The traps were effective for as long as 5 weeks in monitoring lesser grain borer populations.

However, infestations of lesser grain borers often develop deep within the bin, and, therefore, present a special problem in monitoring since traps are usually placed in the headspace of a bin. To surmount this problem, I advocate the use of perforated grain probes similar to those developed by Loschiavo and Atkinson (1967). These could carry the aggregating pheromone to the infestation site so the insect could be monitored and subsequently controlled. For example, *Bacillus thuringiensis* Berliner (BT), or some other pathogen, could be placed in the monitoring probes. This use of pheromone lures for pathogen dissemination would supplement the monitoring efforts and give some protection of the grain from insect damage. Qayyam (1979) presented evidence that BT was effective in suppressing the lesser grain borer. However, care must be taken in the selection of the pathogen strain, because some strains of BT are more effective than others against the lesser grain borer.

I also suggest that insect growth regulators (IGRs) could be used with monitoring devices as a backup measure to exert some control of the lesser grain borer and perhaps other insects in the same environment. Indeed, though monitoring is my primary and present concern here, I believe we should not ignore the major concept of pest management to combine all of our best monitoring and suppression procedures in a practical and effective plan of insect control. Previous research has indicated that the treatment of corn, wheat, and peanuts with IGRs is disadvantageous in that insect larvae inside the grain may not be directly exposed to an effective concentration (Strong and Diekman, 1973; Hoppe, 1974). Also, McGregor and Kramer (1975) demonstrated that IGRs reduced F_1 populations of the lesser grain borer, but not of *Sitophilus* spp. Nickle (1979) suggested that *Sitophilus* spp. are less susceptible to IGRs. Nevertheless, I suggest that there are major differences in insect behavior that may be responsible for the observed variation in mortality. *Sitophilus* spp. females deposit their eggs directly into the kernels of grain; lesser grain borer females deposit the eggs loosely among the kernels. Thus, the lesser grain borer larvae must crawl over the outside of the kernel some distance to find a suitable place (usually a crack or the softer germ area) to bore inside. In that exposed state, the larva may come in contact with sufficient IGR or other deleterious agents that could have an adverse effect on development. This idea is an extension of the earlier suggestion of Burkholder and Dicke (1966), advancing the use of pheromones combined with pathogens for dermestid control.

Other Grain and Pulse Beetles and Weevils

Pheromones are being extracted, identified, and developed for a variety of other grain-infesting beetles and weevils, and these compounds will probably be available within a few years. The most promising ones are those for *Tribolium* spp., *Sitophilus* spp., *Oryzaephilus* spp., *Stegobium paniceum* (L.), *Lasioderma serricorne* (F.), *Acanthoscelides obtectus* (Say), the bean weevil, and *Callosobruchus* spp., seed weevils.

Bin probes and bin or floor surface-traps containing pheromones or attractants appear to be a promising monitoring tool for many of these insects, especially those that infest grain and pulse. In addition, several of these beetles (*Sitophilus* spp., *Tribolium* spp., *Oryzaephilus* spp.) respond dramatically to such disturbances as vibration or shaking of the substrate or probing: the insects scramble to the surface of the grain or fall into the perforated probes (Burkholder, 1979). Again, I would propose an integrated approach to these particular pest problems. Monitoring by traps and probes baited with pheromones could be combined with physical disturbance such as mechanically vibrating probes which seem to motivate insects to trapping sites. All these approaches should be combined with additional materials like pathogens, IGRs or insecticides; that is, the integrated system would provide the needed monitoring and have built-in controls, or monitoring could be initiated first and the control method added later as needed.

The pheromones of *S. paniceum* (drugstore beetle), and *L. serricorne* (cigarette beetle), and of some other insects may be used effectively as an aid in monitoring infestations in combination with light traps, devices used for many years in monitoring cigarette beetles. For example, preliminary tests have shown that the synthetic pheromone of the cigarette beetle is highly effective in attracting males. Therefore, an integrated approach could involve such possible measures as pheromones, food attractants, physical traps, IGRs, and insecticides.

LEPIDOPTERA

Reichmuth et al. (1976) studied the use of pheromone traps for moth control in storage. Their experiments were carried out in small rooms, large warehouses, and a chocolate factory in West Berlin during 1975, and involved (*Z*,*E*)-9,12-tetradecadien-1-ol acetate vaporized from plastic capsules in sticky traps. Results showed the value of such pheromone traps for monitoring infestations of *Plodia interpunctella* (Hübner) and *Ephestia elutella* (Walker).

Reichmuth et al. (1978) were also able to monitor the seasonal fluctuations of *E. elutella* populations in grain stores in West Berlin, but pheromone traps were somewhat more effective in granaries with light infestations. Also, their studies in Greece during 1977-78 pinpointed seasonal emergence cycles of *Anagasta kuehniella* (Zeller), *E. cautella* (Walker), and *P. interpunctella*. Also, germane to this report are the evaluations of 8 trap designs and the combined usage of *P. interpunctella* and *Sitotroga cerealella* (Oliver) pheromone in those traps by Vick et al. (1979). Baiting a single trap with pheromone from both species had no adverse effect on capture of *P. interpunctella*, but the presence of *P. interpunctella* pheromone may have reduced captures of *S. cerealella*. A most recent report in this area comes from Curtis (unpublished data), who reported on the field attraction of the male navel orangeworm, *Amyelois transitella* (Walker), to its synthetic pheromone.

In the future, attractant pheromones will undoubtedly be used to monitor moth populations in warehouses. Then, when a population is detected, increased monitoring could be phased into a control program that would involve mass-trapping. Such a sequence appears to be theoretically feasible if the traps are highly efficient in removing males, and if a supplemental control program is also initiated. Supplemental control measures may consist of increased inspection of the commodity, improved sanitation, spot treatment with pesticides, and the use of approved pathogens such as *Bacillus thuringiensis* applied directly or dispersed using a pheromone-luring technique previously discussed. Combinations of this type would provide a promising new integrated control program.

SUMMARY

Biomonitoring for stored-product insects is an important part of pest management programs. With the increased availability of pheromones, it will soon be possible to initiate many new insect monitoring programs. These programs could be combined into an overall plan for integrated control involving novel insect suppression schemes that include IGRs, pathogens, and conventional pesticides.

We need to strive for a reduction in the normal growth of an insect population to a level below the economic threshold. The use of monitoring by pheromones combined with biological and/or limited chemical control may make it possible to manage an insect population in order to prevent it from reaching the threshold level that would result in a major chemical control program. I propose the term, "managed growth curve" (Fig. 1), to describe population suppression to a level below the usual economic threshold. A combination of intensive pheromone monitoring, the use of pathogens, or parasites, and predators, along with physical or limited chemical controls, may reduce the need for a major chemical treatment. Pilot studies are needed to evaluate these combinations. Also, it will be necessary to demonstrate the multiple control techniques proposed here are more efficient, economical, and safe than the present control methods.

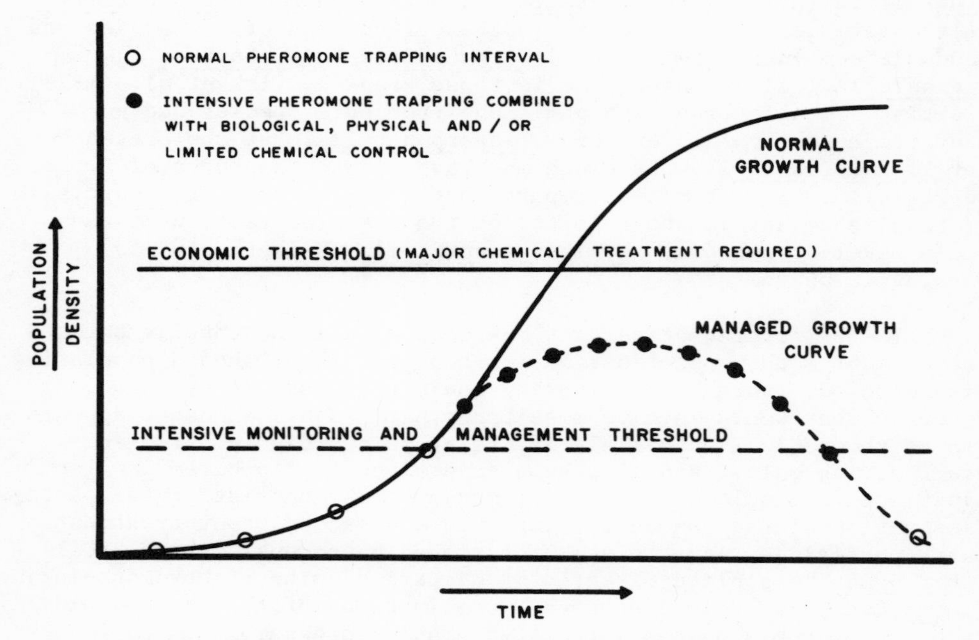

Fig. 1. Monitoring and management model for warehouse and grain storage insects

ACKNOWLEDGMENT

Research was supported by the College of Agriculture and Life Sciences, University of Wisconsin, Madison, and by a cooperative agreement between the University of Wisconsin and Agricultural Research, Science and Education Administration, U. S. Department of Agriculture.

REFERENCES

Barak, A. V., and Burkholder, W. E., 1976, Trapping studies with dermestid sex pheromones, Environ. Entomol., 5: 111.

Barak, A. V., and Burkholder, W. E., 1977a, Behavior and pheromone studies with *Attagenus elongatulus* Casey (Coleoptera: Dermestidae), J. Chem. Ecol., 3: 219.

Barak, A. V., and Burkholder, W. E., 1977b, Studies on the biology of *Attagenus elongatulus* Casey (Coleoptera: Dermestidae) and the effects of larval crowding on pupation and life cycle, *J. Stored-Prod. Res.*, 13: 169.

Brady, V. E., 1973, Isolation, identification and stimulatory activity of a second component of the sex pheromone system (complex) of the female almond moth, *Cadra cautella* (Walker), *Life Sci.*, 13: 227.

Brady, V. E., and Daley, R. C., 1972, Identification of a sex pheromone from the female raisin moth, *Cadra figulilella*, *Ann. Entomol. Soc. Am.*, 65: 1356.

Brady, V. E., and Nordlund, D. A., 1971, *Cis*-9,*trans*-12 tetradecadien-1-yl acetate in the female tobacco moth *Ephestia elutella* (Hübner) and evidence for an additional component of the sex pheromone, *Life Sci.*, 10: 797.

Brady, V. E., Nordlund, D. A., and Daley, R. C., 1971a, The sex stimulant of the Mediterranean flour moth *Anagasta kuehniella*, *J. Ga. Entomol. Soc.*, 6: 215.

Brady, V. E., Tumlinson, J. H., III, Brownlee, R. B., and Silverstein, R. M., 1971b, Sex stimulant and attractant in the Indian meal moth and in the almond moth, *Science*, 171: 802.

Burkholder, W. E., 1979, Application of Pheromones and Behavior-Modifying Techniques in Detection and Control of Stored-Product Insects, Proceedings of the Second International Working Conference on Stored-Product Entomology, Ibadan, Nigeria, Sept. 1978, pp. 56-65.

Burkholder, W. E., and Dicke, R. J., 1966, Evidence of sex pheromones in females of several species of Dermestidae, *J. Econ. Entomol.*, 59: 540.

Burkholder, W. E., Ma, M., Kuwahara, Y., and Matsumura, F., 1974, Sex pheromone of the furniture carpet beetle, *Anthrenus flavipes* (Coleoptera: Dermestidae), *Can. Entomol.*, 106: 835.

Chuman, T., Kato, K., and Noguchi, M., 1979a, Synthesis of (±)-serricornin,4,6-dimethyl-7-hydroxy-nonan-3-one, a sex pheromone of cigarette beetle (*Lasioderma serricorne* F.), *Agric. Biol. Chem.*, 43: 2005.

Chuman, T., Kohno, M., Kato, K., and Noguchi, M., 1979b, 4,6-Dimethyl-7-hydroxy-nonan-3-one, a sex pheromone of the cigarette beetle (*Lasioderma serricorne* F.), *Tetrahedron Lett.*, 25: 2361.

Coffelt, J. A., and Burkholder, W. E., 1972, Reproductive biology of the cigarette beetle, *Lasioderma serricorne*. 1. Quantitative laboratory bioassay of the female sex pheromone from females of different ages, *Ann. Entomol. Soc. Am.*, 65: 447.

Coffelt, J. A., Vick, K. W., Sower, L. L., and McClellan, W. T., 1979a, Sex pheromone mediated behavior of the navel orangeworm, *Amyelois transitella*, *Environ. Entomol.*, 8: 587.

Coffelt, J. A., Vick, K. W., Sonnet, P. E., and Doolittle, R. E., 1979b, Isolation, identification, and synthesis of a female sex pheromone of the navel orangeworm, Amyelois transitella (Lepidoptera: Pyralidae), J. Chem. Ecol., 5: 955.

Cross, J. H., Byler, R. C., Cassidy, R. F., Jr., Silverstein, R. M., Greenblatt, R. E., Burkholder, W. E., Levinson, A. R., and Levinson, H. Z., 1976, Porapak-Q collection of pheromone components and isolation of (Z)- and (E)-14-methyl-8-hexadecenal, potent sex attracting components, from females of four species of Trogoderma (Coleoptera: Dermestidae), J. Chem. Ecol., 2: 457.

Fukui, H., Matsumura, F., Barak, A. V., and Burkholder, W. E., 1977, Isolation and identification of a major sex-attracting component of Attagenus elongatulus (Casey) (Coleoptera: Dermestidae), J. Chem. Ecol., 3: 539.

Fukui, H., Matsumura, F., Ma, M. C., and Burkholder, W. E., 1974, Identification of the sex pheromone of the furniture carpet beetle, Anthrenus flavipes LeConte, Tetrahedron Lett. 40: 3563.

Hope, J. A., Horler, D. F., and Rowlands, D. G., 1967, A possible pheromone of the bruchid, Acanthoscelides obtectus (Say), J. Stored-Prod. Res., 3: 387.

Hoppe, T., 1974, Effect of a juvenile hormone analogue on Mediterranean flour moth in stored grains, J. Econ. Entomol., 67: 789.

Horler, D. F., 1970, An allenic ester produced by the male dried bean beetle, Acanthoscelides obtectus (Say), J. Chem. Soc., (c): 859.

Keys, R. E., and Mills, R. B., 1968. Demonstration and extraction of a sex attractant from female Angoumois grain moths, J. Econ. Entomol., 61: 46.

Khorramshahi, A., and Burkholder, W. E., 1980, Behavior of the lesser grain borer Rhyzopertha dominica (Coleoptera: Bostrichidae): Male-produced aggregation pheromone attracts both sexes, J. Chem. Ecol. (In press).

Kuwahara, Y., and Casida, J. E., 1973, Quantitative analysis of the sex pheromone of several physitid moths by electron-capture gas chromatography, Agric. Biol. Chem., 37: 681.

Kuwahara, Y., Kitamura, C., Takahashi, S., Hara, H., Ishii, S., and Fukami, H., 1971a, Sex pheromone of the almond moth and the Indian meal moth: Cis-9,trans-12, tetradecadienyl acetate, Science, 171: 801.

Kuwahara, Y., Hara, H., Ishii, S., and Fukami, H., 1971b, The sex pheromone of the Mediterranean flour moth, Agric. Biol. Chem., 35: 447.

Kuwahara, Y., Fukami, H., Ishii, S., Matsumura, F., and Burkholder, W. E., 1975, Studies on the isolation and bioassay of the sex pheromone of the drugstore beetle Stegobium paniceum (Coleoptera: Anobiidae), J. Chem. Ecol., 1: 413.

Kuwahara, Y., Fukami, H., Howard, R., Ishii, S., Matsumura, F., and Burkholder, W. E., 1978, Chemical studies on the Anobiidae: Sex pheromone of the drugstore beetle, Stegobium paniceum (L.) (Coleoptera), Tetrahedron Lett., 34: 1769.

Levinson, H. Z., and Barilan, A. R., 1967, Function and properties of an assembling scent in the khapra beetle Trogoderma granarium, Riv. Parassitol. 28: 27.

Levinson, H. Z., and Levinson, A. R., 1979, Trapping of storage insects by sex and food attractants as a tool of integrated control, in: "Chemical Ecology: Odour Communication in Animals," F. J. Ritter, ed., Elsevier/North Holland Biomedical Press, Amsterdam.

Loschiavo, S. R., and Atkinson, J. M., 1967, A trap for the detection and recovery of insects in stored grain, Can. Entomol., 99: 1160.

McGregor, H. E., and Kramer, K. J., 1975, Activity of insect growth regulators, hydropene and methoprene, on wheat and corn against several stored grain insects, J. Econ. Entomol., 68: 668.

Nickle, D. A., 1979, Insect growth regulators: New protectants against the almond moth in stored inshell peanuts, J. Econ. Entomol., 72: 816.

Qayyum, H. A., 1979, Insect Pests on Stored Cereal Grains and Their Control in Pakistan, Research Report, April 30, 1979, Dept. of Entomol., University of Agriculture, Faisalabad, 40 pp.

Reichmuth, C. von, Wohlgemuth, R., Levinson, A. R., and Levinson, H. Z., 1976, Untersuchungen uber den Einsatz von pheromonbekoderten Klebefallen zur Bekampfung von Motten in Vorratsschutz, Z. Angew. Entomol., 82: 95.

Reichmuth, C. von, Schmidt, H. V., Levinson, A. R., and Levinson, H. Z., 1978, Die Fangigkeit Pheromonbekoderter Klebefallen fur Speichermotten (Ephestia elutella Hbn.) in unterschiedlich dicht befallenen Getreidelagern, Z. Angew. Entomol., 86: 205.

Rodin, J. O., Silverstein, R. M., Burkholder, W. E., and Gorman, J. E., 1969, Sex attractant of female dermestid beetle Trogoderma inclusum LeConte, Science, 165: 904.

Shapas, T. J., and Burkholder, W. E., 1978, Patterns of sex pheromone release from adult females, and effects of air velocity and pheromone release rates on theoretical communication distances, J. Chem. Ecol., 4: 395.

Shapas, T. J., Burkholder, W. E., and Boush, G. M., 1977, Population suppression of Trogoderma glabrum by using pheromone luring for protozoan pathogen dissemination, J. Econ. Entomol., 70: 469.

Silverstein, R. M., Rodin, J. O., Burkholder, W. E., and Gorman, J. E., 1967, Sex attractant of the black carpet beetle, Science, 157: 85.

Silverstein, R. M., Cassidy, R. F., Burkholder, W. E., Shapas, T. J., Levinson, H. Z., Levinson, A. R., and Mori, K., 1980, Perception by Trogoderma species of chirality and methyl branching at a site far removed from a functional group in a pheromone component, J. Chem. Ecol. (In press).

Strong, R. G., and Diekman, J., 1973, Comparative effectiveness of fifteen insect growth regulators against several pests of stored products, J. Econ. Entomol., 66: 1167.

Vick, K. W., Su, H. C. F., Sower, L. L., Mahany, P. G., and Drummond, P. C., 1974, (Z,E)-7,11-Hexadecadien-1-ol acetate: The sex pheromone of the Angoumois grain moth, Sitotroga cerealella, Experientia, 30: 17.

Vick, K. W., Kvenberg, J., Coffelt, J. A., and Steward, C., 1979, Investigation of sex pheromone traps for simultaneous detection of Indian meal moths and Angoumois grain moths, J. Econ. Entomol., 72: 245.

Williams, H. J., Silverstein, R. M., Burkholder, W. E., and Khorramshahi, A., 1980, Dominicalure 1 and 2, the components of the aggregation pheromone from the male lesser grain borer, Rhyzopertha dominica (F.) (Coleoptera: Bostrichidae), J. Chem. Ecol., (In press).

Yarger, R. G., Silverstein, R. M., and Burkholder, W. E., 1975, Sex pheromone of the female dermestid beetle Trogoderma glabrum (Herbst), J. Chem. Ecol., 1: 323.

THE USE OF PHEROMONE TRAPS TO MONITOR DISTRIBUTION AND POPULATION TRENDS OF THE GYPSY MOTH

Joseph S. Elkinton

Department of Entomology
University of Massachusetts
Amherst, MA 01003 USA

Ring T. Cardé

Department of Entomology and
Pesticide Research Center
Michigan State University
East Lansing, MI 48824 USA

INTRODUCTION

Goals of a Gypsy Moth Pheromone-Trapping Program

Traps baited with synthetic pheromones are in widespread use in the survey and detection of the gypsy moth, *Lymantria dispar* (L.) and other important insect pests in many parts of the world. In the United States approximately 95,000 pheromone traps were deployed to monitor the gypsy moth in 1979. Pheromone traps offer several advantages over other methods of sampling populations. First, such traps are relatively inexpensive and easy to deploy. Gypsy moth traps can be set in place early in the summer and left untended until early fall. Second, pheromone traps generally catch only the target species. The time-consuming sorting to species required for other kinds of traps is avoided. Third, pheromone traps are effective at extremely low population densities. Other sampling methods for the gypsy moth, such as egg mass surveys or collecting pupae in burlap bands wrapped around tree trunks, are impractical when density is low or the area to be surveyed is large.

The use of pheromone traps for the gypsy moth differs from that employed for other major insect pests. In most agricultural crops pheromone traps are used primarily to monitor the phenological timing of adult activity so that control efforts can be applied at the most effective moment (e.g., Welch et al., 1978). The single flight of the male gypsy moth occurs several weeks after the larval stage, which would be the target of most insecticide applications. Detection of the flight of male gypsy moths, therefore, cannot be used to time the control activities.

The primary goal of the gypsy moth pheromone-trapping program has been the detection of population centers in areas where population densities are sparse. Our ability to locate population centers from the distribution of male captures in a grid of pheromone traps is aided by the fact that males generally do not disperse very far (Schwalbe, 1980; Elkinton and Cardé, 1980).

A second goal of gypsy moth pheromone-trapping systems is to monitor population trends and to estimate density. The number of males captured in traps has been correlated with moderate to dense pupal populations (over 200/ha) (Granett, 1974). Mark-recapture studies have shown that, depending on the weather, consistent proportions of males are captured in trapping grids of a given density (Elkinton and Cardé, 1980). This fact permits a straightforward estimate of adult male density from the numbers captured in traps. When traps are used to monitor population trends, they can be utilized to evaluate the effectiveness of various control procedures.

HISTORY OF GYPSY MOTH PHEROMONE-TRAPPING

The use of pheromone traps for the gypsy moth began in the 1890's during the first years of outbreak following the accidental release of the gypsy moth in Medford, MA in 1869. C. H. Fernald attracted males to experimental traps baited with caged females (Forbush and Fernald, 1896). A barrier zone to prevent further spread of the gypsy moth beyond New England was created in 1922 from the Canadian border south along the Hudson River Valley (Burgess, 1940). Collins and Potts (1932) demonstrated that benzene or xylene extracts of female abdominal tips could be used instead of live females to attract males to traps. This discovery made feasible the subsequent widespread use of pheromone traps to detect incipient populations beyond the barrier zone. By 1948, thousands of traps baited with female tip extracts were deployed in New England and the Mid-Atlantic States (reviewed in Schwalbe, 1980). After World War II approximately 30,000 such traps were deployed each year in conjunction with the widespread application of DDT in an effort to halt the spread of this insect (Holbrook et al., 1960).

Observations that abdominal extracts had short-lived activity led to efforts to stabilize the extracts chemically (Haller et al., 1944) and to isolate the active ingredients (Acree, 1953). Jacobson et al. in 1960 proposed d-10-acetoxy-*cis*-7-hexadecen-1-ol (gyptol) as the active component. Gyplure, an 18-carbon analogue of gyptol, was also reported to be highly attractive (Jacobson et al., 1960; Jacobson and Jones, 1962). The widespread pheromone-trapping program initiated by the USDA in the 1960's utilized gyplure which was readily synthesized in large quantities. It was soon apparent, however, that gyplure did not attract male gypsy moths (Doane, 1961). Reinvestigation (Jacobson et al., 1970) later confirmed that neither gyptol nor gyplure was attractive and these authors suggested that the supposed activity of the earlier preparations was due to an unidentified contaminant. Definitive field tests demonstrating any attractancy in these lures have never been published.

Consequently, the USDA returned to the use of female abdominal tip extracts in the trapping program and reinitiated the effort to isolate and identify the sex pheromone. This research culminated in the identification of disparlure, <u>cis</u>-7,8-epoxy-2-methyloctadecane, as the sex pheromone of the gypsy moth (Bierl et al., 1970). All subsequently deployed gypsy moth pheromone traps were baited with this material.

Disparlure is a chiral compound and its (+) enantiomer has been found to elicit trap catch 10-fold above the racemate (Cardé et al., 1977b; Plimmer et al., 1977). Several synthetic pathways for (+) di disparlure have been reported (Iwaki et al., 1974; Mori et al., 1976; Farnum et al., 1977; Pirkle and Rinaldi, 1979), but the material remains expensive. In current trapping for survey and detection of the gypsy moth, (+) disparlure is gradually replacing (±) disparlure as the former becomes available in sufficient quantities.

Efforts to design an efficient and low cost gypsy moth trap (reviewed in Holbrook et al., 1960; Schwalbe, 1980) proceeded apace with the efforts to identify the sex attractant. Currently, the USDA and various state agencies use a triangular-shaped "delta" trap coated on the inside with adhesive and stapled to a tree. Such traps have a very limited capacity. As moths become stuck in the sticky surface, its adhesive qualities decline and subsequent moths can escape more easily. Therefore, such traps cannot be used to monitor population trends, except perhaps at the lowest densities. Consequently, efforts have been made to develop large capacity, non-sticky traps (Granett, 1973; Cardé et al., 1977a).

The factors that affect trap efficiency are complex (Cardé, 1979). The size of the pheromone active space downwind of the trap and the concentrations of pheromone near the trap are determined by the rate of pheromone release and the speed and turbulence of the wind. The shape and concentration of the pheromone plume near the trap is also affected by the shape of the trap (Lewis and Macauley, 1976). Unless the trap is radially symmetrical, the trap efficiency may vary with the wind direction. The visual image presented by the trap and its surroundings, the proximity of trees or underbrush and the height of trap placement (Granett, 1974; Cardé et al., 1977a) are all factors of importance.

The emission characteristics of the pheromone dispenser are also important. An ideal dispenser would release pheromone at an optimal rate, independent of temperature, over long periods. The release rate of many dispensers such as cotton wicks, however, declines exponentially with time. The laminated plastic Heron® dispenser currently used in gypsy moth survey traps releases attractant at a relatively constant rate over the whole summer flight period.

Although efforts to develop more efficient traps are important, it is also essential that the same traps be used consistently from year-to-year (Minks, 1977). Otherwise, between-year comparisons of the numbers caught in traps will have no meaning in terms of population trends. Year-to-year comparisons also require a consistent methodology and density of trap placement. Current USDA recommendations are summarized by Schwalbe (1979). Since 1972, most of the USA (31-40 states) where susceptible forests exist has been surveyed systematically for gypsy moths with pheromone-baited traps. New infestations have been identified in North Carolina, Florida, Ohio, Wisconsin, California, Michigan, Illinois and Virginia (Schwalbe, 1980).

ESTIMATION OF DENSITY AND DISTRIBUTION FROM TRAP CATCH

Estimates of Dispersal

Efforts to pinpoint population centers are aided by the fact that the adult male does not disperse very far. Mark-recapture studies by Schwalbe (1980) and Elkinton and Cardé (1980) have shown that $<5\%$ of released males disperse further than 800 m in a grid of pheromone traps with a density of 1 trap/0.65 km^2 or 4 traps/mi^2 These findings appear consistent with earlier studies by Holbrook et al. (1960). At higher trap densities the average distance of dispersal before capture is much lower. Consequently, our ability to delineate population centers depends upon trap density. A large-scale effort conducted in Michigan in 1979 used 1 trap/0.08 km^2 or 32 traps/mi^2 over several counties. From the number and distribution of males caught in the grid of traps, the location of population centers can be deduced. At any one trap, the presence of males (especially just one male) may or may not indicate the presence of a breeding population. Males may be carried far from the breeding site on human transportation. While the vast majority of adult males appear to disperse at most a few hundred meters (Schwalbe, 1980; Elkinton and Cardé, 1980), some may disperse much further under particular environmental or behavioral conditions. A single male caught in a trap may indicate a small nearby infestation or it may represent a "rogue" individual that dispersed as a first instar larva or an adult many miles from a large infestation. Nevertheless, the pattern of trap catch in the grid as a whole should reveal the location of population centers.

Estimates of Density

In addition to detection and delimitation of gypsy moth populations, pheromone traps have great potential as a tool for monitoring population trends and estimating population density. Several workers have demonstrated the correlation of gypsy moth pheromone trap catch (using the less effective racemic disparlure) with the other methods of sampling the population. Granett (1974) found, in relatively dense populations, a good correlation between male trap catch and the number of pupae collected beneath burlap bands placed around oak trees on the study site. Maksimović (1958, 1965) used the correlation between male trap catch and the number of egg masses to identify those areas in Yugoslavia where egg masses had reached a "critical density" indicative of an incipient population outbreak.

Mark-recapture techniques offer another powerful method for relating the numbers caught in traps to population density. These techniques are widely used in estimating densities of fish and other vertebrates but are rare in studies of insect populations, primarily because insects are very short-lived and the proportions recaptured are usually very low. However, the high recapture efficiencies of pheromone traps make mark-recapture studies with gypsy moths and other insects feasible. In most studies it is necessary to release laboratory-reared animals because a sufficient number of wild organisms would not be available. A key assumption in such a substitution is that both laboratory and feral organisms have similar dispersal and pheromone response characteristics (Elkinton and Cardé, 1980). This hypothesis can be tested most conveniently by simultaneous releases of laboratory-reared and field-collected populations.

Most population estimates derived from mark-recapture studies are derivations of the so-called Petersen estimate (Petersen, 1896). This estimate asserts that, after the release of marked individuals into a population, the ratio of marked and ummarked individuals in the population as a whole is equal to the ratio of marked to unmarked individuals in samples taken from that population. Since the number of marked individuals released into the population is known, we can then estimate the total population size if we assume no birth or immigration. More recent theoretical treatments (Bailey, 1951; Leslie, 1952; Darroch, 1959; Seber, 1962; Jolly, 1963, 1965; Manly and Parr, 1968) elaborate upon the Petersen estimate by relaxing certain assumptions such as no birth or immigration.

In addition, they estimate additional factors such as the rate of survival. Most of these treatments involve multiple releases and recaptures. With pheromone traps, however, moths cannot be released a second time following capture. Even if they are captured alive, continuous exposure to the pheromone in the traps may radically and permanently alter their subsequent behavior and thus invalidate any estimate based upon second- and third-time recaptures. Several theoretical treatments, however, are based upon sequential releases of differently marked groups and single recaptures (Seber, 1962; Jackson, 1937; Ricker, 1975). These techniques can be utilized in mark-capture studies with pheromone traps.

Various studies have utilized the recapture of marked male gypsy moths not to estimate population parameters but to examine some aspect of male behavior or trap efficiency (e.g., Richerson and Cameron, 1974; Cardé and Webster, 1979). Schwalbe (1980), however, reported a series of experiments in which laboratory and field-collected males were released from a point at the center of a grid of pheromone traps. At a density of 1 trap/0.65 km^2 or 4 traps/mi^2, he discovered that approximately 1% of such moths were captured. The great majority of captures occurred at the nearest traps which were within 400 m of the release site. In further studies, Schwalbe (1980) compared the capture of males at 3 different trap densities greater than the densities used in the 1st study. At the greatest trap density (traps separated by 88 m), the total proportion of moths captured was highest and the average distance of dispersal before catpure was lowest. At all densities, trap catch was highest near the point of release and dropped off precipitously further away.

In similar studies, Elkinton and Cardé (1980) demonstrated how one might extrapolate such mark-recapture results to the captures of wild males in a survey situation to obtain estimates of population size and the rate of adult male survival. As in Schwalbe's first experiment, males were released at the center of a 65 km^2 (25 mi^2) grid of pheromone traps at a density of 1 trap/0.65 km^2. In contrast to Schwalbe's experiment, however, 2 different release strategies were used. One group of males was released from a single point at the center of the grid. Simultaneously, a 2nd group of males was released from 100 points spaced evenly across an 800 x 800 m area at the center of the grid of traps. As in Schwalbe's study, an average of 0.9% of the males released from the single point at the center of the grid was captured. Of the males released uniformly across the 800 x 800 m area, 3.9% were captured (Figure 1). In nature, males will originate from many points in the area surrounding the nearest pheromone traps as approximated by the uniform release strategy. The dispersion of males in nature is undoubtedly not uniform. At any given site, males may be clumped nearer or farther away than average from the nearest pheromone traps. However, over the trapping grid as a whole, the average distances between a given moth and the nearest traps will be the same for a uniform, clumped or random distribution provided there is no systematic relationship between the location of traps and the location of population centers. Consequently, one might expect the 3.9% recapture to approximate the average recapture rate of males in nature provided one can account for several factors discussed below that will cause trap catch to vary. In contrast, the single point release at grid center represents the maximal possible clumping of males away from the nearest traps. Consequently, one might expect the 0.9% recapture of males from this release strategy to approximate a lower limit of the proportion of wild males captured in a survey grid of this trap density.

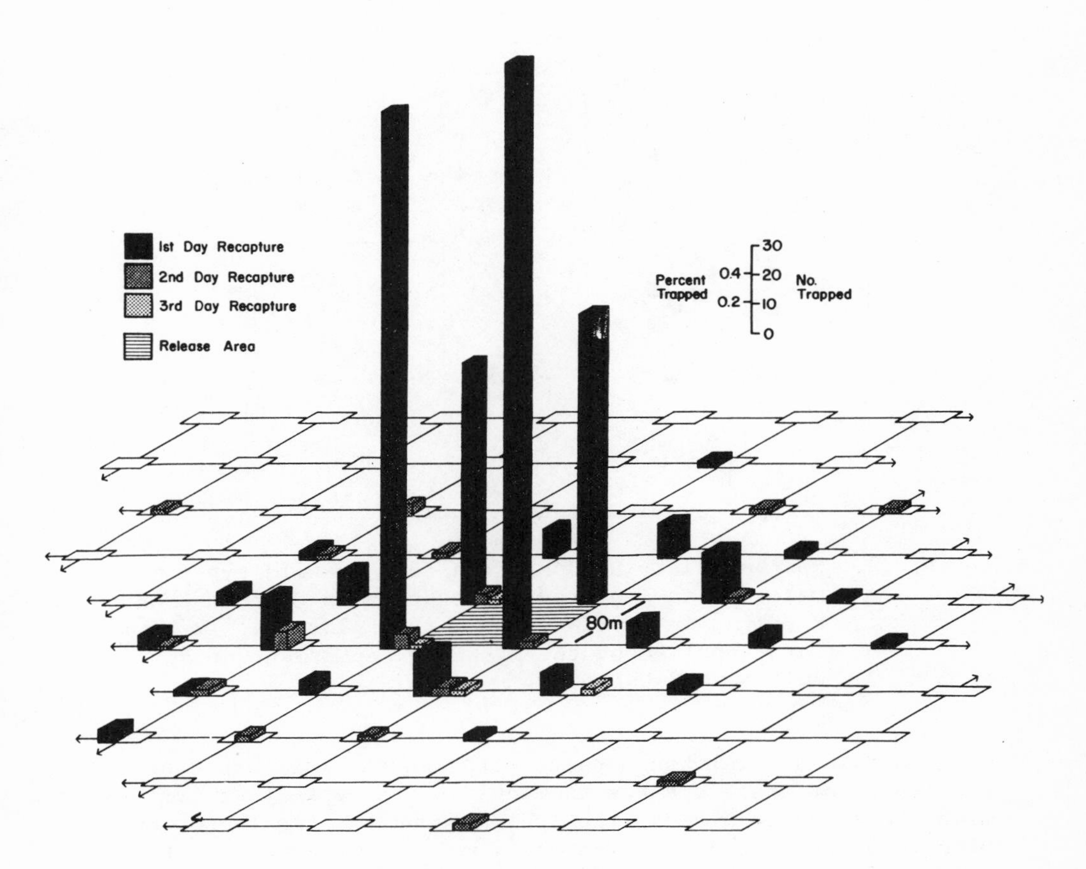

Fig. 1 Spatial distribution of capture of male, laboratory-reared gypsy moths. Marked males were released from a single point in the grid center and uniformly from 100 points in the 800 x 800 m center of a 8 x 8 km grid of pheromone traps spaced 800 m apart.

These estimates cannot be extrapolated to the survey situation with natural populations until we understand how various factors affect trap catch. First of all, these estimates would only apply to those survey situations where trap capacity is not a limiting factor. The delta trap, now widely used in gypsy moth survey, has an upper limit of capacity of ca. 25-40 males, and the efficiency of capture undoubtedly declines as the trap fills. When actual grid captures in survey situations using these are examined (Figures 2 and 3), the relevance of trap capacity to distribution estimates

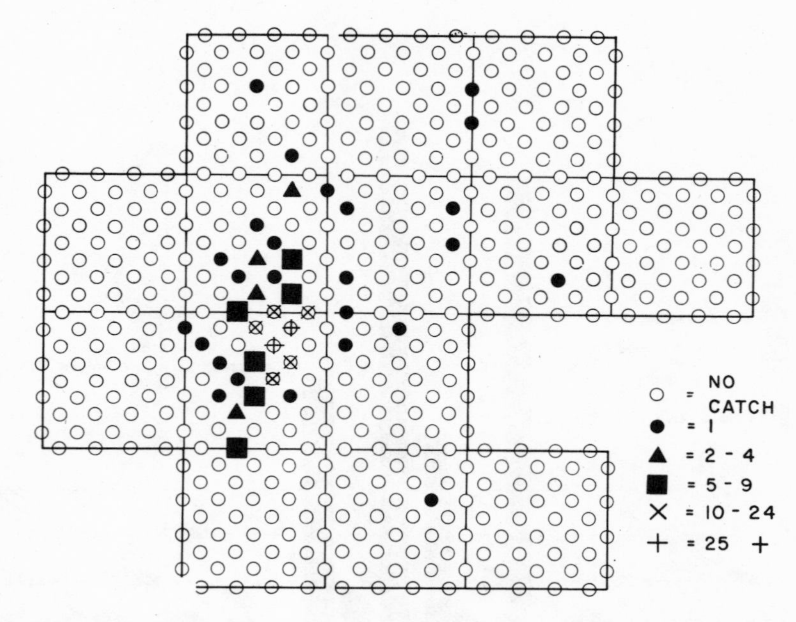

Fig. 2 Spatial distribution of capture of wild gypsy moth males in an isolated infestation in Calhoun County Michigan in 1979. Traps were deployed at 12/km^2. Data supplied by the Michigan Department of Agriculture and USDA-PPQ.

becomes evident. Although there were substantial trap captures in the center of these 2 sites in Michigan, the potential numbers that might have been captured with a high capacity trap cannot be determined. If the actual population of males in the center of this array was substantially higher than suggested by the numbers captured in delta traps, then it is likely that the low numbers/trap caught on the edge represent individuals dispersing from the central area. Conversely, if we accept the numbers caught as being linearly related to actual density, then the distribution of the population is likely to be less clumped and to extend at least to the boundary of trap catches. Interpreting distribution in such situations using data from the limited capacity traps cannot be simply resolved. Thus, decisions as to precisely where insecticide applications should be directed if the entire breeding population is to be eliminated are problematic when limited capacity traps are employed.

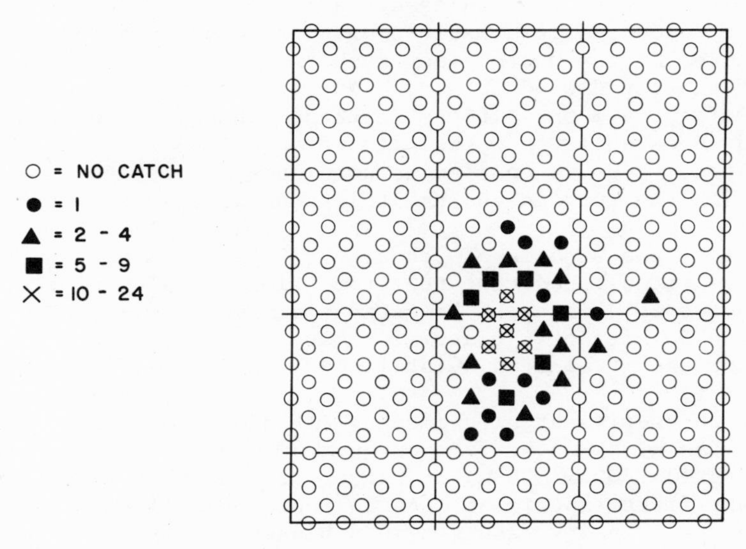

Fig. 3 Spatial distribution of capture of wild gypsy moth males in an isolated infestation in Oakland County Michigan in 1979. Traps were deployed at $12/km^2$. Data supplied by the Michigan Department of Agriculture and USDA-PPQ.

Even with traps of unlimited capacity, we do not know if trap catch is a linear function of population density. The mark-recapture studies cited above were conducted in the absence of a natural population of gypsy moths. There is evidence with several other Lepidoptera (reviewed in Cardé, 1979) that the presence of females in the population producing pheromone in "competition" with the traps will reduce the proportion of males captured as population increases. In other studies (Baker et al., 1980) there was no evidence for such a "competition effect." Theoretical models to describe the competition effect (Knipling and McGuire, 1966; Howell, 1974; Nakamura and Oyama, 1978) suffer from the lack of adequate quantitative evidence. Even if the proportion of males captured declines as population increases, it may still be possible to arrive at an estimate of density by successive approximation to the expected proportion recovered for a particular density.

Weather is another important factor that affects trap catch. Elkinton and Cardé (1980) showed that the proportion of male gypsy moths captured increased with increasing temperature. The effect of other variables such as humidity or rainfall is unknown. Any population estimates based upon trap catch may have to be adjusted to the prevailing weather conditions.

The mark-recapture studies cited above (Elkinton and Cardé, 1980) were conducted in a habitat of continuous woodlands. In much of the United States where gypsy moth traps are deployed, the habitat is one of scattered wood lots. We do not know how the dispersal behavior of gypsy moths differs in these habitats. Further study is needed to establish how the proportion of males captured changes with habitat. Even in continuous woodlands, trap catch is likely to vary with the terrain and the thickness of the vegetation.

Finally, the studies of Schwalbe (1980) and Elkinton and Cardé (1980) have established that the proportion of males captured is a function of trap density. The estimates given above are appropriate only to the density of 4 traps/mi^2. Separate estimates must be derived for each trap density utilized in survey situations.

In a 2nd series of experiments, Elkinton and Cardé (1980) released male gypsy moths from an 80 x 80 m area at the center of a smaller grid (800 x 800 m) of higher trap density (400 traps/mi^2) than in the first experiment. As in Schwalbe's (1980) study, the increased trap density resulted in a higher proportion of males recaptured (an average of 18%) and a shorter average distance of dispersal before recapture. As in the earlier studies, most moths were caught in the trap immediately surrounding the site of release (Figure 4). This trend was especially apparent for those moths captured on the same day after release. Capture on the 2nd- and 3rd-day after release was also concentrated at grid center, but the effect was not as pronounced as on the 1st day.

A major purpose of this 2nd series of experiments was to examine the effects of moth age on the probability of capture and to estimate the moths' daily rate of survival. Moths were held outdoors in emergence cages for 1, 2 or 3 days following eclosion. Each day the 3 age groups were marked a different color and released. The captures of each group were monitored daily for several subsequent days. The average capture rate of 3-day-old males was 24.8% compared to 15.9% and 13.7%, respectively, for 2- or 1-day-old males. The capture rates of all ages were high on the 1st day after release and then declined precipitously on subsequent days, indicative of a high rate of mortality and/or emigration. The daily rate of mortality plus emigration was estimated to be in excess of 95% according to the method of Ricker (1975). However, this estimate assumes that on any given day all moths of the same age had an equal probability of being caught in traps. A large degree of pheromone response variability between males may have elevated this estimate. Most of the "good responders" may have been caught on the 1st day, leaving a pool of "poor responders," only a few of which would be caught on subsequent days.

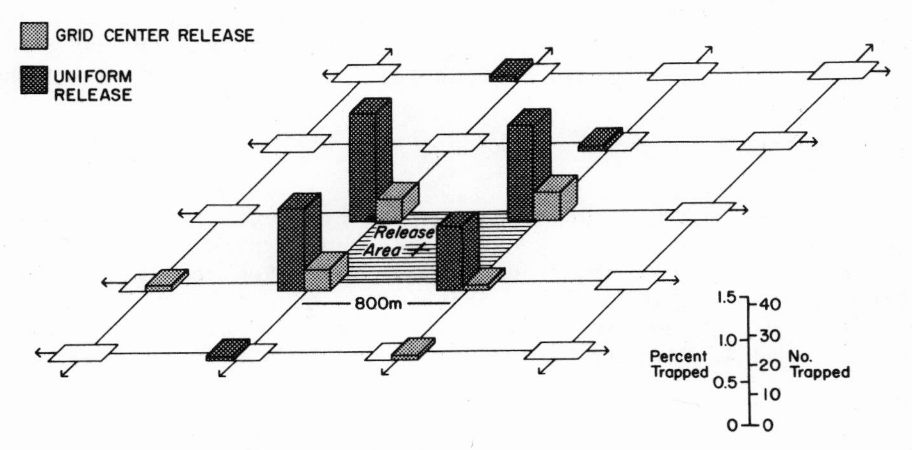

Fig. 4 Daily spatial distribution of captured male, laboratory-reared gypsy moths. Marked males were released from 25 points in the 80 x 80 m center of a 800 x 800 m grid of pheromone traps spaced 80 m apart.

It is clear from the foregoing discussion that the factors that affect trap catch are complex. Simulation models may offer a useful approach in understanding the complex interactions of these factors. Elkinton and Cardé (unpublished data) have developed a simulation model that predicts trap catch in a given area with given trap and population densities and given weather conditions. The model incorporates current knowledge about male dispersal flight, the effect of wind velocity and turbulence on the size and shape of a pheromone plume and the probability of males inside the plume locating the pheromone source from various distances downwind. The model also includes what we know of the diel rhythms of gypsy moth eclosion (ODell, 1978) and female pheromone emission (Richerson and Cameron, 1974). At present, our understanding of all these components is meager. Consequently, we cannot expect the model in its present form to accurately predict the trap catch measured in the field. However, the construction of such a model aids in the conceptual process of putting together the many interactions that affect trap catch and in suggesting the appropriate experiments to unravel these factors.

CONCLUSION

In this essay we have traced the history of pheromone survey trapping for the gypsy moth in the United States. Such traps are primarily used for detection purposes in areas where populations are sparse. However, recent studies have provided the experimental groundwork for the use of pheromone traps for monitoring population trends and for estimating population density and dispersion. Pheromone traps may thus play a more important role in the future guidance and evaluation of management or control activities directed at the gypsy moth.

ACKNOWLEDGMENT

This research was supported in part by grants from the Michigan Department of Agriculture, a USDA-sponsored "Expanded Gypsy Moth Research and Applications Program," the NE-84 Regional Project and NSF Grant BCM-7912014. We are grateful to Dean Lovitt of the Michigan Department of Agriculture and Richard Moore of USDA-PPQ for supplying the data used in Figures 2 and 3.

REFERENCES

Acree, F., Jr., 1953. The isolation of gyptol, the sex attractant of the female gypsy moth, J. Econ. Entomol., 46: 313.

Bailey, N. T. J., 1951, On estimating the size of mobile populations from recapture data, Biometrika, 38: 293.

Baker, T. C., Cardé, R. T., and Croft, B. A., 1980, Relationship between pheromone trap capture and emergence of adult Oriental fruit moths, Grapholitha molesta (Lepidoptera: Tortricidae), Can. Entomol., 112: 11.

Bierl, B. A., Beroza, M., and Collier, C. W., 1970, Potent sex attractant of the gypsy moth, Porthetria dispar (L): Its isolation, identification, and synthesis, Science, 170: 87.

Burgess, A. F., 1940, The value to uninfested states of gypsy moth control and extermination, J. Econ. Entomol., 33: 558.

Cardé, R. T., 1979, Behavioral responses of moths to female-produced pheromones and the utilization of attractant-baited traps for population monitoring, in: "Movement of Highly Mobile Insects: Concepts and Methodology in Research," R. L. Rabb and G. G. Kennedy, eds., North Carolina State University Press, Raleigh.

Cardé, R. T., Doane, C. C., Granett, J., Hill, A. S., Kochansky, J., and Roelofs, W. L., 1977a, Attractancy of racemic disparlure and certain analogues to male gypsy moths and the effect of trap placement, Environ. Entomol., 6: 765.

Cardé, R. T., Doane, C. C., Baker, T. C., Iwaki, S., and Marumo, S., 1977b, Attractancy of optically active pheromone for male gypsy moths, Environ. Entomol., 6: 768.

Cardé, R. T., and Webster, R., 1979, Variation in attraction of individual gypsy moths to (+) and (-)-disparlure, J. Chem. Ecol., 5: 935.

Collins, C. W., and Potts, S. F., 1932, Attractants for the flying gipsy [sic] moth as an aid in locating new infestations, USDA Tech. Bull., 336.

Darroch, J. N., 1959, The multiple-capture census. II. Estimation where there is immigration or death, Biometrika, 46: 336.

Doane, C. C., 1961, Field tests with gyplure, 1961, Conn. Agric. Exp. Stn. Prog. Rept., 1, 3 pp.

Elkinton, J. S., and Cardé, R. T., 1980, Distribution, dispersal and apparent survival of male gypsy moths as determined by capture in pheromone-baited traps, Environ. Entomol., 9: (In press).

Farnum, D. G., Veysoglu, T., Cardé, A. M., Duhl-Emsweiler, B., Pancoast, T. A., Reitz, T. J., and Cardé, R. T., 1977, A stereospecific synthesis of (+)-disparlure, sex attractant of the gypsy moth, Tetrahedron Lett., 46: 4009.

Forbush, E. H., and Fernald, C. H., 1896, "The Gypsy Moth," Wright & Potter Printing Co., Boston.

Granett, J., 1973, A disparlure-baited box trap for capturing large numbers of gypsy moths, J. Econ. Entomol., 66: 359.

Granett, J., 1974, Estimation of male mating potential of gypsy moths with disparlure baited traps, Environ. Entomol., 3: 383.

Haller, H. L., Acree, F., Jr., and Potts, S. F., 1944, The nature of the sex attractant of the female gypsy moth, J. Am. Chem. Soc., 66: 1659.

Holbrook, R. F., Beroza, M., and Burgess, E. D., 1960, Gypsy moth (Porthetria dispar) detection with the natural female sex lure, J. Econ. Entomol., 53: 751.

Howell, J. F., 1974, The competitive effect of field populations of codling moth on sex attractant trap efficiency, Environ. Entomol. 3: 803.

Iwaki, S., Marumo, S., Saito, T., Yamada, M., and Katagiri, K., 1974, Synthesis and activity of optically active disparlure, J. Am. Chem. Soc., 96: 7842.

Jackson, C. H. N., 1937, Some new methods in the study of Glossina morsitans, Proc. Zool. Soc. London, 1936, 811.

Jacobson, M., and Jones, W. A., 1962, Insect sex attractants. II. The synthesis of a highly potent gypsy moth sex attractant and some related compounds, J. Org. Chem., 27: 2523.

Jacobson, M., Beroza, M., and Jones, W. A., 1960, Isolation, identification, and synthesis of the gypsy moth sex attractant, Science, 132: 1011.

Jacobson, M., Schwarz, M., and Waters, R. M., 1970, Gypsy moth sex attractants: A reinvestigation, J. Econ. Entomol., 63: 943.

Jolly, G. M., 1963, Estimates of population parameters from multiple recapture data with both death and dilution-deterministic model, Biometrika, 50: 113.

Jolly, G. M., 1965, Explicit estimates from capture-recapture data with both death and immigration-stochastic model, Biometrika, 52: 225.
Knipling, E. F., and McGuire, J. U., 1966, Population models to test theoretical effects of sex attractants used for insect control, USDA Agric. Info. Bull. 308.
Leslie, P. H., 1952, The estimation of population parameters from data obtained by means of the capture-recapture method. II. The estimation of total numbers, Biometrika, 39: 363.
Lewis, T., and Macaulay, E. D. M., 1976, Design and evaluation of sex-attractant traps for pea moth Cydia nigricana (Steph.) and the effect of plume shape on catches, Ecol. Entomol., 1: 175.
Maksimović, M., 1958, A contribution to the investigation of the numerousness of the gypsy moth by means of the trap method [in Serbo-Croation, English summary], Zast. Bilja, 49/50: 41.
Maksimović, M., 1965, Sex attractant traps with female odor of the gypsy moth used for forecasting the increase of population of gypsy moth, Proc. 12th Int. Congr. Entomol. 1964, 398.
Manly, B. F. J., and Parr, M. J., 1968, A new method of estimating population size, survivorship, and birth-rate from capture-recapture data, Trans. Soc. Brit. Entomol., 18: 81.
Minks, A. K., 1977, Trapping with behavior-modifying chemicals: Feasibility and limitations, in: "Chemical Control of Insect Behavior," H. H. Shorey and J. J. McKelvey, eds., Wiley-Interscience, New York.
Mori, K., Takigawa, T., and Matsui, M., 1976, Stereoselective synthesis of optically active disparlure, the pheromone of the gypsy moth (Porthetria dispar L.), Tetrahedron Lett., 44: 3953.
Nakamura, K., and Oyama, M., 1978, An equation for the competition between pheromone traps and adult females for adult males, Appl. Entomol. Zool., 13: 176.
ODell, T. M., 1978, Periodicity of eclosion and pre-mating behavior of gypsy moth, Ann. Entomol. Soc. Am., 71: 748.
Petersen, C. G. J., 1896, The yearly immigration of young plaice into Limfjord from the German sea, etc., Rept. Danish Biol. Stn., 6:1.
Pirkle, W. H., and Rinaldi, P. L., 1979, Synthesis and enantiomeric purity determination of the optically active epoxide disparlure, sex pheromone of the gypsy moth, J. Org. Chem., 44: 1025.
Plimmer, J. R., Schwalbe, C. P., Paszek, E. C., Bierl, B. A., Webb, R. E., Marumo, S., and Iwaki, S., 1977, Contrasting effectiveness of (+) and (-) enantiomers of disparlure for trapping native populations of the gypsy moth in Massachusetts, Environ. Entomol., 6: 518.
Richerson, J. V., and Cameron, E. A., 1974, Differences in pheromone release and sexual behavior between laboratory-reared and wild gypsy moth adults, Environ. Entomol., 3: 475.
Ricker, W. E., 1975, Computations and interpretation of biological statistics of fish populations, Bull. Fish. Res. Board Can. 191, 382 pp.

Schwalbe, C. P., 1979, Using pheromone traps to detect and evaluate populations of the gypsy moth, USDA Agric. Handbook 544, 11 pp.

Schwalbe, C. P., 1980, Disparlure-baited traps for survey and detection, in: "The Gypsy Moth: Research Toward Integrated Pest Management," USDA Tech. Bull. 1584, (In Press).

Seber, G. A. F., 1962, The multi-sampling single recapture census, Biometrika, 49: 339.

Welch, S. M., Croft, B. A., Brunner, J. F., and Michels, M. F., 1978, PETE: An extension phenology modeling system for management of multi-species pest complex, Environ. Entomol., 7: 487.

MONITORING CODLING MOTH POPULATIONS IN BRITISH COLUMBIA APPLE ORCHARDS

Harold F. Madsen

Agriculture Canada Research Station
Summerland, British Columbia VOH 1ZO
Canada

INTRODUCTION

In British Columbia apple orchards, a program of pest management is well established and administered by professional managers holding a master's degree in pestology. The key to the management program is a monitoring system for codling moth, *Laspeyresia pomonella* (L.), based upon male captures in sex pheromone traps. The monitoring program has been developed for conditions in British Columbia orchards and it should be emphasized that although principles have application elsewhere, details of a monitoring system must be determined for each area.

TRAPS AND LURE

The standard lure consists of a rubber cap stopper impregnated with 1 mg of codlemone (*trans*-8,*trans*-10,dodecadien-1-ol) the codling moth sex pheromone. The lure can be obtained commercially, or it may be made up by pipetting 1 ml of a 1% (W/V) methylene chloride solution of pure pheromone into the concavity of the rubber cap. A number of trap types are available, but the one preferred in British Columbia is the Pherocon 1C® manufactured by Zoecon Corporation (Palo Alto, California). The trap is easy to service and relatively simple to maintain in good condition. Traps should be replaced if the sticky surface loses effectiveness because of contamination by dust, debris or accumulation of insects. The lure may be placed upon the sticky surface of the trap bottom or suspended by a pin from the top of the trap.

TRAP INSTALLATION AND MAINTENANCE

Field studies in South Africa (Myburgh et al., 1974) and in British Columbia (Vakenti and Madsen, 1976) established that 1 trap per ha was sufficient to monitor codling moth populations in apple orchards. British Columbia apple orchards are relatively small (average 4 ha) therefore it is necessary to install sex pheromone traps in neighboring orchards to minimize the immigration of male moths from these sources. Trap captures are difficult to interpret if there is an influx of moths from areas outside of the monitored orchard. Our experience indicates that a trap every 500 ft, 5 tree rows into the neighboring orchard will effectively reduce the immigration problem.

Riedl (personal communication) has shown that traps placed in the upper third of an apple tree capture more moths than those at head height, but for monitoring purposes, traps installed at the lower level have been adequate. There are conflicting data on the effect of aging on the attractiveness of lures. Culver and Barnes (1977) state that pheromone impregnated rubber caps rapidly lose their attractiveness when exposed in the field. On the other hand, Gaunce and Madsen (unpublished data) found little difference in attractiveness between freshly prepared lures and those used for as long as 8 weeks. To be on the safe side, we recommend replacing the lures every 6 weeks.

INTERPRETATION OF MOTH CAPTURES

The critical part of a monitoring program is interpretation of the moth captures and the decision whether chemical treatment is needed. Under British Columbia conditions, a moth capture of 2 per trap in 2 consecutive weeks is considered a treatment level. Traps should be installed in orchards before codling moth emergence, otherwise the traps will be sampling an accumulation of moths and may indicate a treatment level when one does not actually exist. Codling moth pheromone traps rarely attract other moth species. One exception is *Thiodia latens* Heinrich, an Olethreutid species on native shrubs. It is frequently taken in codling moth traps in the early season, but is smaller and lacks the copper spot typical of codling moth.

One might expect a marked decline in codling moth captures after an application of organophosphate insecticide but in most cases this does not occur. Male captures are frequently as high or higher following a spray than they were prior to the application. In all probability, moths are attracted to the traps shortly after emergence and before a toxic dose of pesticide is accumulated. In such cases, it is emphasized that the fruit is protected from larval attack and the pesticide residue should be effective for a 3-week period even though trap captures exceed the treatment level.

When an orchard is monitored with sex pheromone traps at a density of 1 per ha, the average moth catch per trap may be below a treatment level, but one area of the orchard may have a high percentage of the total catch. If this situation is encountered, it may be necessary to treat only a portion of the orchard. When codling moth populations are low, infestation is frequently localized and spot treatment, when the traps so indicate, can eliminate a potential problem.

Riedl and Croft (1978) have discussed the value of predictive models for managing codling moth in Michigan. There is no question that predictive models and computer processing of data provides an accurate means of managing codling moth as well as other pests. However, it is possible to have an efficient codling moth monitoring system without sophisticated models. This is the situation in British Columbia apple orchards. In our small acreages, a grower can monitor his codling moth population and determine his own spray program or he can hand the decision-making to a professional pest manager. A predictive model and computer system would eliminate considerable leg work by pest managers and simplify the management program. Until such procedures are available, the present system of codling moth monitoring provides adequate information to the orchardist and has resulted in a considerable saving in pesticide costs.

Use and Current Status of Codling Moth Management in British Columbia

Codling moth monitoring with sex pheromone traps has been very successful in British Columbia apple orchards and in most cases has reduced the number of pesticide applications necessary to control this key apple pest. The data from one of the pest managed orchards will illustrate the value of codling moth monitoring. Figure 1 shows the 1978 and 1979 codling moth captures from a pest managed orchard and the spray program used to obtain control.

In 1978, the average moth capture for the orchard (12 traps in 14 ha) during first brood was less than the treatment level, but on May 29 and June 12, nearly 80% of the total catch was from one section of the orchard. The orchardist applied a spray to this single section and did not treat the remainder of the orchard. A treatment level was reached in September, but a spray was not applied because harvest would be completed before damage could occur. Codling moth damaged fruit was 0.2%, well below an economic threshold. The figure also illustrates the importance of placing traps in neighboring orchards as the outside traps caught high numbers of moths which would have been captured by the traps inside the orchard.

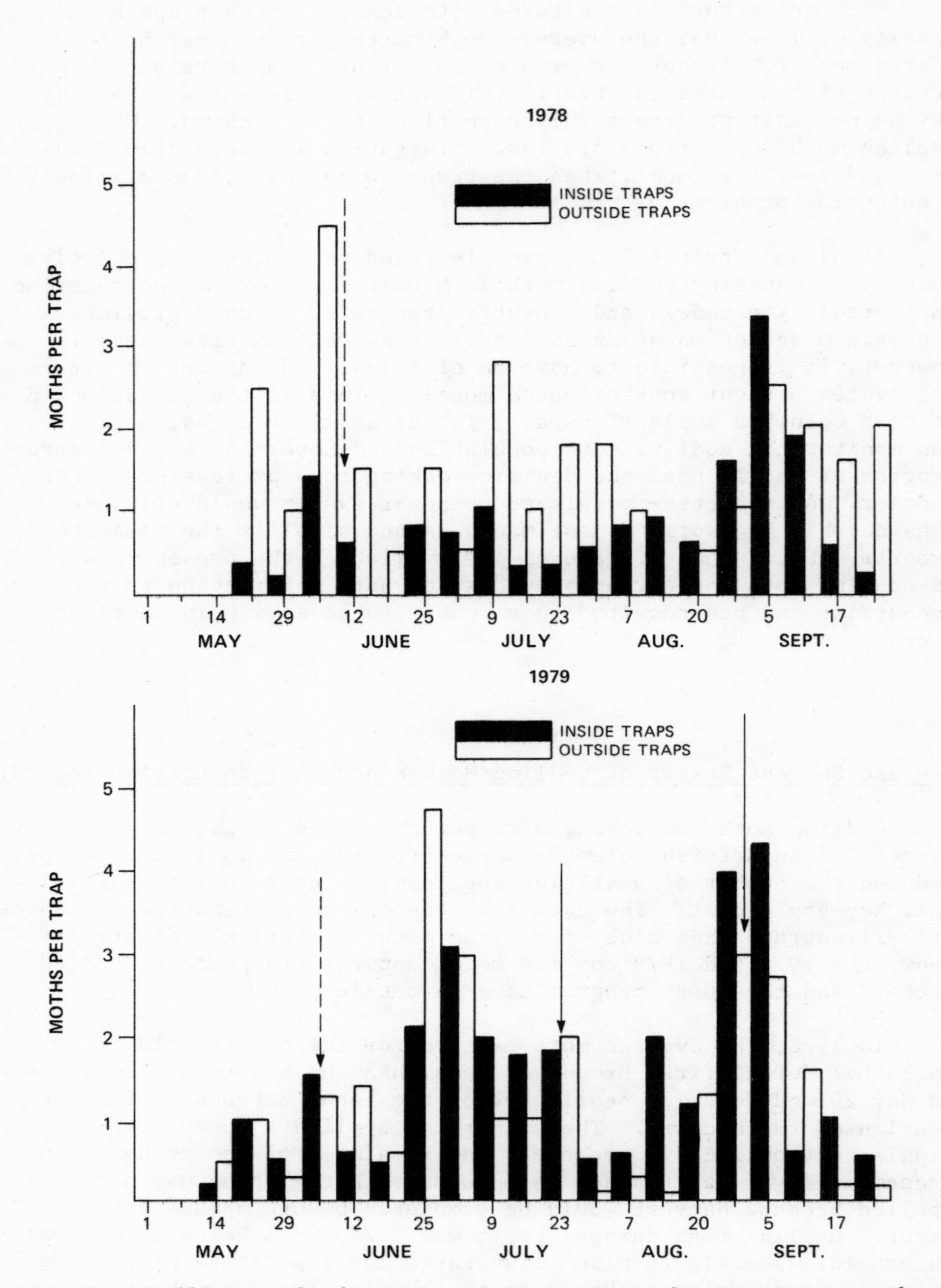

Fig. 1. Codling moth pheromone trap captures in a pest managed orchard (arrows indicate a spray application).

A calendar-based spray schedule for codling moth control in British Columbia apple orchards is 2 sprays for the first brood and 1 or 2 sprays for second brood depending upon the district and severity of the previous year's infestation. By using sex pheromone traps to monitor codling moth, the orchardist applied only one partial spray compared to a routine program of 3 or 4 sprays.

The situation was quite different in 1979, a season with above normal temperatures and near ideal conditions for codling moth survival and reproduction. Again, a spray on a portion of the orchard was applied in early June, and a spray to the entire orchard in July. Trap captures were not greatly above a treatment level in July but were at or near this level for 4 consecutive weeks. Because of favorable weather conditions for codling moth survival, it was decided to treat the entire orchard. In late August, the traps indicated a definite treatment level and a second full spray was applied. At harvest, the codling moth infestation was 0.1%. Therefore in 1979, 2 1/2 treatments were required to obtain codling moth control compared to only one partial spray in 1978. The important point is not the difference between the 2 years, but that the codling moth monitoring program provided the data on which the control program was based. In one year a considerable saving in pesticide costs was realized, and in the other nearly a full spray program was required.

In 1979, ca. 600 ha of apples were under a pest management program and sex pheromone traps were used to monitor codling moth populations. A wide variety of population densities was encountered, but in all instances, a spray program based on the monitoring system provided satisfactory control. No reports were received that indicated the pheromone traps failed to provide adequate information on which to base control decisions.

Although the present monitoring system is satisfactory, there is always room for improvement. A computer based predictive model might simplify the program and reduce the need for monitoring each individual orchard. The current treatment levels are probably on the safe side and it might be possible to raise the levels in areas where climatic conditions are less suitable for codling moth reproduction and survival. One of the basic problems in population monitoring with sex pheromones is that only males are attracted to lures. We need to know considerably more about female activity within an orchard and at present there is no specific lure to attract female codling moths. These are problems that will require research input in future years in order to provide a useful and accurate method of monitoring codling moth activity.

REFERENCES

Culver, D. J., and Barnes, M. M., 1977, Contribution to the use of the synthetic pheromone in monitoring moth populations, J. Econ. Entomol., 70: 489.
Myburgh, A. C., Madsen, H. F., Bosman, I. P., and Rust, D. J., 1974, Codling moth (Lepidoptera: Olethreutidae): studies on the placement of sex attractant traps in South African orchards, Phytophylactica, 6: 189.
Riedl, Helmut, and Croft, B. A., 1978, Management of the codling moth in Michigan, Mich. State Univ. Res. Report, 337: 1.
Vakenti, Jerry M., and Madsen, Harold F., 1976, Codling moth (Lepidoptera: Olethreutidae): monitoring populations in apple orchards with sex pheromone traps, Can. Entomol., 108: 433.

MONITORING FOR THE ANGOUMOIS GRAIN MOTH IN CORN

J. Stockel

F. Sureau

INRA Station de Zoologie
Centre de Recherches de Bordeaux
33140 Pont de la Maye
France

INTRODUCTION

The technique of trapping males with synthetic pheromone is currently used to monitor numerous insect pests. However, pheromones have other uses. One new trend, disrupting communication between sexually mature adults by confusion, was hypothesized by Beroza (1960) and Brown (1961). The difference between attraction and confusion is the amount of sex attractant diffused into the atmosphere.

Many authors studying insect sexual attraction consider correlation between synthetic pheromone dose and the number of males trapped, but their observations differ greatly. Roelofs et al. (1977) baited traps with 0.03 to 5 mg pheromone in polyethylene caps and captured the most *Lithocolletis blancardella* (F.) at the highest dosages. Kraemer et al. (1979) made a similar observation with *Neodiprion pinetum* (Norton): dispensers loaded with 0.1 up to 100 μg doses of 3,7-dimethylpentadecan-2-ol were more attractive at the highest dose (100 μg).

On the other hand, after comparing 1 and 6 mg codlemone doses using different trap designs for *Laspeyresia pomonella* (L.), Charmillot et al. (1975) found that 6 mg did poorly at the beginning of the 1st flight -- probably because high initial concentration inhibited captures. Using another lepidopteran, *Plutella xylostella* (L.), Chisholm et al. (1979) observed that by using between 4 and 2000 μg doses, the best result occurred while using a 100 μg bait dose. In 3 consecutive years (1975, 1976, and 1977) of trapping *Synanthedon exitiosa* (Say), Yonce and Pate (1979) established that a 100 μg dose (when compared to doses from 20 to 500 μg) documented their seasonal flights most successfully.

Lastly, Vick et al. (1979), using separate tests conducted in rooms with the Indian meal moth (IMM) and Angoumois grain moth (AGM), studied the relationship between the amount of synthetic pheromone and the number of trapped males. When dispensers with release rates of 40, 320 and 2750 ng/h for IMM and 77, 310 and 700 ng/h for AGM were used, they noted that the best results were obtained with baits giving release rates of 2750 ng/h for IMM and 77 ng/h for AGM. These authors concluded that an excess of pheromone causes a reduction in trap catch for *Sitotroga cerealella* (Olivier), (as Charmillot et al., 1975 do for *L. pomonella*).

In a previous work (Stockel and Huguet, 1977) we compared attractivity of captive females of *S. cerealella* and increasing amounts of synthetic pheromone for the males of this species. Under these experimental conditions, our results showed that female attractiveness increased up to a maximum of about 10 females before decreasing, probably because of a group effect. With synthetic pheromone the numbers of trapped males were proportional to the decimal logarithm of the quantity of pheromone, between 100 and 2700 μg, loaded in rubber septa dispensers.

This paper contains data on the optimum dose of pheromone for sex trapping applications and the threshold of sexual inhibition for *S. cerealella*.

MATERIALS AND METHODS

From June 8 to November 13, 1979, in order to take into account the significant effect that host plants have on the number of male *S. cerealella* trapped (Stockel, 1973), we placed traps successively in a rectangular wheat field of 150 x 120 m (June 12-July 17) and in an adjoining cornfield of the same size (July 17-November 13) at Preignac (Gironde).

Sticky traps of INRA design were placed at 50 m intervals 5 m in from the field's edge 1.8 m high. These traps were baited with 1 of 9 concentrations of (*Z*,*E*)-7,11-hexadecadienyl acetate as follows: 1, 11, 33, 100, 300, 900, 2700, 8100 and 24300 μg.

Hoping to reduce the traps' positional effect and possible trap interaction, we alternated a high pheromone concentration trap with a low one (Fig. 1) and then rotated traps one position twice/wk. counting males captured each time.

Synthetic attractants were loaded in rubber septa dispensers which were replaced every 6 weeks. Pheromone release from dispensers was estimated by measuring the weight loss of loaded dispensers set in a "diffusiometre," an apparatus described by Carles et al. (1979a). A constant air flow of 0.3 m/sec at 20°C and 80% RH was maintained over the dispensers and vented to the outside. Pheromone loss from dispensers was computed after comparison with unloaded control dispensers. Weights were taken at regular intervals using a Mettler ME30® microbalance with an accuracy of 1 µg, but balance sensitivity was insufficient to obtain release rate data for the 1 and 11 µg pheromone loads.

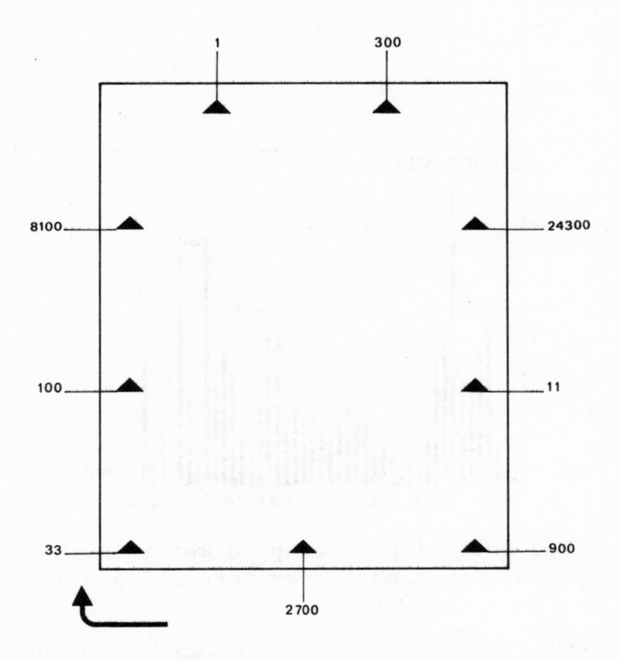

Fig. 1. Experimental field showing arrangement of sex traps schematized by black triangles. Number indicate the amount of pheromone bait (µg). The direction of circular permutation of traps is symbolized by black dart.

RESULTS

The 6,855 male moths trapped from June 8 to November 13 in the 9 traps are shown in Figure 2. The 3 annual generations usually seen in southwest France (Cangardel and Stockel, 1972) are easily distinguished. Their 3 flights peak between July 6-10 for overwintering generation (WG) adults, about September 15 for the next generation (G_1) and October 15 for the last generation (G_2). No correlation seems to exist between males trapped and mean temperature or rainfall.

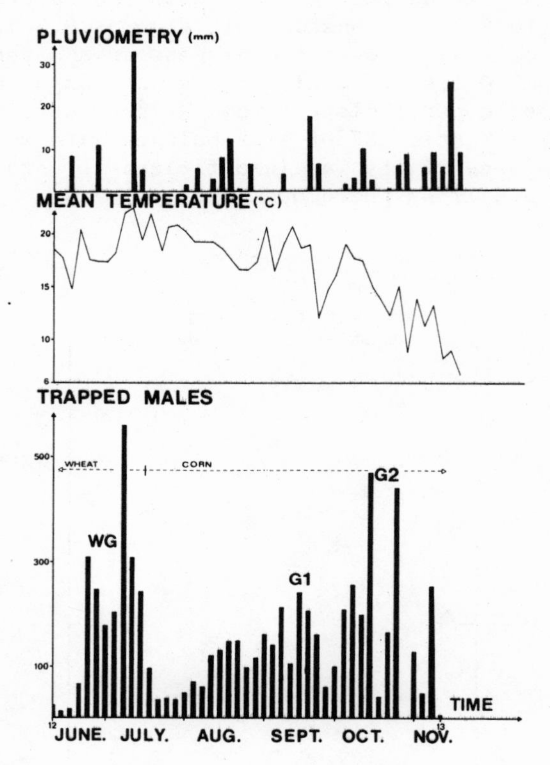

Fig. 2 General evolution of trapped males in the 9 sex traps, compared with mean temperature and pluviometry.

Males trapped in each of the 2 fields are shown in Table 1 (11 sampling days and 34 sampling days). Our goal was to find the easiest way to show correlation between the number of trapped males and pheromone amount used as bait. Our previous data showed that trapped males increased logarithmically. Therefore, we searched for correlations between males trapped and the logarithm of pheromone dose.

We found a linear correlation coefficient between the 9 pairs of variables for each studied period (R_1 = 0.909, R_2 = 0.937, R_3 = 0.937). These correlation coefficients are highly significant (P: 0.001 = 0.798) and indicate to us that within the limits of the experimental doses (1 and 24300 μg) that this correlation explains at least 87% of the observed variability for the whole period (R_2 = 0.8788).

Table 1. Number of trapped males in each sex trap during the 2 successive periods.

Dispenser load	Wheat 8/6-17/7	Corn 11/7-13/11	Total 8/6-13/11
1	3	9	12
11	13	63	76
33	59	159	218
100	175	371	546
300	299	756	1055
900	415	617	1032
2700	413	936	1349
8100	424	918	1342
24300	354	971	1225
Total	2155	4700	6855
Coefficient correlation	0.909	0.937	0.937

Then, we analyzed data using a polynomial adjustment for each of the 9 pairs of variables to determine if the degree of correlation was improved. We carried the analysis out to the 3rd order, but if the correlation coefficient was increased less than 10% above the correlation coefficient of the previous order, we used the regression for the previous order. Table 2 gives this correlation coefficient with the order of the studied polynomial for each period.

Table 2. Improvement of correlation according to the degree of polynomial.

	Wheat 8 Jun-17 Jul	Corn 17 Jul-13 Nov	Total 8 Jun-13 Nov
Correlation coefficient	0.909 +++	0.937 +++	0.937 +++
Corr. coeff.^{2} x 100 (1)			
1st degree	82.69	87.81	87.88
2nd degree	83.72	87.91	88.17
3rd degree	99.08*	95.25*	97.13*

(1) Percentage explanation of the regression.
*Best polynomial adjustment.

The best polynomial regression for each of the 2 periods and for the entire period (June to November 13) is the 3rd order regression (*) shown by the calculated curve in Figure 3.

$$Y_1 = 962.05 - 1596.57 X + 721.61 X^2 - 77.35 X^3$$

This curve has a maximum at X = (log dose + 1) = 4.750, that corresponds to pheromone amount between 8100 and 24300 µg, which should be the higher threshold for sexual attraction and the lower threshold of sexual inhibition for the synthetic pheromone of the Angoumois grain moth.

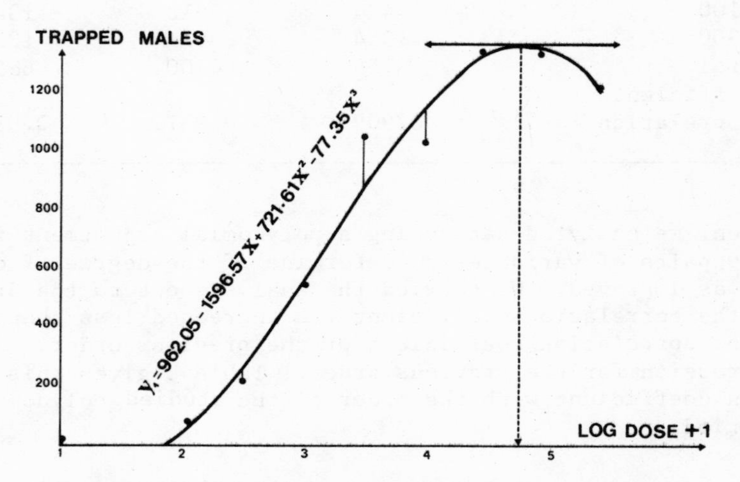

Fig. 3. Polynomial adjustment with 9 sex traps (3rd degree).

On August 17, we added 2 supplementary traps baited with 72,900 and 218,700 µg of the same pheromone. Distribution of the 6,225 trapped males by 11 traps from August 17 to November 13 is given in Table 3, along with sex attractant dose.

Under these conditions, correlation between the 2 studied variables is about the same (correlation coefficient = 0.949 instead of 0.937). We determined the best polynomial adjustment for this regression:

1st order $R^2 \times 100 = 90.13$
2nd order $R^2 \times 100 = 90.41$
3rd order $R^2 \times 100 = 91.22$

Table 3. Number of males trapped in response to dose.

Z7,E11-HDDA (μg)	No. males trapped
1	9
11	59
33	143
100	346
300	712
900	542
2700	844
8100	796
24300	749
72900	946
218700	1079
Total	6225

Using this extensive range of pheromone doses, the correlation seems to be linear and no improvement is given for comparing a 2nd or 3rd polynomial order regression. The calculated curve equation becomes:

$$Y_2 = 211.06\ X - 257.61 \text{ (Fig. 4)}$$

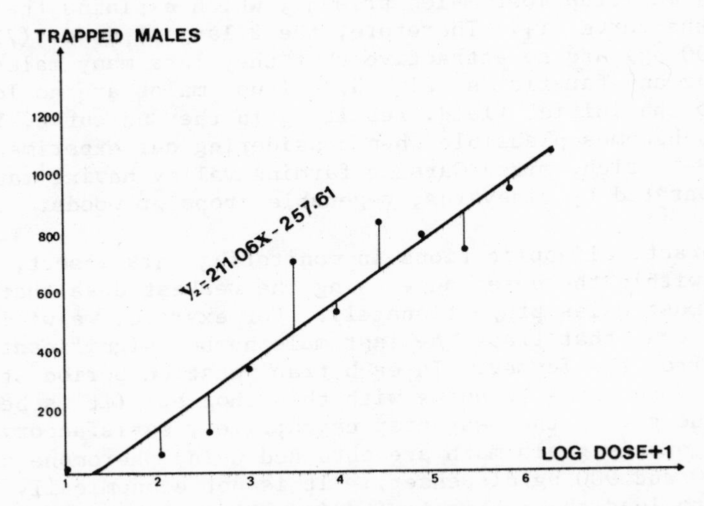

Fig. 4. Logarithmic relation between amount of pheromone and number of trapped males.

The diffusiometre experiment shows that (Z,E)-7,11-hexadecadienyl acetate is released very slowly when compared to the synthetic pheromone of Lobesia botrana (Schiff.), (E,Z)-7,9-acetoxydodecadiene (Carles et al., 1979b). During 30 days under the former conditions, the evaporation rate for higher doses is about 3-4% of the initial load. Under similar conditions the mean release of pheromone for L. botrana was between 50-60% of the initial load for a 39-day-period (Carles et al., 1979b).

DISCUSSION

This study underlines the very low release rate of Z,E-7,11-HDA that confirms the work of Carles et al. (1979a) and Butler and McDonough (1979) on evaporation rates of acetates from natural rubber septa. Our trapping results under natural conditions seem to be of interest for ethological and applied research studies.

The number of insects captured is logarithmically related to pheromone dosage within doses ranging from 1×10^{-6} to 2×10^{-1} g as shown previously by Stockel and Huguet (1977) using a more limited range. Unlike the results of Vick et al. (1979), these show no diminution of male response to high pheromone doses, further showing that the mechanism of male perception and its attraction behavior are neither modified nor disturbed when pheromone concentration increases.

The curve's shape in Figure 3 might be explained if AGM population is limited in one or several contiguous cereal fields. Under these conditions the highest bait doses (2,700, 8,100 and 24,300 μg) may trap most males present, which explains the maximum point of the curve, Y_1. Therefore, the 2 largest doses (72,300 and 218,900 μg) are so attractive that they lure many males from neighboring and far fields (Fig. 5). Thus, males are no longer limited to the initial field, resulting in the 2nd curve, Y_2. This hypothesis becomes plausible when considering our experimental field location -- a rich, mixed Garonne farming valley having many cereal fields separated by vineyards, vegetable crops or woods.

For practical applications in monitoring this insect, we will stay within the dose range using the weakest dose that captures the most males proportionately. For example, we will choose the lowest dose that traps the last male number significantly different from the former. In each trap rotation period studied, the 300 μg dose seems to agree with this choice. (It is best to use the dose giving the best trap catch; i.e., satisfactory results for the Angoumois grain moth are obtained using pheromone amounts between 300 and 900 μg/dispenser). It is not economically necessary to load the rubber septa dispensers with sufficient attractant when capturing maximum insects for monitoring.

Finally, regarding control of this insect by mating disruption, we wish to review our previous ethological observations. There does not seem to be any threshold for male sexual inhibition of this species; each male is able to discover a female and mate. To prevent the male's finding of the female, the amount of synthetic pheromone released from each source must always be higher

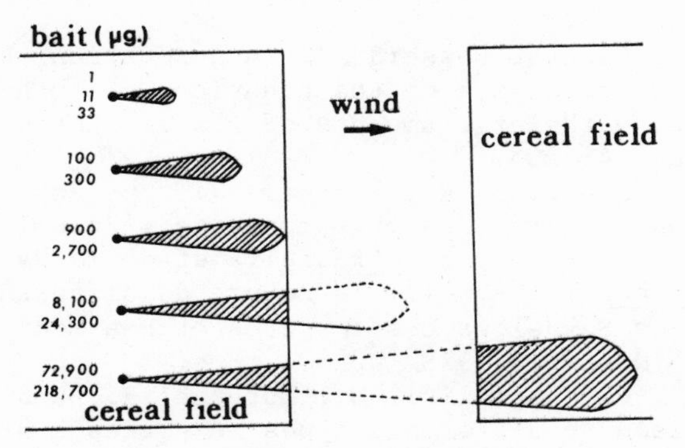

Fig. 5. Tentative ecological explanation of distance attraction:
+ Feathers symbolize attraction areas
+ Males stay in hachured zones.

than pheromone amount from one female at the same time. For these insects the control strategy is to maintain an antagonistic situation between the synthetic pheromone source and the female. This competition must always agree with the synthetic substance to maintain disruption.

ACKNOWLEDGMENT

We thank Miss F. Bruzau and Mr. J. G. Huguet for skilled statistical assistance and Dr. R. Roehrich and Mr. P. Anglade for helpful discussion.

REFERENCES

Beroza, M., 1960, Insect attractants are taking hold. Agric. Chem., 15(7): 37.

Brown, W. L., Jr., 1961, Mass insect control programs: Four case histories, Psyche., 68: 75.

Butler, L. I., and McDonough, L. M., 1979, Insect sex pheromones: Evaporation rates of acetates from natural rubber septa. J. Chem. Ecol., 5: 825.

Cangardel, H., and Stockel, J., 1972, Recherches par l'élevage en insectarium et par piégeage sexuel, sur le cycle annuel de l'Alucite des Céréales Sitotroga cerealella Oliv. (Lep. Gelechiidae) et sur le rôle des cultures de blé et de maïs dans le maintien de l'espèce en Aquitaine (France), Ann. Zool. Ecol. Anim., 4: 311.

Carles, J. P., Fleurat Lessard, F., and Roehrich, R., 1979a, A device for bioassays on the behaviour of moths in air flows permeated with large amounts of sexual pheromones, Biol. of Behaviour, 4: 205.

Carles, J. P., Roehrich, R., Stockel, J., and Sureau, F., 1979b, Diffusion comparée des phéromones de l'Eudémis Lobesia botrana, de l'Alucite Sitotroga cerealella et de la Pyrale du Maïs Ostrinia nubilalis selon la nature du diffuseur, in "Les Phéromones Sexuelles Des Insectes et Les Médiateurs Chimiques," Antibes, November 1978, pp 80.

Charmillot, J. P., Bagiolini, M., Murbach, R., and Arn, H., 1975, Comparaison de différents types de pièges à attractif sexuel synthétique pour le contrôle du vol du Carpocapse (Laspeyresia pomonella L.). Rech. Agron. Suisse, 14(1): 57.

Chisholm, M. O., Underhill, E. W., and Steck, W. F., 1979, Field trapping of the diamond back moth Plutella xylostella using synthetic sex attractants, Environ. Entomol., 8(3): 516.

Kraemer, M., Coppel, H. C., Matsumara, F., Kikukawa, T., and Mori, K., 1979, Field responses of the white pine sawfly Neodiprion pinetum, to optical isomers of sawfly sex pheromones, Environ. Entomol., 8(3): 519.

Roelofs, W. L., Reissig, W. H., Weires, R. W., 1977, Sex attractant for the spotted tentiform leaf miner moth Lithocolletis blancardella, Environ. Entomol., 6(3): 373.

Sower, L. L., Fish, J. C., 1975, Rate of release of the sex pheromone of the female Indian meal moth, Environ. Entomol., 4(1): 168.

Stockel, J., 1973, Influence des relations sexuelles et du milieu trophique de l'adulte sur la reproduction de Sitotroga cerealella Oliv. (Lép. Gelechiidae). Conséquences écologiques. Thèse Docteur-Ingénieur n° 3 Tours.

Stockel, J., and Huguet, J. G., 1977, Relations entre la dose de phéromone de synthèse ou le nombre de femelles vierges utilisées comme appât et le taux de capture des mâles chez

Sitotroga cerealella Oliv. (Lep. Gelechiidae). Biol. of Behaviour, 3(2): 273.

Vick, K. W., Kvenberg, J., Coffelt, J. A., Steward, C., 1979, Investigation of sex pheromone traps for simultaneous detection of Indian meal moth and Angoumois grain moth, J. Econ. Entomol., 72(2): 245.

Yonce, C. E., and Pate, R. R., 1979, Seasonal distribution of Synanthedon exitiosa in the Georgia Peach Belt monitored by pheromone trapping, Environ. Entomol., 8(1): 32.

SEX ATTRACTANT TRAPS: THEIR ROLE IN THE MANAGEMENT OF SPRUCE BUDWORM

C. J. Sanders

Canadian Forestry Service
Great Lakes Forest Research Centre
Post Office Box 490
Sault Ste. Marie, Ontario

INTRODUCTION

The use of sex attractant traps for monitoring population densities of spruce budworm [*Choristoneura fumiferana* (Clemens)] has to be viewed within the context of the techniques currently being used to estimate density. Present control strategies depend largely upon the aerial application of insecticides on high density populations. Such populations may number 50 million or more budworm/ha, and they are easily detected by the associated defoliation which can be spotted by aerial surveys. Once detected, the necessary information for the planning of control operations can be collected by egg sampling in the fall of the previous year and by larval sampling in the spring before the control operations begin. The egg sampling permits decisions to be made on the need for protection and on the broad areas to be treated; and since it can be carried out in the fall, it provides sufficient lead time for planning and for the ordering of materials and equipment. The spring larval sampling permits final adjustments to be made in the delineation of the spray boundaries. Within such a strategy there is little scope for the use of sex attractant traps. Relatively few samples are required to provide the necessary accuracy of egg and larval numbers at high densities. Also at high densities dispersal of female moths becomes a major factor in the redistribution of egg masses and hence in the larvae and the damage they cause. Investigations in the Maritime Provinces suggest that a higher proportion of females disperse than males; thus, catches of male moths in sex attractant traps may not be a good indication of the abundance of females ovipositing in the same area.

Hopefully, in the future, the strategy used in forest management will be directed toward a reduction in the possibility of severe outbreaks. To do this it will be necessary to detect population surges well below the densities at which defoliation is detectable. Available evidence suggests that the critical density at which populations escape the constraints holding them at endemic levels is in the order of 0.1-1 larva/45 cm branch tip--between 10,000 and 100,000 larvae/ha, several orders of magnitude below outbreak population densities. This is about the lower limit at which conventional larval sampling is practical in terms of manpower (a single population estimate accurate to within ±10% of the mean at these densities would involve 6 or more man-days). Below this level there is only one technique that has been used extensively for monitoring population densities: the beating of foliage with long sticks to dislodge large larvae onto cloth trays where they can be counted. This technique has been used widely across eastern Canada in the past; it is a cheap and fast method, and in detecting population trends over large areas its usefulness has been demonstrated in New Brunswick (Miller et al., 1968) and in Ontario (A. H. Rose, personal communication). But when population densities are very low, unless the area of foliage beaten is large, the incidence of zero counts presents problems in interpreting the data; yet any increase in sample size reduces the number of areas that can be sampled, undermining the usefulness of this technique. A relatively new technique has potential as a monitoring tool for low density populations -- the extraction of overwintering larvae from foliage using sodium or potassium hydroxide solution (Miller and McDougall, 1968). This technique, however, has not gained the popularity which might have been expected. Although work in New Brunswick and Quebec has shown that it gives population estimates of reasonable accuracy, recent work by Blais (1979) has cast some doubt on the validity of the data. Personal experience with the technique for monitoring low density populations in Ontario has led us to seriously question its use for monitoring sub-outbreak population densities. Comparison with later larval samples suggested that many of the overwintering larvae were missed. Also, it was difficult to distinguish spruce budworm larvae from the larvae of other species of Lepidoptera, which (particularly on white spruce) may outnumber the spruce budworm.

Therefore, sampling of low density populations presents problems which as yet have not been resolved. It is within this area that a role is being considered for the use of sex attractant traps. Already the traps have demonstrated that they are extremely efficient; anywhere that baited traps have been placed out during the budworm flight period in host forests within the range of the spruce budworm, some male budworms have been captured.

Two strategies for use of the traps are possible, either to monitor rates of population change -- large increases in trap catch from one year to the next heralding large surges in population density -- or to identify threshold catches which would indicate that population densities had reached the transition stage from endemic to epidemic. Which of these 2 strategies would be more appropriate will depend upon the accuracy with which catches can be correlated with population density.

In either event, it will be necessary during the development stage to calibrate the catches and/or rates of change in catches with population densities. As already mentioned, the population densities of concern are 100,000 larvae/ha and lower (1 larva/45 cm branch tip in an average stand) which translates into about 10,000 male moths/ha or less.

CHARACTERISTICS OF THE TRAPPING SYSTEM

An essential feature of the trapping system is comparability among catches in any one location from year to year. Reasoning that commercial quality control would be more likely to produce a consistent product than "homemade" traps, we evaluated 4 types of commercial traps. Of the 4 designs, Pherocon® 1C, Pherocon 1CP, Sectar® XC-26, and Sectar 1, the Pherocon 1CP trap appeared to suit this need best (Sanders, 1978). While being relatively cheap and easy to handle, the Pherocon 1CP trap is sturdy, satisfactorily surviving a full 12-month exposure in the forest, and its design makes it specific to the target species, excluding extraneous debris and errant insects of other species.

The second component of the system is the lure. Here the critical factors are stability of the synthetic sex attractant and a uniform release rate. During the development of the system, we have concentrated on a PVC formulation (Fitzgerald et al., 1973; Daterman, 1974). This can be easily made up in the laboratory, allowing experimentation with different concentrations and additives. Once the required release rate has been determined, then any formulation providing that release rate throughout the flight period of the budworm could be used.

Sanders and Weatherston (1976) found that the pheromone emitted by female spruce budworm was 96:4, (*E*:*Z*)-11-tetradecenal. More recently, Silk et al. (unpublished data) found the ratio to be closer to 95:5 (*E*,*Z*). They also detected trace amounts of the corresponding acetate and tetradecanal, although the biological significance of these 2 compounds, if any, is unknown. Field trapping data (Table 1) show conclusively that the most attractive blend is 97:3 (*E*,*Z*), but differences in catch over the range 92.5 -99% (*E*) are frequently not significant (Sanders and Weatherston, 1976; Table 1). Therefore, we are confident that a blend of 95-97% (*E*) will give maximum catches.

Release rates of this blend (for which the name "fulure" has been coined) measured by weight loss from various concentrations in PVC, and a comparison between release rates from PVC and polyethylene vial stoppers is shown in Fig. 1. Clearly, the PVC provides a more uniform rate of release than the vial stoppers. Before use the PVC pellets are aged for 10 days in a fume hood and since the flight period of the spruce budworm lasts up to 4 weeks, release rates from 10 to 40 days bracket the period of concern.

Table 1. Average catches of male spruce budworms in Pherocon 1C traps baited with different blends of (E,Z)-11-tetradecenal in PVC lures (attractant by weight = 4.4 mg) or polyethylene vial stoppers (each containing 1 mg of each attractant). Each treatment replicated 5 times, Ontario 1976. (n.b. Catches with PVC and stoppers are not comparable since they were tested on different days).

% E Isomer	PVC formulation (4.4 mg attractant)*	Polyethylene stoppers (1.0 mg attractant)**
100	12.2 cd	37.6 d
99	26.6 a	124.8 b
97	28.6 a	147.0 a
95	26.4 a	112.6 b
92.5	22.0 ab	73.2 c
90	15.6 b	71.2 c
85	7.8 de	28.2 d
80	6.6 de	-
70	3.4 de	-
60	4.2 e	-
Blank	1.6 e	7.6 e

*June 28-29. **June 29 - July 1. Numbers followed by the same letters in each column not significantly different (Duncan's New Multiple Range Test P = .05).

Using PVC pellets 4 mm diam. x 10 mm long, a concentration of 3% (ca. 4.5 mg) gave release rates in the order of 60 μg/day at 10 days and 28 μg/day at 40 days. For 0.3% the figures were 8.0 and 2.3, and for 0.03%, 2.3 and 0.16, respectively. Allowing for inaccuracies in the weighing of such small quantities, it is probable that 10-fold increases in concentration are associated with 10-fold increases in release rate.

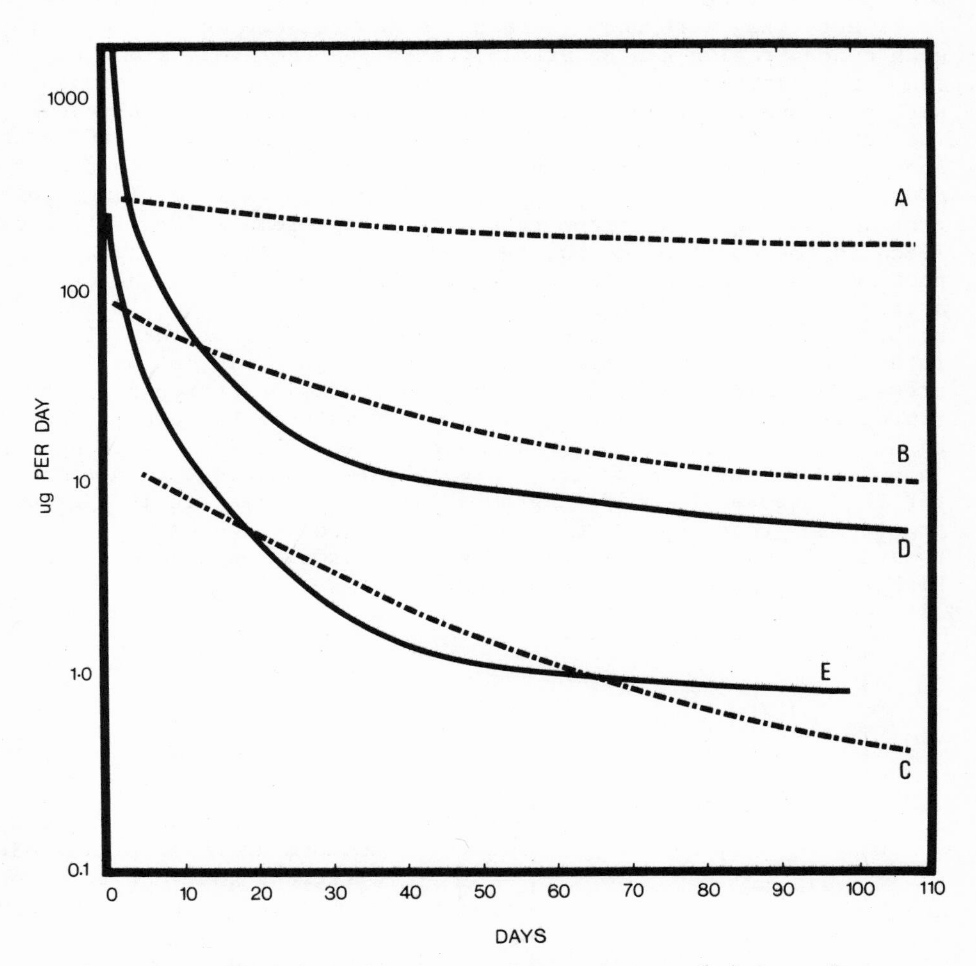

Fig. 1 Comparison between release rates of fulure from PVC and polyethylene stoppers.
A 82.5 mg fulure in PVC 10 mm diam x 10 mm long
B 4.35 mg " " " 4 mm " " "
C 0.51 mg " " " 2.5 mm " " "
D 10 mg fulure in polyethylene vial stoppers
E 1 mg " " " " "

Catches of moths in traps baited with PVC containing the different concentrations of fulure and aged for different lengths of time, all exposed at the same time, are shown in Fig. 2. Differences between the treatments were significant, but within treatments they were not. A rather wide variance in catches within each concentration can be attributed to the fact that since the traps were in a grid, 20 m between traps, the lures were competing with each other for captures; hence, catch depended greatly upon the potency of the neighboring lures. Experiments using different concentrations of fulure in PVC lures have demonstrated that a 0.03% concentration in a PVC pellet, 4 mm diam. x 10 mm long, most closely parallels the catch rate of a virgin female (Fig. 3). This leads to the conclusion that females must be releasing their pheromone at a comparable rate to the PVC containing 0.03% of attractant; i.e., between 2 and 0.2 μg/24 h, or between 10 and 100 ng/h, with the lower figure more probable. Estimates of release rates from females vary from 10 to 40 ng/night (Silk et al., in press). Females call under laboratory conditions throughout the scotophase (an 8-h period) giving an estimate of between 1 and 5 ng/h; i.e., within the order of magnitude of the PVC. Although it is not necessary to utilize a release rate equivalent to a female for the purpose of monitoring population densities, it is reassuring to know that the 2 are in agreement, and it avoids possible behavioral abnormalities which might be associated with abnormally high release rates of the attractant.

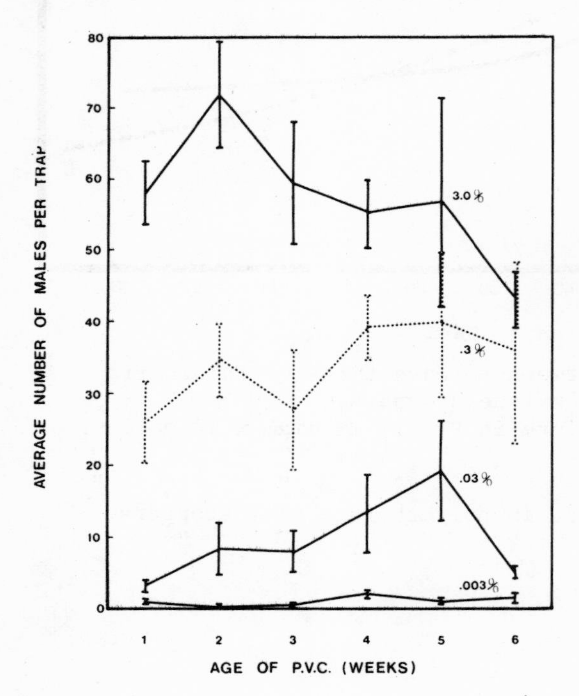

Fig. 2. Catches of male spruce budworm in traps baited with different concentrations of fulu[re] formulated in PVC, pre-exposed f[or] different lengths of time. Verti[cal] bars indicate ± 1 SE.

POTENTIAL PROBLEMS

The use of sticky traps have a number of limitations.

(1) The stickiness of the adhesive material deteriorates with age, which manifests itself as a decreasing number of the attracted insects being captured as the traps age.

(2) Catches vary with weather. Data from very low density populations in northwestern Ontario in the 1960s (Fig. 4) where virgin females were used as lures, indicated a significant correlation between catch and temperature for 2 out of 4 years' of data, and a significant inverse relationship with humidity during one of the 4 years. There was also evidence that catches peaked during periods of decreasing atmospheric pressure (Fig. 5).

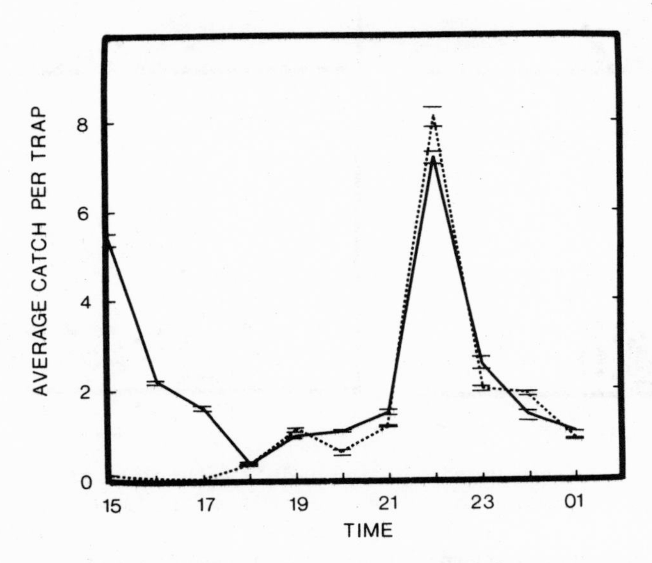

Fig. 3 Comparison between catches of male spruce budworm in traps baited with individual virgin females (dotted line) or PVC pellets containing .03% fulure by weight, ca. 44 μg (solid line). Vertical bars indicate ± 1 SE.

(3) The traps have a limited capacity, functionally saturating at about 50 moths/trap (Fig. 6).

(4) Traps placed too close together interfere with each others catches. Catches in a single trap, surrounded by a circle of 6 other traps at various distances, were lower than in a single

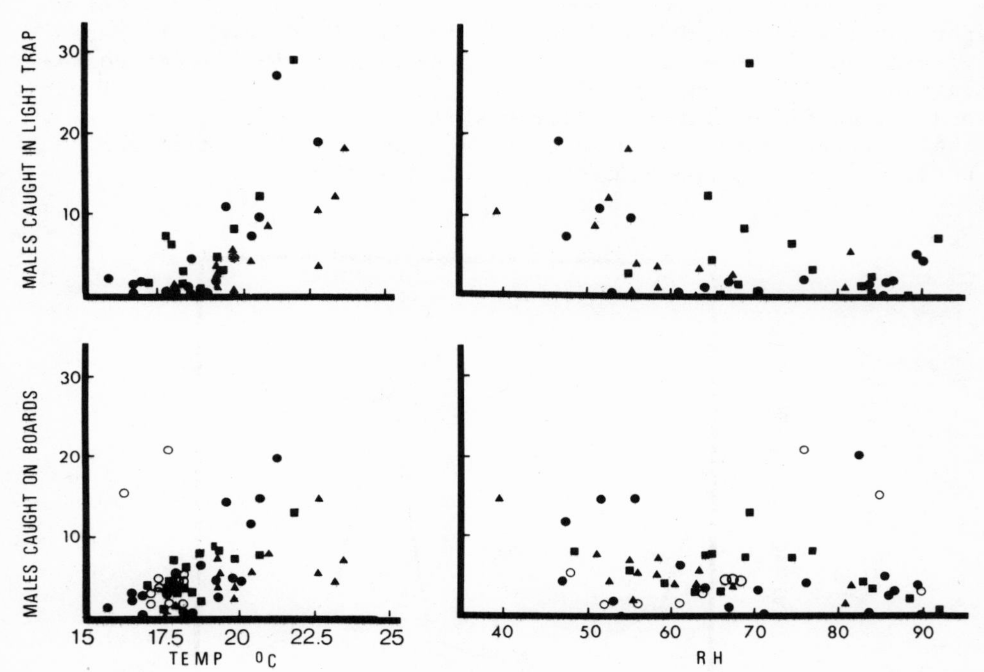

Fig. 4 Relationship between daily catches of male spruce budworm moths (as percent of season total) in a black light-trap, or on sticky trap boards baited with virgin females, and temperature (C at 2000 hr) or relative humidity

$\left(\frac{\text{RH max.} + \text{RH @ 2000 hr}}{2}\right)$

(o - 1965, ▲- 1966, ■- 1967, ●- 1968)

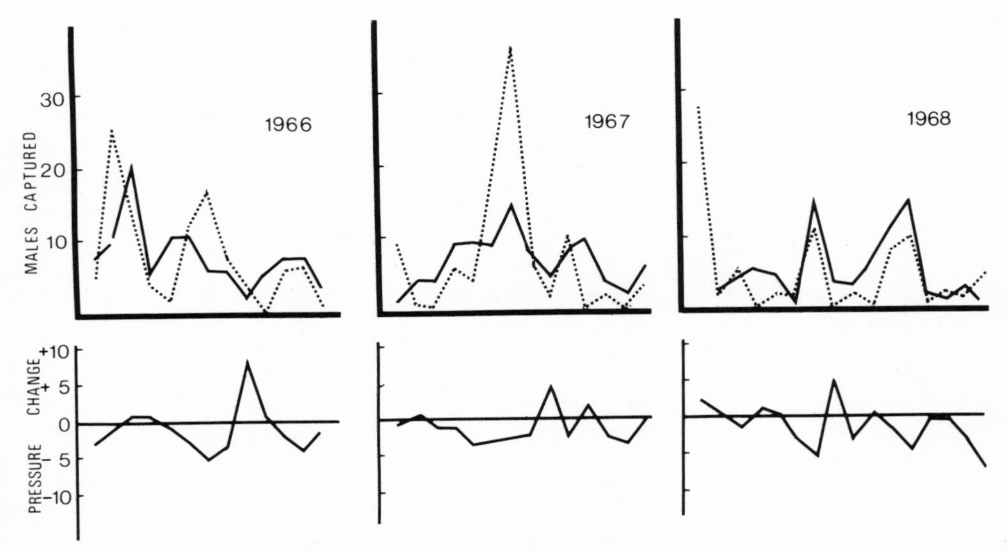

Fig. 5 Daily catches of male spruce budworm moths (as percent of season total) in a 15 watt black light trap (dotted line) or on trap-boards baited with virgin female budworm (solid line), compared with changes in barometric pressure (pressure at 1200 - pressure at 2100).

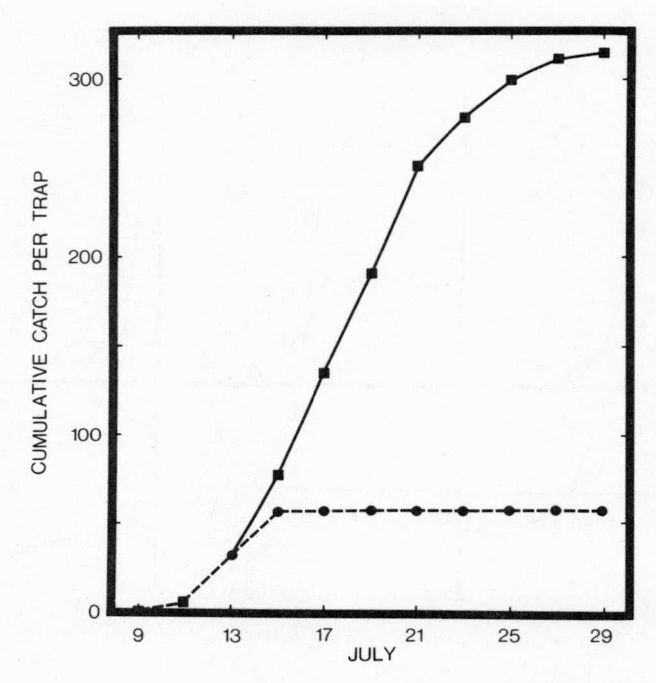

Fig. 6 Catches of male spruce budworm moths in Pherocon 1CP traps: A trap bottoms changed every 2 days (solid line).

B trap bottoms unchanged throughout (dashed line).

isolated trap, and those in the 6 peripheral traps at distances up to 40 m in low density populations (Sanders, unpublished data).

(5) Catch varies with both location and height in the stand, maximum catches occurring in traps placed in the stand canopy (Miller and McDougall, 1973; also Table 2).

Table 2. Catches of male spruce budworm moths in traps baited with virgin females placed at different heights in 2 types of balsam fir/white spruce--hardwood stand in northwestern Ontario: (a) an overmature stand of scattered white spruce 25 m high and a balsam fir/spruce understory 10 m high, and (b) an immature stand 10 m high. Note that maximum catches occur within the continuous canopy.

	Trap height (m)	Year				
		1965	1966	1967	1968	Total
Overmature stand	24	9	34	36	20	99
	17	32	83	32	51	198
	10	51	54	59	72	236
	3	59	72	56	40	227
Immature stand	9.5	-	-	63	14	77
	7			60	15	75
	4.5	-	-	39	11	50
	2			23	21	44

DEALING WITH THE PROBLEMS

These problems can be resolved as follows: Stickiness of the adhesive can be maintained by regularly changing the bottoms of the traps every 6-8 days, the point where there is evidence of serious differences between fresh and old traps. However, this would entail considerable work, which would counteract the usefulness and simplicity of the trap system. An alternative approach is to assume that the deterioration of the sticky surface proceeds at the same rate each year, and to place traps out at the same time each year relative to the flight period so that this factor can be treated as a constant from year to year.

The effects of weather are more problematical and may represent one of the more important factors affecting correlations between catch and population density in any one year. But, if we assume that over the moth flight period conditions will more or less even themselves out, then we can again treat this factor as a constant. The degree of validity is not known. But it can be argued that a bad season for catches probably represents a bad season for mating and oviposition. Therefore, lowered catches may be a better indicator of next year's population density than the previous larval population density would have been.

While limited capacity of the traps could be solved by changing the bottoms more frequently, this means more visits to each trapping location which again reduces the utility of the trapping system. A more reasonable approach is to ensure that potency of the traps is appropriate to the population density with which we are concerned. Potency can be reduced either by using less optimum blends or by decreasing the release rates. However, catches with less optimum blends can be misleading. In a situation where traps were competing for males (e.g. Table 1) an 80:20 (E,Z) blend caught only 23% of the catch by the optimum blend. However, in a noncompetitive situation where the different blends were 1 km apart, the 97:3 averaged 44.2/trap; the 80:20, 41.8; a nonsignificant difference. Therefore, a more predictable method for reducing potency appears to be the reducing of release rates, giving effects similar to those in Fig. 2. If our interest is solely in the detection of a critical population density (i.e., a threshold), then only that concentration which gives reasonable catches at that density would be required--if lower populations give zero catches and higher populations saturate the traps, this would not affect the conclusions. However, if we are interested in rates of change coupled with some idea of fluctuations and trends, then trapping over a wide range of low population densities would be desirable. This can be achieved by using several concentrations in one sampling location to cover the possible range of densities. Table 3 shows the results of using different concentrations of synthetic pheromone at different population densities in Ontario in 1979. None of the populations were as low a density as those experienced in northwestern Ontario and northern New Brunswick in the early 1960s when estimates were as low as 0.001 larvae/45 cm branch tip. But it would appear that 2 concentrations, 0.0003 and 0.03, would cover the possible ranges of populations we are concerned with.

If traps are placed out in groups, then interference between them must be allowed for. Although tests have not been conducted with the lower concentrations, it is probable that interference will occur when traps are placed within 10 m of each other. One solution is to use only a single trap in each location, but that may be risky and may introduce errors because of slight variations in trap-stickiness or lure-potency. An alternative is to separate the traps sufficiently to avoid interference but this is inconvenient; we have found that 10 m between traps is most convenient--one trap can be seen from another--and facilitates placement and collection. Therefore, our solution is to treat the catch from a group of 5 traps placed in a circle 10 m apart as a single sample, and not to express catches as those for the average trap.

Table 3. Total catches of male spruce budworm in clusters of 5 Pherocon 1CP traps baited with PVC lures containing different concentrations of synthetic attractant (% weight), deployed in 5 different population densities in Ontario, 1979. (PVC lures were 4 mm diam. x 10 mm long, and averaged 0.13 g).

Concentration of attractant (% weight)	Population density (larvae/45 cm branch tip)				
	0.1	0.5	3.0	4.8	17.2
.03	+	+	+	+	+
.003	134	28	+	+	+
.0003	28	12	231	+	+
.00003	9	0	100	+	+
0 (check)	0	0	20	112	92

+ = Traps saturated (i.e., >50 males/trap).

The problem of differences in trap catch because of position can be avoided quite simply by ensuring that the traps are placed in the same location each year. This will not prevent differences among stands in the relationship between catch and population density, but this factor is not a problem if our concern is with rates of change rather than absolute density estimates.

A number of these problems could be avoided with the design of a nonsaturating, nonsticky trap (i.e., some form of high capacity, nonexit trap). Certainly, this would remove the problems of trap saturation and deterioration in stickiness, but all the other problems will remain. Ultimately, the decision as to whether an improved trap design is warranted will depend upon costs and convenience of handling.

FIELD APPLICATION

Evaluation of the traps for operational use has been carried out in northwestern Ontario since the 1960s. Two aspects have been under consideration: (a) the practicality of such a system and (b) the establishment of correlations between trap catch and population density obtained by intensive sampling of larval populations. Unfortunately, during this period records have not been kept continuously because of a shortage of manpower. The picture is further complicated by the fact that different techniques for trapping the male budworm moths have been used, making comparisons among the data difficult. However, allowing for these shortcomings, it is evident (Fig. 7) that the trap catches have paralleled population densities throughout the period, thus demonstrating the potential of this technique.

In this same area of northwestern Ontario, traps have been placed out in 29 locations annually since 1973. Synthetic attractant (containing 97% E-isomer) incorporated at a rate of 3% in PVC (4.4 mg) were used throughout this period. For the first 3 years (1973-1975) Sectar 1 traps were used; thereafter, Pherocon 1CP traps. No sampling to determine population densities has been carried out in any of these areas. The plots were distributed along a triangular route 80 x 80 x 64 km and were selected to provide a range of stand types, including a 6,000 ha area burned in 1969, a mature black spruce stand, immature balsam fir thickets, and mature spruce/fir mixed-wood stands. The catches from 7 such plots, spaced at 30-km intervals along this route, including the catches from the center of the burned area, are shown in Fig. 8. Even allowing for the fact that the jump in catches from 1975 to 1976 may be attributable to the changes from Sectar 1 to Pherocon 1CP traps, the point of interest is that the same trend in catches is evident from all the plots. Hopefully, this is an indication that population densities over the whole area were rising and falling synchronously. However, the same effect could be explained by postulating years of good climatic conditions and years of bad climatic conditions for male flight and orientation. Since the climatic factors fluctuate on a macro scale, this effect would be similar over large areas. In any event, the important point is that a few trapping locations could be used to indicate general trends over large areas.

FUTURE STRATEGIES

A possible design for the use of the pheromone traps when monitoring spruce budworm populations, therefore, is as follows: Sampling plots would be located along accessible routes in stands of susceptible host trees at intervals of up to 50 km. Ideally, the intervals should be less than this, perhaps every 10 km, but in reality, the number of locations that can be sampled will be limited by practical considerations of manpower and time. The stands selected should be principally those which, according to our current knowledge, will support the most rapid spruce budworm population growth rates; i.e., mature spruce/fir stands where tree crowns are well exposed to the sun. Two clusters of 5 traps each would be placed out in each location, the clusters separated by at least 200 m to avoid interference and the traps within each cluster separated by 10 m. Both clusters would be baited with PVC lures, 4 mm diam. x 10 mm long, one cluster containing .03% fulure by weight, the other .0003%. Traps would be deployed at a height of 2 m on marked trees about 10 days before the start of moth flight. Traps would be collected after the moth flight. Deployment and

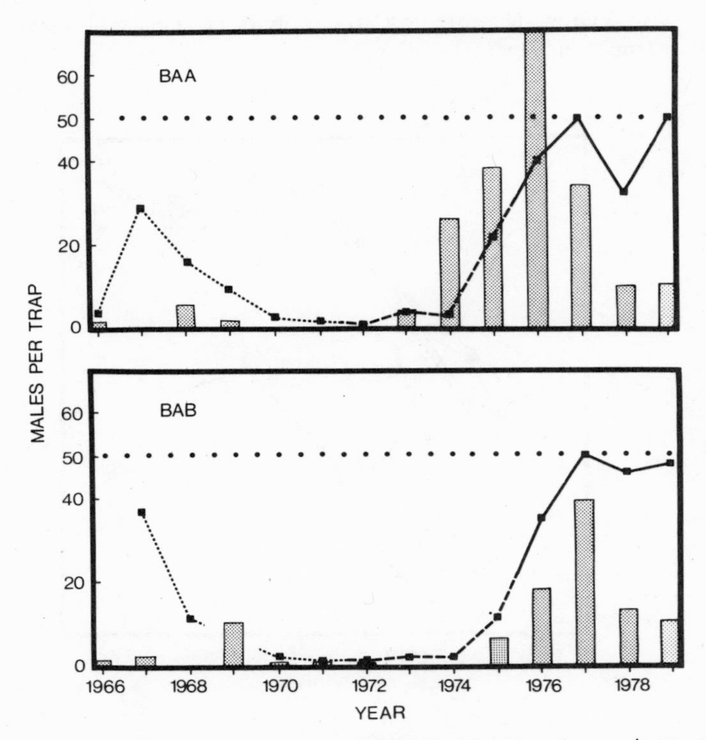

Fig. 7 Density of spruce budworm larvae (nos./100 branch tips; histograms) and catches of male budworm in sex attractant traps. 1966 through 1971, virgin females as lures in 30 x 30 cm trap-boards; 1972 through 1975 PVC lures containing 4.5 mg fulure in Sectar 1 traps; 1976 through 1979 PVC lures (as above) in Pherocon 1 CP traps.

collection of traps could be timed to coincide with the sampling of large larvae and egg populations, respectively, if they are being carried out in the same areas to save the number of visits to the sampling plots. Trapping would be carried out annually. When catches showed significant population increases or had exceeded the prescribed threshold, then a more intensive sampling program would be conducted in the area using techniques which provide accurate population estimates. This, of course, could include sex attractant traps if catches are found to be well correlated with population density.

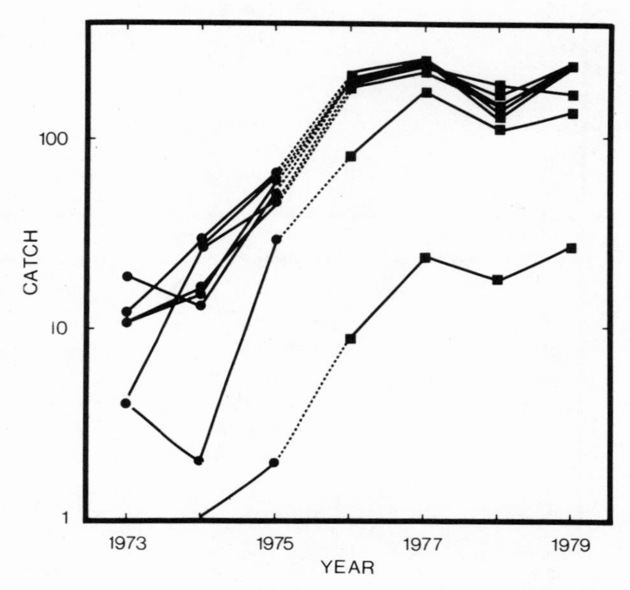

Fig. 8 Catches of male spruce budworm at 7 locations approximately 30 km apart along a triangular route in northwestern Ontario, and in the center of a 6000 ha area burned in 1969 (lower line). All traps baited with PVC containing 4.5 mg fulure. From 1973 through 1975 Sectar 1 traps; 1975 through 1979 Pherocon 1CP traps.

The most important step remaining before such a program can be implemented is to establish correlations between catch and population density over a wide range of stand types and geographic locations. This will enable us to calibrate the changes in catch with rates of change in population density and to establish thresholds indicating that populations have reached a threatening level.

REFERENCES

Blais, J. R., 1979, Comparison of spruce budworm overwintering populations with emerged populations (peak L_2) in the lower St. Lawrence region of Quebec, 1978, Can. Dept. Environ. Bi-Mo. Res. Notes, 35: 33.

Daterman, G. E., 1974, Synthetic sex pheromone for detection survey of European pine shoot moth, USDA For. Serv. Res. Paper PNW-180, 12 pp.

Fitzgerald, T. D., St. Clair, A. D., Daterman, G. E., and Smith, R. G., 1973, Slow release plastic formulation of the cabbage looper pheromone cis-7-dodecenyl acetate: Release rate and biological activity. Environ. Entomol., 2: 607.

Miller, C. A., and McDougall, G. A., 1968, A new sampling technique for spruce budworm larvae. Can. Dept. For. & Rural Develop., Bi-Mo. Res. Notes, 24: 30.

Miller, C. A., and McDougall, G. A., 1973, Spruce budworm moth trapping using virgin females, Can. J. Zool., 51: 853.

Miller, C. A., Forbes, R. S., and Dobson, C. M., 1968. Studies of the spruce budworm in the Maritimes Region by the Forest Insect & Disease Survey. I. A comparison of larval surveys, 1957-1967 with two independent sources of population data, Can. Dept. For. & Rural Develop. Internal Rept. M-34, 15 pp.

Sanders, C. J., 1978, Evaluation of sex attractant traps for monitoring spruce budworm populations (Lepidoptera: Tortricidae). Can. Entomol., 110: 43.

Sanders, C. J., and Weatherston, J., 1976, Sex pheromone of the eastern spruce budworm: Optimum blend of trans and cis-11-tetradecenal, Can. Entomol., 108: 1285.

Silk, P. J., Tan, S. H., Wiesner, C. J., and Ross, R. J., 1980, Sex pheromone chemistry of the eastern spruce budworm, Choristoneura fumiferana, Environ. Entomol., (In press).

FUTURE THRUSTS FOR DEVELOPMENT OF INSECT SEX PHEROMONES AS MONITORING TOOLS

Monitoring pest populations with pheromone-baited traps has long been recognized as one of the major benefits of sex pheromone research. At first glance this technique seems simple enough. After all we have powerful synthetic pheromones, manufactured traps, and reasonably effective formulation systems. The problem lies in interpreting the trap catches. We find it difficult to estimate population sizes from the number of trapped insects. The season of the year, population level, plant host, weather conditions, and many other conditions affect our ability to extrapolate from our trapping sample to the whole population.

Situations where precise population size estimates are not required have proved thus far to be more amenable to pheromone monitoring. Biomonitoring in quarantine situations and in food processing plants usually require only a "yes" or "no" answer. The question to be answered is not how many insects but whether there are <u>any</u> insects. Most monitoring situations are going to require a more quantitative answer.

It seems obvious that the major bottleneck to more fully developing pheromone monitoring systems is in the area of correlating trap catch to field populations and to subsequent damage levels. Progress toward solving this problem is slow. In fact, few people are even working in this area.

We feel that government will have to continue to bear the major expense for the development of pheromone monitoring. There just does not appear to be sufficient monitoring incentives to entice commercial concerns into any concerted effort in this field as there is in pheromone disruption. So what can be done?

(1) Recruitment of additional people in this field, particularly ecologists, is a primary concern if the work is to be accelerated.

(2) Increased government support. This support should not be difficult to justify. Reduced pesticide usage, improved pesticide usage, greater use of natural controls, and extended life of pesticides are well documented benefits of biomonitoring.

(3) Give high priority to the identification of sex pheromones for key pest species.

PANEL: K. W. Vick, USDA; W. E. Burkholder, USDA; R. T. Cardé, Michigan State University; H. F. Madsen, Canada; J. Stockel, France; D. D. Hardee, Mississippi; C. J. Sanders, Canada; J. W. Kennedy, Maryland.

THE "3-BODY" PROBLEM ANALOGY IN MASS-TRAPPING PROGRAMS

Roy T. Cunningham

Tropical Fruit and Vegetable
Research Laboratory
AR-SEA, USDA
Honolulu, HI 96804 USA

INTRODUCTION

Every mass-trapping program has to deal with decisions as to the pattern of allocation of the attractant in space and time. Is it better to have many traps each with a little lure or to have a few traps with a lot of evaporating surface and lure? If a trap geometry or formulation slows the evaporation rate, is this an advantage or disadvantage? How do we strike a balance to achieve the most efficient use? That is, how do we get the greatest kill per unit or lure used?

Usually, these 3 variables; spacing, dose/trap, and vaporization rate, are somewhat subject to our control (the last, however, much less controllable than we would like). With a given amount of lure available, changing one of these variables affects the amount available for the other 2, and I have chosen to look upon our situation as a kind of analogy to the 3-body problem in Newtonian mechanics--that is the determination of the vectors of 3 bodies moving under mutual gravitational influence. The conservation of the momentum of the system corresponds to the fixed amount of lure we have to work with, and the interdependence of the 3 variables in trapping operations is analogous to the interdependent movement of the 3 bodies. However, the mechanics problem can be readily solved by partial differentials plugged into a computer program, and the system is much more exact and simple than the biological system we are dealing with. The complexities and uncertainties introduced by unknown, uncontrolled, and/or ill-measured variables and the nonlinear interrelationships in trapping programs are not easily solved. Nor, unfortunately, for the most part, has the entomological research been systematically directed (and I am including our own) toward gathering the data necessary for mathematical optimization procedures. Notable exceptions to this have been the work of Wolf et al. (1971), Hartstack et al. (1971), and McClendon et al. (1976). Their work concentrated on trap spacing. I would like to review some of the work in our laboratory that is in a similar vein to theirs, but from a somewhat different approach.

We are fortunate to have 3 good male lures, one for each of the 3 species of tropical tephritid fruit flies with which we work. Methyl eugenol (Howlett, 1915) for the Oriental fruit fly, *Dacus dorsalis* Hendel; trimedlure (Beroza et al., 1961) for the Mediterranean fruit fly, *Ceratitis capitata* (Wiedemann); and cue-lure (Alexander et al., 1962) for the melon fly, *D. cucurbitae* Coquillett. (None of these are pheromones, and their biological fit is not really clear.) Methyl eugenol has been used in mass-trapping experiments for over 20 years (Steiner and Lee, 1955; Steiner et al., 1970), and for the last 8 years has been used on a routine basis in mass-trapping programs developed to protect California from the periodic introduction of this fly from untreated, infested fruit smuggled in from Hawaii (Chambers et al., 1974).

Our flies are highly mobile beasts, and feeding, and breeding sites (though more limited with regard to the melon fly) are usually scattered over a wide area and include both wild and cultivated plants. Some of our best information on the effect of trap spacing has come from large-area test plots over several square kilometers in extent and, in this fashion, differs from the work of previously cited authors who have worked on the opposite scale in relatively small-field tests. Some of our future work will be patterned after their methods because both scales should be investigated.

METHODS AND RESULTS

In one of our large-scale tests in a 5-km^2 macadamia nut orchard (Cunningham and Steiner, 1972) in which the melon fly was breeding in *Momordica charantia* L., a weedy vine; we hung 5-cm^2 fiberboard blocks from the trees at regular spacing intervals. Each block had 25 g of a solution 5% naled in cue-lure. The first distribution rate was at 0.5 blocks/ha. Since these blocks retain undiminished effectiveness for 7 months, subsequent monthly distributions spaced between prior distributions were additive and we could follow the decrease in population with increase in trap density. The proportion of the population of males killed was a hyperbolic function of the number of blocks/ha (Fig. 1). That is, survival probability was in inverse relation to the number of traps/ha. It is an asymptotic function so that the number of additional captures becomes less and less on a per trap basis as trap density increases. Beyond a certain point, increasing density becomes uneconomical.

A similar hyperbolic curve was obtained in a test against the Oriental fruit fly, with 25% malathion in methyl eugenol soaked on cigarette filter tips (Hardee et al., 1972) and distributed by aircraft over a 7-km^2 plot of guava scrub and forest on the island of Lanai (Fig. 1). (These data are also well-fitted on an asymptotic regression curve). The asymptote was well below the maximum possible. We suspected this was because of poor distribution patterns on the ground. Although we were increasing the average number of tips/ha, there was very little side scatter beneath the flight path. To test this effect, we did the following small-scale test: In a 5-ha plot in a mixed tropical fruit orchard in repeated trials, we changed spacings between lines of filter tips dropped on the ground while maintaining the total number of tips constant. That is, as the space between rows of tips increased, the space between tips in the row decreased. The rows became functionally like a solid line of tips and the functional variable was the spacing between lines of tips. Our purpose was to investigate the effect of flight-line spacing variation. Again (Fig. 1), we got a hyperbolic relationship so that survival was an inverse function of the number of drop lines per plot; the spacing--not simply the average number of tips/ha--was an important variable.

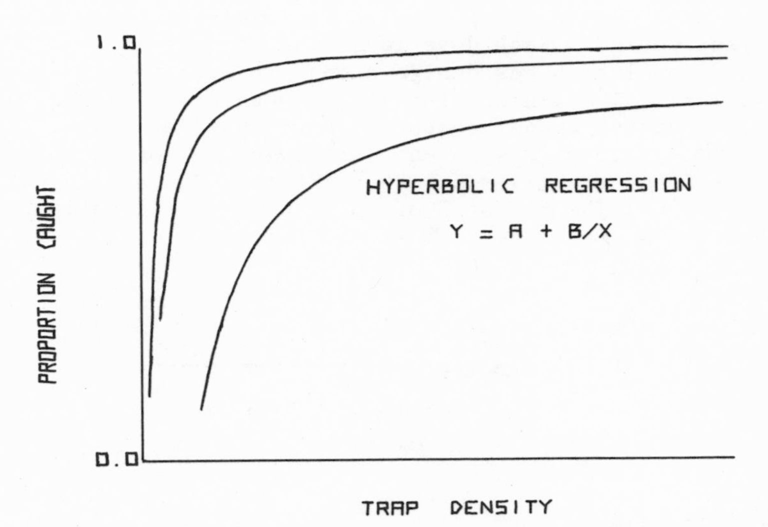

Fig. 1. Influence of trap density on proportion caught. Upper curve is cue-lure for melon fly on 5 km^2. Middle curve is methyl eugenol for Oriental fruit fly with various spacings between lines of "tips" on 5 ha. Lower curve is methyl eugenol on filter tips by aerial distribution over 7 km^2.

Obviously, from the shape of these curves with their steep initial slopes, we can get very good return for an increased investment when the trap density is in the low end of the range.

In other tests we studied the effect of varying the dosage of lure/trap site. We increased the number of methyl eugenol-soaked filter tips from 0.5 to 128/trap site (a trap being 1 or more topless 4-liter plastic buckets) so that the same surface to volume ratio was maintained throughout the series. That is, at the highest dose, the evaporating surface was 256 times larger than that of the lowest dose. This, of course, is an indirect measure of the vapor output/trap site.

I have found that there is a tendency for people with limited experience with good lures in field situations to overestimate their effectiveness. I refer to these people as belonging to the "open-bottle" school of thought. All you have to do is open the lure bottle and all the little insect automatons from nearby hectares will fly into the bottle. I am sure all of you have encountered similar recommendations in mass-trapping programs.

From the data we have just seen, we can say that at least 2 bottles would be a lot better; but even aside from the coverage and spacing problem, there is an inefficiency in placing too much lure at 1 trap site. We ran 3 separate tests of dosage response curves over a 1-year period. Each dose level was replicated 10 times in a randomized complete block design along a single 8-km-long semicircular trap line. The linear distance between traps in the line was 75 m so that between traps, competition was not important. Again (Fig. 2), an asymptote is apparent, although the more complex asymptotic regression curve gives a better fit in this case than the simple hyperbola. The fit of 2 of these curves is excellent as you

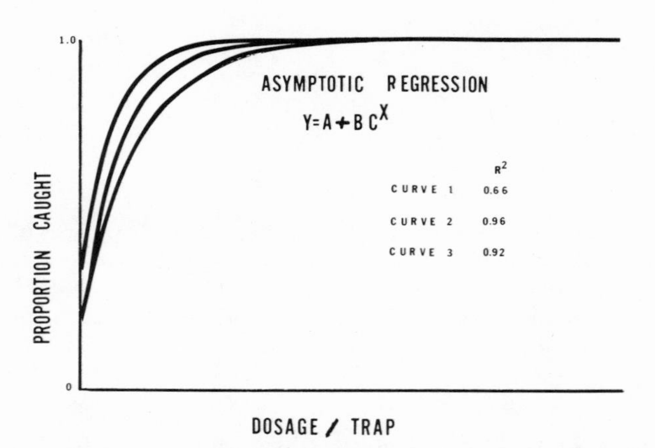

Fig. 2. Influence of amount of lure/trap on proportion caught. Methyl eugenol for Oriental fruit fly.

can see from the high coefficients of determination. This asymptotic limit to the dose-response curve cannot be accounted for by the physical aspects of diffusion but must result from the increasing insect searching error with distance from the point source. The open-bottle school might have some justification if we were dealing solely with physical diffusion because increasing vapor output should give an increasing diffusion area and increasing catch.

Although curves have a pleasingly close fit, we must insert a caveat here. They were generated by the accumulative catch during the first 2 weeks. If we were to graph the catches from the 4th plus 5th weeks, the slopes would be much shallower and the asymptotic limits would not be so evident (Fig. 3). This brings us to the 3rd variable usually somewhat, though imperfectly, controlled by the experimenter--the lure vaporization rate-lure allocation in time. This is usually imperfectly controlled by the formulation, and the trap and/or lure reservoir geometries.

Both this variable, the vaporization rate and the previous one, lure dosage, are indirect means in practice of controlling the true independent variable, vapor concentration at the trap site. In one experiment with methyl eugenol on cotton wicks in Steiner traps (Steiner, 1957), we enclosed a set of wicks (10 replicates) in glass tubes so that the lure could only evaporate from the ends. This set was never retreated to saturation during the 8 or so months of the experiment. A 2nd set of glass-enclosed wicks was retreated to saturation periodically to measure lure loss and relative performance in the untreated wick. The encased wick maintained undiminished effectiveness for 7 months after which it started to decline below the performance of the set which was kept saturated. We also had a set of unencased wicks. The encased wicks caught about only 20% of the catch rate of the unencased wicks. In 2 different periods where the rate of lure loss of the unencased wick was measured, we found good correspondence between catch rate and lure loss. The unencased wick evaporated 6 and 4 times as much lure and caught 6 and 4 times as many flies in these 2 periods. Obviously, we were working on the low end of our asymptotic dose-response curve.

DISCUSSION

If we were faced with making decisions in an expensive large-area, mass-trapping program over hundreds or thousands of square kilometers, we need to answer the question as to what is the best strategy--many traps with low rates of output retreated every 6 months, or fewer traps with higher rates of output that are retreated every month?

I have prepared a simple model of the interaction of 2 of our variables (Fig. 4). This model is still theoretical because we have not actually tested the interactions, but I think it is a reasonable guess as to what we will find for this particular system,

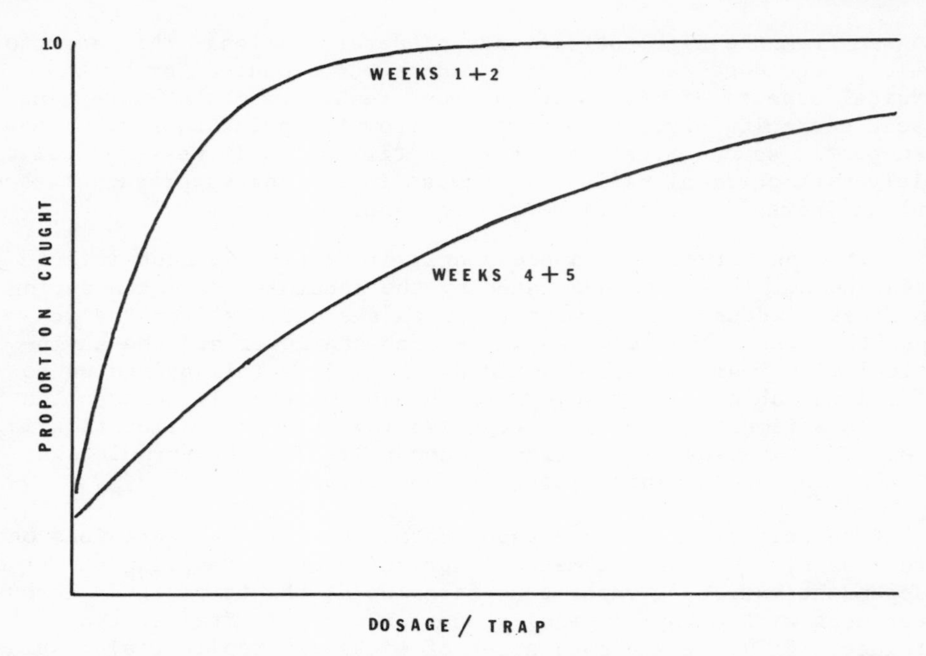

Fig. 3. Influence of amount of lure/trap on proportion caught with changing time. Methyl eugenol for oriental fruit fly.

methyl eugenol on filter tips. This 3-dimensional model generates a smoothly curving response surface. Because of the asymptotic nature of both curves, there is a large relatively flat area which gives poor return on increased investment. This model, although a step in the right direction, is far too simple. The 4th dimension on this should be time. And, of course, in any real program we should plug in costs to get the 5th through the nth dimensions.

The basic tools for mass-trapping programs were developed by entomologists teaming with biochemists. If your experience has been like mine, and I think it has, you will have applied as much art and experimental intuition as science in getting them to work on a large scale. Our 3-body problem is, in reality, a multibody, a multidimensional problem.

If we wish to see the good groundwork develop to its full fruition on routine large-area, mass-trapping programs extending over hundreds or thousands of square kilometers, then, we are going to have to solicit aid from colleagues in other specialties, especially mathematical optimization procedures. We will have to design our research to address the questions as formulated by them to replace most, but never all, of the art with science.

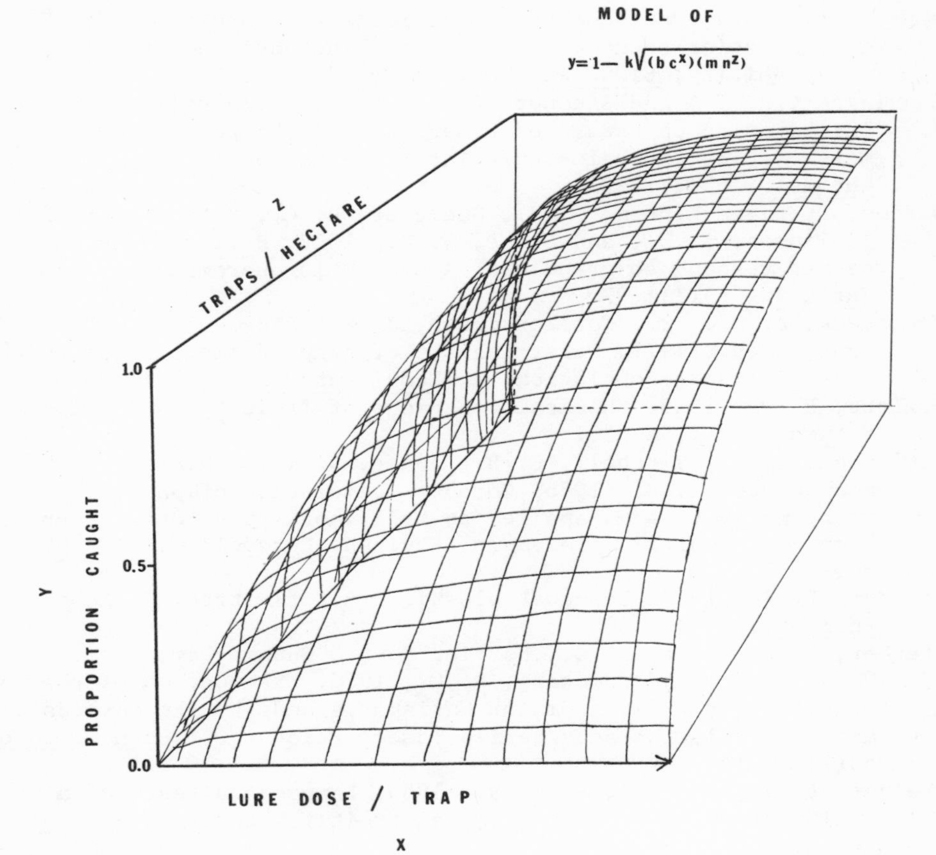

Fig. 4. Response surface model of proportion caught as a function of trap density and lure dosage/trap.

ACKNOWLEDGMENT

I wish to especially acknowledge the aid of David Y. Suda, Agricultural Research Technician, for his valuable help over the many-years that these studies encompassed.

REFERENCES

Alexander, B. H., Beroza, M., Oda, T. A., Steiner, L. F., Miyashita, D. H., and Mitchell, W. C., 1962, The development of male melon fly attractants, J. Agric. Food Chem., 10(4): 270.

Beroza, M., Green, N., Gertler, S. I., Steiner, L. F., and Miyashita, D. H., 1961, New attractants for the Mediterranean fruit fly, J. Agric. Food Chem., 9(5): 361.

Chambers, D. L., Cunningham, R. T., Lichty, R. W., and Thrailkill, R. B., 1974, Pest control by attractants: A case study demonstrating economy, specificity, and environmental acceptability, BioScience 24(3): 150.

Cunningham, R. T., and Steiner, L. F., 1972, Field trial of cue-lure + nalad on saturated fiberboard blocks for control of the melon fly by the male-annihilation technique, J. Econ. Entomol., 65(2): 505.

Hardee, D. D., McKibben, G. H., Gueldner, R. C., Mitchell, E. B., Tumlinson, J. H., and Cross, W. H., 1972, Boll weevils in nature respond to grandlure, a synthetic pheromone, J. Econ. Entomol., 65(1): 97.

Hartstack, A. W., Jr., Hollingsworth, J. P., Ridgway, , R. L., and Hunt, H. H., 1971, Determination of trap spacings required to control insect population, J. Econ. Entomol., 64(5): 1090.

Howlett, F. M., 1915, Chemical reactions of fruit flies, Bull. Entomol. Res., 6: 297.

McClendon, R. W., Mitchell, E. B., Jones, J. W., McKinion, J. M., and Hardee, D. D., 1976, Computer simulation of pheromone trapping systems as applied to boll weevil population suppression: A theoretical example, Environ. Entomol., 5(5): 799. 806.

Steiner, L. F., 1957, Low-cost plastic fruit fly trap, J. Econ. Entomol., 50(4): 508.

Steiner, L. F., Hart, W. G., Harris, E. J., Cunningham, R. T., Ohinata, K., and Kamakahi, D. C., 1970, Eradication of the Oriental fruit fly from the Mariana Islands by the methods of male annihilation and sterile insect release, J. Econ. Entomol., 63(1): 131.

Steiner, L. F., and Lee, R. K. S., 1955, Large-area tests of a male-annihilation method for Oriental fruit fly control, J. Econ. Entomol., 48(2): 311.

Wolf, W. W., Kishaba, A. N., and Toba, H. H., 1971, Proposed method for determining density of traps required to reduce an insect population, J. Econ. Entomol., 64(4): 872.

SUPPRESSION OF *DENDROCTONUS BREVICOMIS* BY USING A MASS-TRAPPING TACTIC

William D. Bedard

Pacific Southwest Forest and Range
Experiment Station
Forest Service, USDA
Berkeley, CA 94701 USA

David L. Wood

Department of Entomological Sciences
University of California
Berkeley, CA 94720 USA

INTRODUCTION

Bark beetles (Scolytidae that feed on phloem and cambium) are a major pest of coniferous forests throughout the northern hemisphere (Rudinsky, 1962). Many bark beetle species use aggregative pheromones to colonize temporary habitats where they feed, mate, and reproduce (Wood, 1972; Borden, 1974). These pheromones offer exciting possibilities for pest suppression because they attract both sexes and because aggregation is essential for reproduction (Wood, 1972; Wood and Bedard, 1977).

The western pine beetle, *Dendroctonus brevicomis* LeConte, kills about a billion board feet of ponderosa pine, *Pinus ponderosa* Dougl. ex. Laws., yearly in the western United States and Canada (Miller and Keen, 1960). Research to develop a mass-trapping suppression tactic over large, contiguous, forested areas has progressed further with this pest than with other bark beetle species.

The research has followed the necessary sequence of: 1) demonstration of a potent attractant from a natural substrate, 2) chemical characterization of the attractant, 3) development of a mass-trapping treatment, 4) determination of the effects of use at an operational scale, and 5) development of guides for applying the treatment in the context of all forest management goals.

This report reviews that research and discusses its relevance to developing a mass-trapping suppression tactic for other bark beetle species. It outlines the progress that has been made to date and work that needs to be done before this new technology is available for use.

Demonstration of Attraction

For mass-trapping to be effective, it must be used with a potent synthetic attractant that mimics the characteristics of the natural attractant (Vité, 1970; Bedard and Wood, 1974). But first it must be demonstrated that a potent natural attractant exists. Then this attractant would serve as a standard for developing bioassays in the process of chemically reproducing it.

Preliminary studies of attraction with *D. brevicomis* were conducted as early as 1916 (Miller and Keen, 1960). By 1933 attraction was reported to be associated with trees showing specific characteristics (e.g. slow growth), the presence of attacking beetles in their entrance tunnels, and the fermention of inner phloem of ponderosa pine (Miller and Keen, 1960). Person (1931) hypothesized that an initial weak attraction originated in the host, and that stronger, secondary attraction was associated with attacking beetles, probably caused by fermentation of inner phloem by yeast introduced by the beetles. Vité and Gara (1962) reported that the mass attack stimulus for *D. brevicomis* was produced by mature females boring in ponderosa pine. Wood and Bushing (1963) demonstrated in a laboratory bioassay that *D. brevicomis* females boring in ponderosa pine were attractive to males. Bedard et al. (1969) found that ponderosa pine bolts freshly infested with female *D. brevicomis* were attractive to both sexes in the field. These experiments were adequate to prove chemically elicited attraction and to develop laboratory and field bioassay procedures to evaluate attraction. But they did not compare the potency of the attractant produced by females boring in cut bolts with the attractant produced in living trees under attack by this bark beetle. Bedard et al. (unpublished data) demonstrated that the attraction produced by boring females was enhanced by the addition of males to female galleries. Thus, the possible origin of the secondary attraction described by Person (1931) for trees under attack by *D. brevicomis* may finally have been determined.

Chemical Characterization of the Attractant

Research leading to the discovery that the mixture of (1R,5S, 7R)-(+)-*exo*-brevicomin, (1S,5R)-(-)-frontalin, and myrcene is a highly potent attractant has been reviewed by Wood (1972) and by Wood and Bedard (1977). *exo*-Brevicomin is released by females, frontalin by males, and myrcene from the host. Browne et al. (1979) demonstrated that *exo*-brevicomin and myrcene are released from female-infested bolts and that the addition of males to female-infested bolts results in release of frontalin.

Verbenone and *trans*-verbenol were also isolated from male and female *D. brevicomis*, respectively (Renwick, 1967; Pitman et al., 1968). Pitman et al. (1968) did not identify a role for these compounds, but Renwick and Vité (1970) postulated that verbenone interrupts (Wood, 1977) the response of *D. brevicomis* to the attractant. A mixture of verbenone and *trans*-verbenol reduced the catch of *D. brevicomis* at a source of the attractive mixture of racemic *exo*-brevicomin, racemic frontalin, and myrcene (Bedard et al., 1980a; Bedard et al., 1980b). Each compound released alone reduced catch at and near a source of the attractive mixture of *exo*-brevicomin, frontalin and myrcene (Bedard et al., 1980b). *trans*-Verbenol may, however, act as a multi-functional pheromone for *D. brevicomis* (Bedard et al., 1980b) because it enhances attraction to the attractive mixture at a low release rate.

In evaluating the relative attractiveness of compounds, Bedard et al. (1980a) found that single compounds and most binary mixtures are either unattractive or very weakly attractive. *exo*-Brevicomin and frontalin combined are highly attractive as are the ternary mixtures containing *exo*-brevicomin, frontalin, and ponderosa pine turpentine or a monoterpene constituent, e.g. myrcene. The variation in daily catch at traps with the same compounds was so large that traps with other compounds with greatly different total catches were not significantly different.

These findings are important to the accurate and complete characterization of other bark beetle attractive pheromones for several reasons. More than one compound is usually involved: different compounds emanate from different sources -- both sexes and the host; beetle response can depend on enantiomeric composition of the compounds; compounds can be multifunctional; some compounds interrupt the response to attractants; compounds that are unattractive or weakly attractive alone may synergize the attractiveness of other compounds or mixtures of compounds; and finally, field bioassays may not be sensitive enough to detect subtle effects (Bedard et al., 1980a).

Development of Technology

Once a potent synthetic attractant is available, some means of using it to mass-trap beetles must be developed; i.e., a trapping system optimized for catch at the attractant source that can be produced in large quantities must be developed. The trap, attractant release rate, and the response of natural enemies and other non-target organisms must all be considered.

Tilden et al. (1979) found that *D. brevicomis* would fly into unbaited sticky traps near a source of attractant. They found that more beetles are caught when the trapping surface was increased. When a simulated tree stem silhouette is present at a source of attractant, the catch is increased at the source, and reduced but not eliminated at surrounding unbaited silhouettes (Tilden, 1976). As attractant release rate was increased, more beetles were caught and they tended to be trapped farther from the source of attractant. Higher attractant release rates also increased the tendency for host trees near a source of attractant to be attacked by *D. brevicomis*.

On the basis of these results, Browne (1978) developed a sticky trap with these attributes: It is easily portable, has a large trapping surface (6 m^2), is relatively inexpensive (ca. $10 ea in 1970), and durable (one flight season under field conditions). It proved adequate for experiments to evaluate the effects of operational scale mass-trapping on tree mortality caused by *D. brevicomis* (Bedard and Wood, 1974; Browne et al., 1979).

Ideally, a trapping system should exclude predators. Bedard et al. (1969) and Vité and Pitman (1969) reported that *Temnochila chlorodia* (Mannerheim) (Coleoptera: Trogositidae), a predator of *D. brevicomis* and other bark beetles and wood boring insects, is attracted to *exo*-brevicomin. We chose to evaluate the effects of operational scale mass-trapping before justifying additional research on trap design to exclude larger-size predators. Meanwhile, such a trap was developed (Moser and Browne, 1978), although its efficacy in catching *D. brevicomis* needs to be established.

In addition to an adequate supply of traps, an adequate supply of attractant must be available. Although myrcene is relatively inexpensive; i.e., dollars/kilogram, *exo*-brevicomin and frontalin are both very expensive; i.e., thousands of dollars/kilogram. Rodin et al. (1971) developed an upscaled synthesis for *exo*-brevicomin. Work is currently underway for developing more economical means of synthesizing *exo*-brevicomin and frontalin (Mundy et al., 1977).

Tilden (1976) and Bedard and co-workers (unpublished data) demonstrated that attractant release rate is critical. A low rate (0.1 mg/24 hrs/compound) resulted in low catch while a high rate (10 mg/24 hrs/compound) resulted in beetle attack on hosts near the attractant source. Thus, a means to release attractant at an appropriate rate consistently, throughout extended periods, is needed. Silverstein and Rodin (personal communication) developed a system of vials, one for each compound. By varying the length and inside diameter, each compound with a different vapor pressure was released at the desired rate, e.g. for *D. brevicomis*, 2 mg/24 hrs, for a month.

The attractive compounds must maintain their authenticity under field conditions. Both *exo*-brevicomin and frontalin are stable in air and sunlight (Silverstein, personal communication), and myrcene is kept from polymerizing by adding 1% by weight of "ethyl" antioxidant 330 and by using an aluminum sun screen (Silverstein and Rodin, unpublished data).

Operationally Scaled Mass-Trapping Experiment

Before mass-trapping can be used operationally, it must accord with current policies and laws on biological and economic efficacy and environmental safety; and it must be used in the context of an integrated pest management system (Waters and Stark, 1980; Wood, 1977, 1979, 1980). The development of such a system is complicated by the need to consider the diversity of forest ecosystems, the benefit and costs of treatment, and differences in management objectives.

Efficacy. To use a treatment, forest managers need to know its end results in terms of protected stand growth and development (Waters and Stark, 1980; Wood, 1977, 1978, 1980). In turn, predicting such outcomes depends on an understanding of what the major direct and indirect treatment effects are, and how they interact within the forest ecosystem to influence trends in stand growth and development. Such major direct and indirect effects of large-scale, mass-trapping experiments with *D. brevicomis* are not easily assessed (Bedard et al., 1979; Wood, 1979). The effect of many traps in an area cannot be determined by trials with a single trap because many traps could elicit different behavior, e.g. attractant released from many traps could result in interruption while the same release rate from a single trap could result in attraction. Moreover, decisions as to how an evaluation should be made; i.e., the evaluation criteria and methodology, and how long it should continue depend on the suspected treatment outcomes, the purpose(s) of the evaluation, and the availability of technology to make the appropriate measurements and estimates.

The potential effects of operational scale mass-trapping include: a) The movement over some distance of D. brevicomis and natural enemies in response to the treatment; b) tree killing by attracted beetles; c) interruption of aggregation on trees so that they are not successfully mass attacked (thus, attacking beetles are killed by the host oleoresin system); d) interruption of response to traps by high release rate of attractant; e) augmentation or reduction of natural enemies (both species and number of individuals) relative to D. brevicomis (Bedard and Wood, 1974). Therefore, an evaluation should cover at least the spatial distribution and abundance of D. brevicomis, its principal natural enemies, trees killed by D. brevicomis, and the incidence of both successfully and unsuccess fully attacked trees near traps studied in a large area before, during and after the treatment.

A large-scale (65 km^2) experiment to suppress D. brevicomis by trapping (Bedard and Wood, 1974; Browne et al., 1979; DeMars et al., 1980) demonstrated that such data are essential. During the period of the treatment, D. brevicomis-caused tree mortality was concentrated in areas where traps were concentrated (4 plots each 2.56 km^2). A large number of D. brevicomis and T. chlorodia were trapped in these plots and in the surrounding area where the same traps were deployed on an 0.8-km grid. Tree mortality caused by D. brevicomis declined in both the suppression plots and the surrounding area during the treatment (283 trees killed in the generation prior to trapping; 91 in the last generation during trapping) and further declined after trapping (about 24 trees the generation after trapping) and remained very low for the next 4 years. It appears that treatment effects occurred far beyond the area where traps were deployed, both in space and time. Preliminary analyses indicate that the sampling methods used to estimate absolute populations (i.e., total number within a defined region and time) of specified species for evaluation of space/time effects will have to be improved before treatments can be adequately assessed at these levels.

In essence, this approach to treatment evaluation as well as development of the capability to predict the course of infestations requires further research in population biology to determine what agents or factors influence change in pest numbers and related damage, and the role of treatments in the context of these population changes. Obviously, this research is complex and requires a large commitment of resources. Even with this approach, uncertainty arises about the generality and practical applicability of the information obtained.

Are there alternative means to evaluate treatment effects and safety and to predict the occurrence and trends of infestations? This question was recently examined in the context of a related pest, the southern pine beetle, D. frontalis Zimmerman, (Coster and Searcy, 1979). The several alternatives examined range from explanatory, quantitative population biology in which simulation modeling is used to evaluate a variety of treatment tactics in a variety of pest situations (Coulson et al., 1979), to an empirical approach (Billings and Pase, 1979) in which an operational tactic is evaluated in terms of the spatial and temporal aspects of beetle-caused tree mortality. The former method provides the potential to adequately understand population biology in order to fully evaluate the effects of treatments, determine their safety with respect to environmental aspects, and develop the capability to predict the course of infestations with or without trapping or other suppression tactics. Its weakness is that it may not reflect reality accurately and may require so much additional work in the iterative process of validation, update, and further validation that it is economically unfeasible. The strength of the empirical method is that it reflects reality. It does not, however, provide much understanding on how the treatment works, and supports prediction with regard to limited aspects of tree mortality and with no guarantee of accuracy at the local infestation level. Also, it considers only a few aspects of biological safety of treatments; i.e., those reflected in tree mortality patterns observed.

These different approaches should be sequential leading toward the goals of evaluating efficacy, determining environmental safety, and developing some of the understanding needed for prediction. The initial evaluation should be extensive and intensive to detect the possible major treatment effects, e.g. the study described earlier for D. brevicomis. In this approach, causal relationships between treatment and effects can be obscured by the complex and interactive nature of the system under study and by the lack of adequate replication dictated by the scope of the study. Results from this approach, however, can then be used to design a replicated study to demonstrate that mass-trapping prevents D. brevicomis-caused tree mortality. With utility thus established, further studies for specific purposes, e.g. trap improvement, treatment optimization for specific pest situations, and prediction of treatment outcome in specific pest situations, can be justified.

Safety. Safety of a treatment may be judged on the basis of its hazard to man in manufacture, formulation, handling and application; its hazard to wildlife and domestic stock as it is operationally formulated and applied; and any possible irreversible damage to the physical, biological, economic of social aspects of the environment (Wood, 1979).

The 3 components of the attractant for D. brevicomis are not toxic as determined in acute oral and dermal, eye and skin irritation, and in inhalation tests with laboratory animals (Bedard, unpublished data). This lack of toxicity combined with trapping where the compounds are held unavailable for direct contact should insure safety to humans in handling and application. The release rates during trapping do not exceed those estimated to occasionally occur in nature (Browne et al., 1979) and the compounds are not released in a way that will contaminate plant, animal, soil or aquatic systems. Thus, the release of the attractant into the environment should be safe.

Non-target flying insects and a few bats and small birds are caught and perish on the sticky traps. When catches of baited and unbaited traps were compared in a later, large-scale mass-trapping experiment, only *D. brevicomis* and *T. chlorodia* were caught in higher numbers on baited traps (Wood, unpublished data). This trapping via attraction would be expected to cause a temporary, local reduction in abundance, while the trapping of apparently unattracted species would be expected to cause the same effect but to a lesser degree. In both cases, trapping probably would not cause irreversible damage and could be avoided if an efficacious, non-sticky trap can be developed.

Costs. Costs of treatments can be both monetary and environmental. Efficacy trials provide an opportunity to determine both costs. Monetary costs are the direct expenditures for the treatment. Environmental costs are the values attached to undesirable effects of treatments. The results of tests to date suggest that environmental costs probably are minimal for mass trapping *D. brevicomis*.

Treatment in Relation to Forest Management Goals

The era of justifying pest suppression on publicly-owned forest land solely on the basis of apparent damage ended with the passage of the National Environmental Protection Act, National Forest Management Act, and Federal Insecticide, Fungicide and Rodenticide Act (Waters and Stark, 1980). Now, suppression must be justified on a benefit/cost basis in the context of other forest management goals. Benefits and costs associated with the physical, biological, economic, and social aspects of the environment must all be considered. The information required for a benefit/cost analysis include the values (e.g. monetary, environmental) to be lost if suppression is not done, the values to be gained by suppression, and the costs (monetary, environmental) of suppression (Wood, 1977, 1979). This analysis depends on a detailed understanding of the population biology of the pest in order to predict the outcome of an infestation with or without suppression (Waters and Stark, 1980; Wood, 1979) and the current and future changes in values caused by the pest's damage.

When we compare what we know about D. brevicomis suppression with what we need to know in order to make informed decisions in the context of integrated pest management, we find that we know too little about the population biology of the pest to predict the natural course of infestations, how to estimate beetle abundance in order to evaluate treatment effects on abundance, the effects of beetle-caused tree mortality on stands to predict stand growth and development with and without that mortality, and the current and future values affected by beetle activity to predict the values lost with no treatment or the benefits of treatment.

Wood's (1977, 1979) conclusions on the use of behavior-modifying chemicals in integrated pest management can be applied to D. brevicomis; i.e., identification of the attractant and development of the technology to support mass-trapping experiments are the initial steps. These first steps can be carried out by biologists, chemists and statisticians on modest budgets. The other steps, the development of methods to evaluate treatment effects and to predict biological and economic outcomes with or without treatment, require the additional expertise of economists, population biologists, silviculturalists, and forest pathologists, and are much more costly. Thus, compliance with legal and political requirements for the use of any pest treatment requires the understanding and support of research administrators. Otherwise, alternatives to the use of toxic chemicals will remain only a promising technology unavailable to forest managers to combat damaging pests.

ACKNOWLEDGMENTS

We thank L. E. Browne, K. Q. Lindahl, Jr., P. A. Rauch, and P. E. Tilden for their substantial contributions in the development of the ideas, and for completing much of the work reported herein. Also, we thank the latter 3 individuals and, E. P. Lloyd and W. E. Waters for helpful reviews of the manuscript.

REFERENCES

Bedard, W. D., Tilden, P. E., Wood, D. L., Silverstein, R. M., Brownlee, R. G., and Rodin, J. O., 1969, Western pine beetle: Field response to its sex pheromone and a synergistic host terpene, myrcene, Science, 164: 1284.

Bedard, W. D., and Wood, D. L., 1974, Programs utilizing pheromones in survey or control: Bark beetles -- the western pine beetle --, in: "Pheromones," M. C. Birch, ed., Elsevier/North-Holland, Amsterdam.

Bedard, W. D., Wood, D. L., and Tilden, P. E., 1979, Using behavior modifying chemicals to reduce western pine beetle-caused tree mortality and protect trees, in: "Current Topics in Forest Entomology: Selected Papers from the XVth International Congress of Entomology, Washington, DC, August 1976," USDA Forest Service General Tech. Report WO-8.

Bedard, W. D., Wood, D. L., Tilden, P. E., Lindahl, K. Q., Silverstein, R. M., and Rodin, J. O., 1980a, Field response of the western pine beetle and one of its predators to host- and beetle-produced compounds, J. Chem. Ecol., 6: 625.

Bedard, W. D., Tilden, P. E., Wood, D. L., Lindahl, K. Q., and Rauch, P. A., 1980b, Effects of verbenone and trans-verbenol on the response of Dendroctonus brevicomis to natural and synthetic attractant in the field, J. Chem. Ecol., (In press).

Billings, R. F., and Pase, H. A., III., 1979, Spot proliferation patterns as a measure of the area-wide effectiveness of southern pine beetle control tactics, in: "Evaluation Control Tactics for the Southern Pine Beetle: Symposium Proceedings, Many, LA, January-February, 1979," USDA Forest Service Tech. Bull. No. 1613.

Borden, J. H., 1974, Aggregation pheromones in the Scolytidae, in: "Pheromones," M. C. Birch, ed., Elsevier/North-Holland, Amsterdam.

Browne, L. E., 1978, A trapping system for the western pine beetle using attractive pheromones, J. Chem. Ecol., 4: 261.

Browne, L. E., Wood, D. L., Bedard, W. D., Silverstein, R. M., and West, J. R., 1979, Quantitative estimates of the attractive pheromone components, exo-brevicomin, frontalin and myrcene, of the western pine beetle in nature, J. Chem. Ecol., 5: 397.

Coster, J. E., and Searcy, J. L., eds., 1979, "Evaluating Control Tactics for the Southern Pine Beetle: Symposium Proceedings, Many, LA, January-February, 1979," USDA Forest Service Tech. Bull. No. 1613.

Coulson, R. N., Feldman, R. M., Fargo, W. S., Sharpe, P. J. H., Curry, G. L., and Pulley, P. E., 1979, Evaluating suppression tactics for Dendroctonus frontalis in infestations, in: "Evaluating Control Tactics for the Southern Pine Beetle: Symposium Proceedings, Many, LA, January-February, 1979," USDA Forest Service Tech. Bull. No. 1613.

DeMars, C. J., Slaughter, G. W., Bedard, W. D., Norick, N. X., and Roettgering, B., 1980, Estimating western pine beetle-caused tree mortality for evaluating an attractive pheromone treatment, J. Chem. Ecol., (In press).

Miller, J. M., and Keen, F. P., 1960, "Biology and Control of the Western Pine Beetle," USDA Misc. Publ. 800.

Moser, J. C., and Browne, L. E., 1978, A nondestructive trap for Dendroctonus frontalis Zimmerman (Coleoptera: Scolytidae), J. Chem. Ecol., 4: 1.

Mundy, B. P., Lipkowitz, K. B., and Dirks, G. W., 1977, Chemistry of the 6,8-dioxabicyclo [3.2.1] octane series sources, synthesis, structures and reactions, Heterocycles, 6: 51.

Person, H. L., 1931, Theory in explanation of the selection of certain trees by the western pine beetle, J. Forestry, 29: 696.

Pitman, G. B., Vité, J. P., Kinzer, G. W., and Fentiman, A. F., Jr., 1968, Bark beetle attractants: trans-Verbenol isolated from Dendroctonus, Nature, 218: 168.

Renwick, J. A. A., 1967, Identification of two oxygenated terpenes from the bark beetles Dendroctonus frontalis and Dendroctonus brevicomis, Contr. Boyce Thompson Inst. Pl. Res., 23: 355.
Renwick, J. A. A., and Vité, J. P., 1970, Systems of chemical communication in Dendroctonus, Contr. Boyce Thompson Inst. Pl. Res., 24: 283.
Rodin, J. O., Reece, C. A., Silverstein, R. M., Brown, V. H., and DeGraw, J. I., 1971, Synthesis of brevicomin, principal sex attractant of western pine beetle, J. Chem. Engin. Data, 16: 380.
Rudinsky, J. A., 1962, Ecology of Scolytidae, Ann. Rev. Entomol., 7: 327.
Tilden, P. E., 1976, Behavior of Dendroctonus brevicomis near sources of synthetic pheromones in the field, MS Thesis, University of California, Berkeley, 66 pp.
Tilden, P. E., Bedard, W. D., Wood, D. L., Lindahl, K. Q., and Rauch, P. A., 1979, Trapping the western pine beetle at and near a source of synthetic attractive pheromone: Effects of trap size and position, J. Chem. Ecol., 5: 519.
Vité, J. P., 1970, Pest management systems using synthetic pheromones, Contr. Boyce Thompson Inst. Pl. Res., 24: 343.
Vité, J. P., and Gara, R. I., 1962, Volatile attractants from ponderosa pine attacked by bark beetles (Coleoptera: Scolytidae), Contr. Boyce Thompson Inst. Pl. Res., 21: 251.
Vité, J. P., and Pitman, G. B., 1969, Insect and host odors in the aggregation of the western pine beetle, Can. Entomol., 101: 113.
Vité, J. P., and Renwick, J. A. A., 1970, Differential diagnosis and isolation of population attractants, Contr. Boyce Thompson Inst. Pl. Res., 24: 323.
Waters, W. E., and Stark, R. W., 1980, Forest pest management: Concept and reality, Ann. Rev. Entomol., 25: 479.
Wood, D. L., 1972, Selection and colonization of ponderosa pine by bark beetles, in: "Insect/Plant Relationships," H. F. van Emden, ed., Blackwell Scientific Publications, Oxford, England.
Wood, D. L., 1977, Manipulation of forest insect pests, in: "Chemical Control of Insect Behavior: Theory and Application, "H. H. Shorey and J. J. McKelvey, Jr., eds., John Wiley and Sons, New York.
Wood, D. L., 1979, Development of behavior modifying chemicals for use in forest pest management in the U.S.A., in: "Chemical Ecology: Odour Communication in Animals," F. J. Ritter, ed., Elsevier/North-Holland Biomedical Press, New York.
Wood, D. L., 1980, Approach to research and forest management for western pine beetle control, in: "New Technology of Pest Control," C. B. Huffaker, ed., John Wiley and Sons, New York.
Wood, D. L., and Bedard, W. D., 1977, The role of pheromones in the population dynamics of the western pine beetle, in: "Proceedings of the XV International Congress of Entomology," D. White, ed., Washington, DC, August 19-27, 1976.

Wood, D. L., and Bushing, R. W., 1963, The olfactory response of *Ips confusus* (LeConte) (Coleoptera: Scolytidae) to the secondary attraction in the laboratory, *Can. Entomol.*, 95: 1066.

PHEROMONE-BAITED TRAPS AND TRAP TREES IN THE INTEGRATED MANAGEMENT OF BARK BEETLES IN URBAN AREAS

Gerald N. Lanier

SUNY College of Environmental
Science and Forestry
Syracuse, NY 13210 USA

INTRODUCTION

Dutch elm disease (DED) has decimated elm populations in eastern North America and threatens to destroy remaining concentrations of elms in Canada and central, southern and western United States. Elm preservation is possible with existing technology, but costs have limited the application of sustained and intensive elm care programs.

The backbone of every DED program is sanitation--the destruction of diseased elms serving as breeding material for bark beetles transmitting DED. However, because of cost, difficult access and erroneous information on flight capacity of bark beetles, sanitation has generally been limited to urban, residential and high-use recreational areas.

Lately, new technology has been widely accepted for therapeutically treating diseased elms and preventing infection by injecting fungicidal chemicals into healthy elms (Smalley, 1978). Unfortunately, expense and technical difficulty of these treatments limit their application to highly valued trees; moreover, these treatments do not solve the problem of managing elm populations and elm bark beetles.

Identification of Aggregation Pheromone

Scolytus multistriatus (Marsh), the principal vector of DED, uses an aggregation pheromone to colonize elm wood (Peacock et al., 1971). Pearce et al. (1975) identified this pheromone as a synergistic blend of 3 compounds: 1) an alcohol, 4-methyl-3-heptanol (H);

2) a bicyclic ketal, α-multistriatin (M); 3) a sesquiterpene, α-cubebene (C). Virgin females produce H and M while C originates in moribund elm wood (Gore et al., 1977). Stereospecific synthesis in the lab and field demonstrated that, of 8 possible isomers in M, only (-)-α-multistriatin is active (Lanier et al., 1977; Elliott et al., 1979). Mori (1976) defined the isomerism of the natural form of H, but no tests of the relative activity of the 4 forms have been reported. Fortunately, a mixture of the racemic synthetic compounds (multilure) proved to be a powerful attractant to both male and female *S. multistriatus* in the field.

Utilization of Multilure

Field trial results from mass-trapping *S. multistriatus* using multilure-baited sticky traps (Lanier et al., 1976; Peacock et al., unpublished data) have been mixed. DED reductions have consistently accompanied mass-trapping near intensively managed elm groves and within a scattered residual elm population (Lanier, 1979). However, large-scale trapping killing millions of *S. multistriatus* did not seem to diminish the incidence of DED or reduce the succeeding population of *S. multistriatus* (Peacock et al., unpublished data). Therefore, operational applications of multilure are currently recommended to aid grove situations (N-Traps, Albany International, Inc., Needham Heights, MA). Meanwhile, research on intricacies of trapping strategies and biological effects of multilure continues.

In 1978 we began a field demonstration using multilure-baited trap trees to eliminate diseased elms and to absorb extant elm bark beetles, thus effectively destroying breeding material that would produce future generations of beetles (O'Callaghan et al., unpublished data).

This paper's purpose is trifold: to examine several aspects of mass-trapping, to outline the philosophy, technology and results of the trap tree approach, and to discuss the potential of mass-trapping and trap trees in the integrated management of elm populations.

EXAMINATION OF MASS-TRAPPING IN 2 AREAS

Detailed studies for 2 of 12 "grove-trapping" areas (Lanier, 1979) for which Peacock et al. (this volume) summarized results are examined in this paper. These areas, Bradford Hills in eastern Syracuse, NY (SYR-E) and Chevy Chase, MD (CC), respectively represent the most and least successful of grove trapping experiments.

Descriptions of Test Areas

SYR-E is an area of 434 ha that contained a base population of 117 amenity-value elms and ca. 8,000 "weed" elms, 4-20 cm in diameter in 1979. Elms were never the dominant shade trees in this area and few streets were elm-lined before DED destroyed most of Syracuse's elms in the mid-1960s (Miller et al., 1969). Trapping here began in 1975 with the intent of assessing various trap and pheromone dosage parameters rather than controlling DED. A total of 104 traps was placed in rows of 4 or 5 on utility poles. Rows of traps were scattered at intervals of ca. 400 m so that interaction between experimental replications (rows) would be minimal. Surveys beginning in 1974 of 100% of this area provided elm population data. Losses have been offset by new discoveries and in-growth so that the base population has hardly changed in 6 years. About a dozen elms that became diseased were saved by pruning or by a combination of pruning and fungicide injecting.

In 1976-78 traps were positioned as they had been in 1975, and a contiguous 300 ha area in DeWitt, NY, was included in the trapping program. In 1979 the total trapping area was expanded to cover about 4,000 ha--ca. 1/3rd of the total Syracuse metropolitan area. Traps were dispersed to ca. 35% of the previous density so that about twice the number of traps previously used in SYR-E and DeWitt covered an area that was ca. 6X larger than the original areas combined. The trap-tree method (explained later) was applied wherever applicable throughout the expanded area. For consistency, DED and trap-catch data refer to the original SYR-E trapping area, although SYR-E DED trends and trap catches were closely parallel in DeWitt and the expanded area.

Chevy Chase (CC) is a residential area on the northwestern boundary of Washington, DC. Within a 5 ha trapping area there were 250 elms in 1976 when the area was surrounded with 19 traps. DED, held at levels between 1-3%, exploded in Washington, DC in the early 1970s. The shock from this release from control was manifest in CC by an increase from 1.4% DED in 1974 to 5.3% in 1975. When trapping began in 1976, DED trees were conspicuous in the section of Washington, DC bordering CC. Although these trees were marked for removal in 1976, most of them were not removed until spring of 1978. Within CC, root graft transmission of DED (that became acute in localized areas) and large populations of bark beetles severely tested its DED program. Eight trees listed for removal by CC during winter of 1977-78 were erroneously left to yield prodigious numbers of bark beetles in spring of 1978. The number and positions of traps remained unchanged from 1976-79, but in July of 1979 most of the DED trees present were converted to trap trees.

Results and Observations

Contrasting physical situations and differing numbers of traps per elm tree probably drove contrasting DED patterns in SYR-E and CC. DED rates plummeted in SYR-E where there was one trap/amenity elm, little possibility for root graft and moderate pressure from immigrating elm bark beetles (Fig. 1). In CC, where elms were densely spaced (13/trap), and adjacent to areas containing abundant

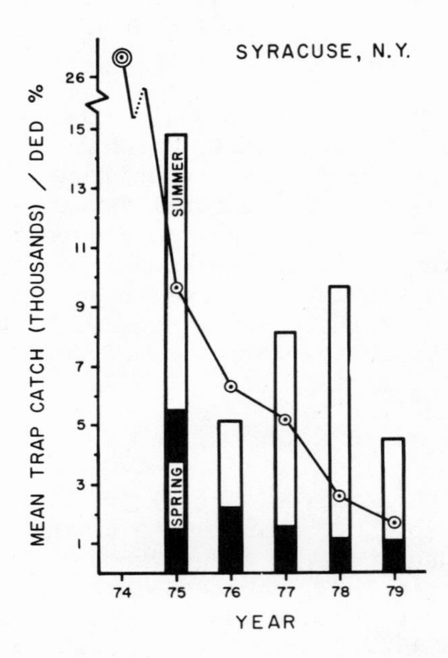

Fig. 1 Dutch elm disease (DED) rates and sticky trap catches in eastern Syracuse, NY. DED is indicated for 1974 through 1977 by circled dots. Histograms indicate mean trap catch in 1,000's during the spring (black) and summer (white) generations, 1975-1979.

elms and brood wood, DED increased from 1976-78 despite catching 20-30 thousand beetles per trap (Fig. 2).

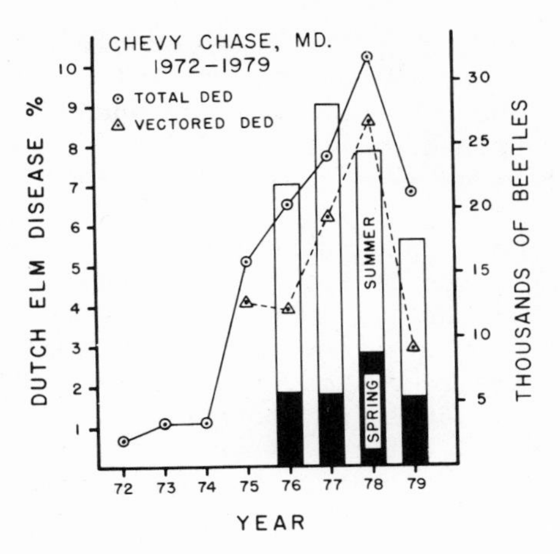

Fig. 2 Dutch elm disease (DED) rates and sticky trap catches in Chevy Chase, MD. Histograms indicate mean trap catch in 1,000's for the spring (black) and summer (white) generations, 1976-1979.

Most beetles trapped probably originated outside of the areas, although numbers emerging inside were high for SYR-E in 1975 and CC in 1978. DED rates followed trends in spring beetle catches. This trend was evident in all of our trapping areas except Richmond, VA, and Raleigh, NC, where the spring flight usually begins before elms are in full-leaf. Thus, the initial appearance of the beetles is not coincident with the susceptibility of elms to infection. In addition, it is possible that traps are most effective in the absence of competing odors from growing elms. Summer catches in both areas decreased markedly in 1979 when trap trees absorbed substantial numbers of elm bark beetles.

Catches of S. multistriatus on multilure-baited traps probably reflect large differences in populations between areas. However, mean yearly trap catches within an area were often incongruous with DED rates and the apparent population, as reflected by the amount of beetle brood wood. This anomaly is manifest in the summer brood catches in SYR-E (Fig. 1) from 1976-78. Each year DED dropped; by 1978 no brood wood could be located within the trapping area. We hypothesized that effectiveness of traps increases as competing pheromone sources (infested elm wood) decrease.

To construct a method more direct than DED rates when measuring the effect of mass-trapping on beetle populations, we developed an index of twig crotch feeding in young elms, 8-35-ft tall

(Rabaglia, 1980). Twig sampling data and elm population estimates were used to project estimates of total twig injuries in juvenile elms within areas. We have not yet developed a system for sampling mature elms so our data do not reflect total twig injuries and our methods should not be applied to stands of mature elms.

Table 1 shows that the index of twig feeding moves with DED rate, but mean trap catch is not clearly related to either index of twig feeding or DED. Most notably, in 1978 DED was epidemic in Fayetteville-Manlius, NY, yet survey traps caught a mean of only 5,170 beetles. At the same time, there were 9,680 beetles/trap in SYR-E, even though all other parameters indicated that the beetle population was very low. It is clear that traps were more efficient in SYR-E than in Fayetteville-Manlius where competing pheromone sources were very abundant.

During 1978-79 (Table 1) conspicuously higher indices of twig feeding, located immediately outside the SYR-E trapping area rather than within the area, corroborate DED rates indicating that mass trapping can reduce populations of *S. multistriatus* and DED transmitted by these beetles. In 1979 beetles trapped in SYR-E was 4.6X greater than estimated twig injuries.

Limitations of Sticky Traps

Our most favorable results for protecting healthy elms by pheromone-baited sticky traps always occurred in areas relatively free from brood wood and newly infested material that acted as competing pheromone sources. Unfortunately, these areas are relatively rare. The following factors in various combinations in most areas depreciate the potential contribution of pheromone-baited sticky traps:

(a) Sanitation of diseased trees in wood-lots and green spaces is inadequate;

(b) Removal of diseased elms from private property is not thoroughly enforced;

(c) Trees scheduled for removal are left beyond the allowable period because of managerial or fiscal problems;

(d) Cut elms hoarded for fuel wood may harbor beetles or become infested;

(e) The native elm bark beetle is not affected by the *S. multistriatus* pheromone.

In order for S. multistriatus pheromone traps to be broadly effective, sanitation programs must be more intensively applied. The problems in overcoming the fiscal and political limitation to wide-scale improvements in sanitation by traditional methods are enormous. New techniques for sanitation are clearly needed if S. multistriatus pheromone is to be broadly used for DED management.

Table 1. Indices of twig crotch feeding compared to trap catches in the Syracuse, NY vicinity. Feeding and catch data expressed in 1000s.

Area	DED rate	Twig feeding Index[a]	Twig feeding Total	Total catch
		1978		
SYR-E				
Trapping area	2.30	6.6	468	1050
Outside area ca.	10.00	22.6	-	-
SYR-SW[b]	4.3-8.00	9.2	1236	605
F-M[c]	ca. 20.00	30.2	2196	269
		1979		
SYR-E				
Trapping area	1.65	2.8	142	656
Outside area	ca. 6.00	16.3	-	-
SYR-SW[b]	3.3-6.00	3.0	611	137

[a] Mean number of injuries by S. multistriatus/100 prime twig crotches (Rabaglia, 1980).

[b] Data combined for treatment, check and buffer zones in the SYR-SW trap tree experiment. Trapping data is for survey traps spaced at 1/4th mile intervals throughout SYR-SW.

[c] Fayetteville and Manlius, villages each ca. 11 miles from Syracuse. DED and elm phloem necrosis were epidemic in this area during 1978.

APPLICATION OF THE TRAP-TREE TECHNIQUE

For at least 3 centuries European foresters have used the aggregating behavior of bark beetles as a means to control them. The strategy for this control was based upon providing felled or girdled trees that bark beetles preferred to attack rather than standing timber. Once infested, trap trees were debarked or burned to destroy beetle broods (Schwerdtfager, 1973). The prevailing thought of foresters was that odors of logs, *per se*, attracted all beetles; the phenomenon of aggregation by bark beetles in response to pheromones apparently was not understood until the mid-20th Century (Anderson, 1948).

Synthetic pheromone has been used to attract conifer-feeding bark beetles to trap logs, living trees and trees killed by treatment with the silvicide, cacodylic acid (Copony and Morris, 1972; Pitman, 1971; Knopf and Pitman, 1972; Pitman, 1973; Coulson et al., 1975). Cacodylic acid injections into pheromone-baited trees further increased their attractiveness to some species (Svihra, 1974), yet brood production in these trees was drastically reduced.

This method is cheap because further action is not necessary. Buffam and Yasinski (1971) found that cacodylic acid treatments of trees to be felled for road and trail construction prevented build-up of the spruce beetle, *Dendroctonus rufipennis* (Kirby), at $3.80/tree as opposed to $20/tree for conventional insecticide treatment.

The destruction of unwanted or diseased elms as a contribution to a DED control program was initially investigated by Whitten (1941) and later refined by Himelick and Neely (1961). The mild advocacy of this principle was insufficient to overcome political difficulties in application of tree-killing for DED control.

Rexrode (1974) reported 100% mortality of *S. multistriatus* in elms pressure-injected with cacodylic acid, but Hostetler and Brewer (1976) did not observe significant mortality in trees killed by topical application of cacodylic acid to an axe-frilled girdle. However, in the former case, trees were infested after they were treated, while in the latter case, beetle broods were probably advanced by early August when trees were treated. Other workers (Chansler and Pierce, 1966; Buffam and Flake, 1971) have reported high rates of bark beetle mortality when recently infested conifers were treated using the axe-frill method. Gardiner (1979) demonstrated that the native elm bark beetle, *Hylurgopinus rufipes* (Eichhoff), was attracted to, and infested elms treated in axe-frills with 0.2 ml cacodylic acid/cm circumference, but brood development succeeded in at least the upper portions of the treated trees.

We thought that with improvement, a trap-tree system might overcome the previously outlined limitations to the effectiveness of pheromone-baited sticky traps in controlling DED.

Development of Treatment Procedures

During 1977 we treated 96 elms with cacodylic acid to assess the feasibility of the trap-tree concept. Variables tested include:

(1) Method of application (axe-frill vs. injection);

(2) Time of year during which treatments are made;

(3) Dosage;

(4) Degree of infestation by DED and beetles at the time of treatment;

(5) Baiting or not baiting trap-trees with multilure.

These preliminary tests demonstrated that diseased trees could be eliminated easily, quickly, and cheaply. Virtually no bark beetle broods would be produced in these trees even though most of them were attacked by the native or European elm bark beetle. Trap-trees baited with multilure were attacked intensely by *Scolytus*, and conditions in the treated tree were altered. The conspicuously dry sapwood apparently was colonized by fungi other than *Ceratocystis ulmi* (Buisman C. Moreau); salmon-colored conidia mats of *Tuberculia vulgaris* (Monilaes) usually were present on the outer bark. Sufficiently clear results enabled us to develop guidelines for application of the trap-tree technique and thus begin operational tests in 1978.

Field Demonstrations

Syracuse-Southwest and Fayetteville-Manlius. During 1978 we began field tests in 2 locations in central New York; the southwestern part of Syracuse (SYR-SW) and Manlius Township, which includes Fayetteville and Manlius (F-M) villages 11 km east of Syracuse. Each location encompassed 10 km^2 and was divided into ca. equal treatment and check areas separated by a "buffer" zone ca. 400 m wide.

To monitor beetle populations, sticky traps baited with multilure were placed on utility poles in a grid pattern at 400-m intervals throughout both locations. The traps were established late in May, replaced in mid-July, and removed late in September.

DED rates were assessed by a survey listing amenity elm trees in each location, first in May and then in August. Elms in wood-lots and saplings less than 10 cm DBH were not listed for survey purposes but were monitored and treated if they became diseased. Red elms

Ulmus rubra Muhl.) and American elms (U. americana L.) were listed but Asian (U. pumila L.) elms were not. Extensive symptoms appearing before July 1 were considered to be previous year infections; symptoms restricted to a single limb system appearing after July 1 were generally considered to be current year infections.

Diseased elms within treatment plots in SYR-SW and F-M experimental areas were treated with cacodylic acid, according to the guideline we developed from our previous experiments. Permission to treat publically owned diseased elms was granted before the experiment, and permission to treat privately owned elms was solicited and usually gained within a week of detection. Only 4 elms were left untreated because of lack of permission.

A sampling scheme was devised to evaluate the success of treatments and the impact on beetle populations (O'Callaghan et al., 1980). An estimate of total brood that would have been produced from the trees, had they not been treated, was extrapolated from X-ray examination of samples from 7 untreated brood trees taken near treatment plots during this study.

In the 1st beetle flight period of 1978 (May 20-July 1), 72 diseased elms were killed in SYR-SW. Six newly infected trees were converted to trap trees during the 2nd flight period (July 1-September 15). These trees absorbed an aggregate of 344,000 attacks accounting for approximately 688,000 beetles (females initiate attack; sex ratio 1:1). Only 94,000 larvae were produced in these trees as compared to an estimated 3,793,000 expected if breeding success had not been affected by the treatment. These data indicate that the maximum actual beetle output from these trees is 7.32X less than beetles absorbed and that the reduction of total output was 97.5% (Table 2).

In 1979 diseased elms and small (4-25-cm diam.) wood-lot weed elms were killed. Brood production in these trees was virtually nil except in 3 trees over 50-cm dbh that were already infested at the time of treatment. One city tree had been missed in the 1978 survey; another had a thin crown but no wilting typical of DED when it was mass-attacked. Owners of a 3rd tree would not permit treatment in

Table 2. Impact of sticky traps and trap trees, Upstate New York. Catch data in 1000s.

Year	DED rate	Beetles caught[a]	% Reduction[b]	Impact[c]
		SYR-SW (treatment)		
1977	7.69	-	-	-
1978	4.30	893	97.5	97
1979	3.30	450	90.8	150
		SYR-SW (check)		
1977	4.98	-	-	-
1978	6.29	-	-	19
1979	6.02	69	-	23
		F-M		
1977	ca. 20	-	-	
1978	ca. 20	6293	88.1	208
		SYR-E + DeWitt[d]		
1977	5.30	1132	-	-
1978	2.80	1371	-	208
1979	1.18	2034	97.0	726

[a]Includes beetles caught on trap trees and sticky traps.

[b]Estimated number of surviving larvae divided by the estimated possible brood (bark surface area of trap-trees x larvae/unit area in untreated checks).

[c]Arbitrary index obtained by dividing total beetles killed by the index of twig feeding.

[d]Data for SYR-E and DeWitt combined for comparisons over the 1979 integrated DED control area.

1978 but allowed treatment in 1979 after the entire crown had wilted. Owing to these trees, larvae produced in SYR-SW exceeded adult beetles absorbed in trap trees during 1979. Nevertheless, the overall reduction in potential brood was 90.8% (Table 2). DED rates declined in treatment areas and remained approximately stable in the check area.

During 1978 assessment of DED rates in F-M was confounded by a concurrent epidemic of elm phloem necrosis (PN). Elms declining from PN were colonized by S. multistriatus that introduced the DED fungus. Beetle populations were prodigious and trees suffering from either or both diseases were too numerous to treat. We abandoned the attempt to manage elms in F-M, but documented absorption of more than 6 million beetles and reduced beetle production by 88% (Table 2) in trap-trees.

SYR-E and DeWitt. Mass trapping had been conducted in SYR-E since 1975 and in a contiguous area of DeWitt since 1976. In both areas DED dropped in successive years. During 1979 we integrated the trap-tree technique using mass-trapping and tree therapy in an expanded 4,000 ha zone that included both trapping areas (see preceding description discussing mass trapping).

Four elms over 50 cm dbh (2 of these were declining from causes other than DED) and several smaller, diseased trees were killed. Most importantly, 6,022 weed elms were sacrificed to make trap trees and to reduce the wood-lot elm population. These trap trees absorbed 1,378,000 beetles; an additional 656,000 beetles were caught with sticky traps while 642,000 larvae survived in trap trees that could have accommodated 21 million beetles. The actual potential disease-carrying and breeding population will be considerably less than larvae because of both preemergence mortality and loss during dispersal. The aggregate of more than 2 million beetles destroyed is 3X larvae produced and more than 14X estimated twig crotch feeding injuries in the area (Table 1).

The known brood wood present during 1979 cannot explain beetles caught. We assumed that most of them immigrated. The very low DED rate (1.18%) and low twig injury index (2.8) (Table 2) in the area indicate that beetle mortality in traps and trap trees more than compensated for immigration stimulated by the use of attractive odorants.

Hennepin County Park Reserve District, MN. In 1979 we began a trial of trap-tree method studies in Lake Rebecca Park, 37 km west of Minneapolis, MN. Baker Park located 9 km east of Lake Rebecca served as a check.

Both parks cover ca. 10,000 ha where very large elms grow on former homesites, wetlands, and areas too rugged for agriculture.

Younger elm trees abound along former roads or fence rows, and saplings grow on land previously farmed. Our random plot survey of 3.3% of the land area in each park indicated that Lake Rebecca Park contained 36,000 elms of which ca. 20% are more than 25 cm dbh and ca. 25% of the 26,600 elms in Baker Park are 25 cm dbh or larger.

In both parks all elms identified as diseased formerly had been removed, but this program did not prevent DED from increasing yearly from 1972-78. Recognizing that sanitation in remote areas was prohibitively expensive as well as damaging, the Hennepin County Park Reserve District (HCPRD) Shade Tree Disease Control Program of 1978 specified that DED sanitation be conducted only in and around high-use areas. The plan also stated that dead trees along rights-of-way be removed. Elms in wooded areas were conceded to DED.

We reasoned that we could quickly and cheaply eliminate identified diseased trees in Lake Rebecca Park by killing them with cacodylic acid; simultaneously, beetles would be destroyed. Gardner's work (1979) and our own research in Syracuse already had found that *H. rufipes* was strongly attracted to elms that had been treated with cacodlic acid.

Multilure-baited traps in Lake Rebecca and Baker Parks established that *S. multistriatus* was rare (<20/trap were caught) and the principal vector of DED in the area was *H. rufipes*. However, root graft transmission was clearly a very significant mode of infection.

Our surveys and treatment records estimate DED rates in Lake Rebecca and Baker Parks to be 1.92% and 1.36%, respectively. Most of these trees became diseased in 1978 or earlier. Estimates of 1979 disease rates will be made following 1980 surveys. In all, 432 of the Lake Rebecca trees were treated with cacodylic acid, 243 were removed and 15 were left for treatment in spring 1980.

Attacks on treated trees were rare. We estimate that the 3,000 (no. attacks X2) beetles (mostly *H. rufipes*) that were absorbed by the treated trees produced 6,000 larvae and even fewer brood adults. However, the eliminated diseased trees could have produced more than 64 million *H. rufipes* if they had been left untreated and had become colonized.

The low number of attacks has 3 principal causes: 1) Most of the trees were treated in June through August--too late to absorb overwintering adults that emerge and attack in May and June; 2) the population of *H. rufipes* was inexplicably low; 3) successively cold winters had virtually eliminated *S. multistriatus* in the area.

To increase the impact of trap-trees on *H. rufipes* during 1980, we will concentrate upon tree killing in the fall and early spring.

Experiences in Syracuse and Vermont showed that trees treated with cacodylic acid before beetle flight were extremely attractive to *H. rufipes*; however, *S. multistriatus* would scarcely attack such trees, even if they were baited with multilure.

APPLICATIONS TO DED MANAGEMENT

The application of multilure to DED management is now commercially feasible. Beginning in 1981, multilure-baited sticky traps will be registered for DED control and will be commercially used in specified situations.

Cacodylic acid and MSMA are registered as aboricides that are commonly used to eliminate unwanted trees for forest improvement. We are currently developing a package to label these materials for conversion from diseased and weed elms to trap-trees to control elm bark beetles and DED.

Mass-trapping *S. multistriatus* with sticky traps is compatible with all other DED control operations. Our data show that a few traps in elm tree groves free of brood wood can reduce risk of DED infections from immigrating beetles.

More intensive trapping may help to bring DED under control in difficult circumstances. In SYR-E, where there was almost 1 trap/amenity value elm, DED plummeted from >20% to <2% after 5 years of trapping.

Sanitation and any other factor reducing the elm wood's susceptibility to colonization will enhance beetle trap effectiveness. Elimination of in-flight beetles further decreases the potential for new inoculation and future brood wood. This, in turn, operates as a positive feedback to improve DED control.

The trap-tree approach may have little applicability when DED is under control, and the trees that become diseased are given therapy or are promptly removed. Unfortunately, intensive, well-managed DED programs are the exception rather than the rule.

Because of its low expense and because vehicle access or special equipment is not a necessity, the trap-tree approach can make feasible DED control in forests, rural areas, and urban green spaces. In urban areas where DED rapidly destroys the elm population, diseased trees can quickly and cheaply be eliminated. If the treatment is properly timed, these trees will absorb prodigious numbers of elm bark beetles that would otherwise spread DED. Instead of requirins immediate removal, trap-trees could be taken down at a convenient schedule, and those in wood lots could be left untouched.

The trap-tree technique also provides the option for municipalities to kill diseased elms on private property rather than subsidizing tree removal or requiring removal to be done at the owner's expense.

Traps and trap-trees baited with multilure can make a major contribution to DED control. Once control is achieved, the application of multilure-baited traps will facilitate DED management.

REFERENCES

Anderson, R. F., 1948, Host selection by the pine engraver, *J. Econ. Entomol.*, 41: 596.

Buffam, P. E., and Flake, H. W., Jr., 1971, Roundheaded pine beetle mortality in cacodylic acid-treated trees, *J. Econ. Entomol.*, 64: 969.

Buffam, P. E., and Yasinski, F. M., 1971, Spruce beetle hazard reduction with cacodylic acid, *J. Econ. Entomol.*, 64: 751.

Copony, J. A., and Morris, C. L., 1972, Southern pine beetle suppression with frontalure and cacodylic acid treatments, *J. Econ. Entomol.*, 65: 754.

Chansler, J. F., and Pierce, D. A., 1966, Bark beetle mortality in trees injected with cacodylic acid, *J. Econ. Entomol.*, 59: 1357.

Coulson, R. N., Foltz, J. L., Mayasi, A. M., and Hain, F. P., 1975, Quantitative evaluation of frontalure and cacodylic acid treatment effects on within-tree populations of southern pine beetle, *J. Econ. Entomol.*, 68: 671.

Elliott, W. J., Hromnak, E., Fried, J., and Lanier, G. N., 1979, Synthesis of multistriatin enantiomers and their action on *Scolytus multistriatus* (Coleoptera: Scolytidae), *J. Chem. Ecol.*, 5: 279.

Gardiner, L. M., 1979, Attraction of *Hylurgopinus rufipes* to cacodylic acid-treated elms, *Bull. Entomol. Soc. Am.*, 25: 102.

Gore, W. E., Pearce, G. T., Lanier, G. N., Simeone, J. B., Silverstein, stein, R. M., Peacock, J. W., and Cuthbert, R. A., 1977, Aggregation attractant of the European elm bark beetle, *Scolytus multistriatus*. Production of individual components and related aggregation behavior, *J. Chem. Ecol.*, 3: 429.

Himelick, E. B., and Neely, D., 1961, Prevention of bark beetle development in undesirable elms for the control of Dutch elm disease. Plant Disease Report, 45: 180.

Hostetler, B. B., and Brewer, J. W., 1976, Translocation of cacodylic acid in Dutch elm-diseased American elms and its effect on *Scolytus multistriatus* (Coleoptera: Scolytidae), *Can. Entomol.*, 108: 893.

Knopf, J. W. E., and Pitman, G. B., 1972, Aggregation pheromone for manipulation of the Douglas-fir beetle, *J. Econ. Entomol.*, 65: 723.

Lanier, G. N., Silverstein, R. M., and Peacock, J. W., 1976,

Attractant pheromone of the European elm bark beetle (Scolytus multistriatus): solation, identification, synthesis, and utilization studies, in: "Perspectives in Forest Entomology," J. F. Anderson and H. K. Kaya, eds., Academic Press, New York.

Lanier, G. N., Gore, W. E., Pearce, G. T., Peacock, J. W., and Silverstein, R. M., 1977, Response of the European elm bark beetle, Scolytus multistriatus (Coleoptera: Scolytidae), to isomers and components of its pheromone, J. Chem. Ecol., 3: 1.

Lanier, G. N., 1979, Protection of elm groves by surrounding them with multilure-baited sticky traps, Bull. Entomol. Soc. Am., 25: 109.

Miller, H. C., Silverborg, S. B., and Campana, R. J., 1969, Dutch elm disease:Relation of spread and intensification to control by sanitation in Syracuse, NY, Plant Disease Report, 53: 551.

Mori, K., 1976, Pheromone synthesis. II. Synthesis of (1S=2R=4S=5R)-(-)-alpha-multistriatin; pheromone in smaller European elm bark beetle, Scolytus multistriatus, Tetrahedron, 32: 1979.

O'Callahan, D. P., Gallagher, E. M., and Lanier, G. N., 1980, Field evaluation of pheromone-baited trap trees to control elm bark beetles, vectors of Dutch elm disease, Environ. Entomol., 9: 181.

Peacock, J. W., Lincoln, A. C., Simeone, J. R., and Silverstein, R. M., 1971, Attraction of Scolytus multistriatus (Coleoptera: Scolytidae) to virgin female produced pheromone in the field, Ann. Entomol. Soc. Am., 64: 1143.

Pearce, G. T., Gore, W. E., Silverstein, R. M., Peacock, J. W., Cuthbert, R. A., Lanier, G. N., and Simeone, J. B., 1975, Chemical attractants for the smaller European elm bark beetle, Scolytus multistriatus (Coleoptera: Scolytidae), J. Chem. Ecol., 1: 115.

Pitman, G. B., 1971, trans-Verbenol and α-pinene: Their utility in manipulation of the mountain pine beetle, J. Econ. Entomol., 64: 426.

Pitman, G. B., 1973, Further observations on douglure in a Dendroctonus pseudotsugae management system, Environ. Entomol., 2: 109.

Rabaglia, R., 1980, Twig crotch feeding by Scolytus multistriatus: Distribution of feedings in the tree crown sampling to estimate populations and evaluation of insecticides, M.S. thesis, SUNY, College of Environmental Science and Forestry, Syracuse, NY.

Rexrode, C. O., 1974, Effect of pressure-injected oxydemeton-methyl, cacodylic acid, and 2,4-D amine on elm bark beetle populations in elms infected with Dutch elm disease, Plant Disease Report, 58: 382.

Schwerdtfager, F., 1973, Forest entomology, in: "History of Entomology," R. F. Smith, T. E. Mittler and C. N. Smith, eds., Annual Reviews, Inc., Palo Alto.

Smalley, E. B., 1978, Systemic chemical treatments of trees for protection and therapy, in:"Dutch Elm Disease: Perspectives After 60 Years," Search Agriculture, 8: 34.

Svihra, P., 1974, The change of the chemical information by bark beetles *Ips typographus* L. in phloem treated with cacodylic acid, *Z. angew. Entomol.*, 75: 247.

Whitten, R. R., 1941, The internal application of chemicals to kill elm trees and prevent beetle attack, *U. S. Dept. Agric. Circ.* 605., 12 pp.

PHEROMONE-BASED SUPPRESSION OF AMBROSIA BEETLES IN INDUSTRIAL TIMBER PROCESSING AREAS

John H. Borden

Pestology Centre
Department of Biological Sciences
Simon Fraser University
Burnaby, B.C., Canada V5A 1S6

John A. McLean

Faculty of Forestry
University of British Columbia
Vancouver, B.C., Canada V6T 1W5

THE PROBLEM

Biology of Ambrosia Beetles

Ambrosia beetles (Coleoptera:Scolytidae) constitute one of the primary breakdown agents in the natural forests of the Pacific Northwest of North America. There are 3 important species, *Trypodendron lineatum* (Olivier), *Gnathotrichus sulcatus* (LeConte) and *G. retusus* (LeConte). Each spring they infest the sapwood of coniferous trees that have died the previous winter. Female *T. lineatum* and male *Gnathotrichus* spp. initiate the attack and produce aggregation pheromones which induce mass attack on suitable hosts (Rudinsky and Daterman, 1964a,b; Chapman, 1966; Borden and Stokkink, 1973; Borden and McLean, 1979). In nature, this process ensures maximal utilization of isolated windthrown, broken or dying hosts (Atkins, 1966). However, in industrial timber-processing operations such as dryland log sorting areas (dryland sorts) (Fig. 1) and sawmills (Fig. 2) the same attack behavior by abnormally high beetle populations constitutes a major problem.

The beetles innoculate the wood with species-specific ambrosia fungi (Francke-Grossman, 1963; Funk, 1965, 1970), on which they feed, and raise their brood in small niches (cradles) (Nijholt, 1978a) off the main trunk of their 1-2 mm-wide galleries (Kinghorn, 1957; McLean and Borden, 1977a).

Differences in the life cycles of the 3 species, as well as habitat and climatic preferences, ensure that attack by at least

Fig. 1. The MacMillan Bloedel, Ltd., Shawnigan Division dryland sort in the rain, showing dense forest margin in which T. lineatum overwinters. Loaded trucks enter the 950-m long perimeter road, pass in the foreground to the weigh scale, and then into the sort. Note truck in right center background being unloaded in single step by a stacker, log sorting in progress in left center background, bays of sorted and decked logs awaiting transport to the mill, and truck in foreground being loaded. (Photo courtesy of MacMillan Bloedel, Ltd.).

one ambrosia beetle species can occur from early March to late October in British Columbia (B.C.).

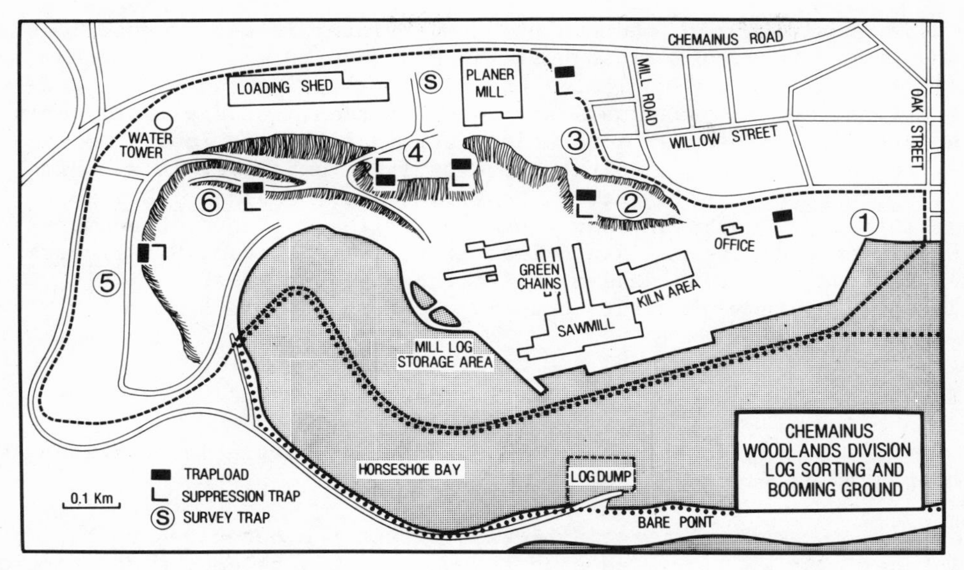

Fig. 2. Map of the MacMillan Bloedel, Chemainus Division Sawmill, Vancouver Island, B.C., showing numbered study locations and approximate position in 1976 of the suppression traps and adjacent trap loads, and the survey trap in location 4. The dashed line indicates the working area of the sawmill including the log storage area, and the dotted line indicates the portion of Horseshoe Bay used for dumping, sorting and booming logs from the Chemainus Woodlands Division. (From McLean and Borden, 1979).

In mid-to-late summer, emergent brood adults of *T. lineatum* fly from the logs in which they matured to their main overwintering site in the litter and duff within the forest margin of a timber processing area (Kinghorn and Chapman, 1959; Dyer and Kinghorn, 1961). Thus, large populations may build up around such industrial areas. *T. lineatum* flies early in the spring, and attacks only aged logs which usually have been felled no later than mid-winter (Prebble and Graham, 1957; Dyer and Chapman, 1965). Anaerobic metabolism in these logs (Graham, 1968) results in the production of ethanol, a primary attractant (Moeck, 1970; 1971) and boring stimulant (MacConnell et al., 1977). There is one generation per year, although vigorous parents may reattack and raise a second, late-summer brood (Nijholt, 1978a). The wide, dark-staining galleries may be very numerous and unsightly.

Gnathotrichus spp. attack logs and stumps within 2 weeks after felling (Mathers, 1935; Cade et al., 1970; McLean and Borden, 1977b). For *G. sulcatus*, ethanol is also known to be a primary attractant

(Cade et al., 1970). Both species overwinter in their brood logs, and emerge to attack fresh hosts in May-June (Prebble and Graham, (1957). In B.C., G. sulcatus has distinct spring and summer generations. G. retusus has a spring generation, and apparently only a partial, late-summer generation (personal observation). We find G. retusus in warmer habitats, preferring stumps over logs, while G. sulcatus occurs in wetter, coastal zones. Gnathotrichus spp. galleries are narrower than those of T. lineatum, are often more widely spaced, and may stain less severely. However, G. sulcatus can attack late into autumn (McLean and Borden, 1977b; Nijholt, 1978a), and is a constant economic problem.

Economic Impact

The damage impact and costs of ambrosia beetle infestations can be placed in 5 categories.

1. Degrade. Ambrosia beetles infest the outer 5-8 cm of logs (Graham et al., 1950), causing the most highly valued, clear portion of the log to produce reduced volumes of high grade lumber and face stock plywood veneer. For example, the loss to number-2 grade western hemlock logs sustaining moderate attack is $8.88/m^3 (Canadian), based on January 1978 prices (Dobie, 1978). The absolute dollar loss to industry caused by ambrosia beetle degrade has never been calculated, although in 1975-76 it was estimated at $7 million in B.C. (Nijholt, 1978a).

2. Export problems. All 3 species may be transported in wood products and dunnage. Their frequent interceptions in countries like New Zealand (Milligan, 1970; Bain, 1974) has resulted in numerous restrictions against exporting products damaged by ambrosia beetles.

3. Remanufacturing and repacking. The cost of such events is mainly in labor and machine time. When one side of a board has severe ambrosia beetle damage, it often needs to be reprocessed (Richmond, 1968). If a board in a load of export-bound lumber becomes infested, the load must be broken down, the offending board replaced, and the load rebuilt.

4. Inconvenience due to the need for rapid inventory turnover. Because the vulnerability of stored logs of unseasoned lumber in the woods and processing areas is high, industry cannot often afford the luxury of retaining such products in storage.

5. Cost of direct control. Any measures taken against the beetles add to production costs, and therefore, the cost of the finished wood product.

PHEROMONES AND THEIR USE

Isolation and Synthesis

In 1967, collaborative research on pheromone isolation was initiated, which culminated in the isolation of 3 species-specific aggregation pheromones (Table 1). Each is chiral in nature and the response by beetles in the field is influenced by the chirality of the pheromone deployed.

The sodium borohydride reduction of the commercially available ketone to racemic sulcatol [(±)-sulcatol] (Byrne et al., 1974) is now routinely performed by organic chemistry students at Simon Fraser University, ensuring a cheap supply of pheromone. A lengthy synthesis for *S*-(+)-sulcatol (Schuler and Slessor, 1977) produced small amounts of pheromone, but a more facile synthesis (Johnston and Slessor, 1979) has assured its ready supply. Because of its complex, tricyclic structure, lineatin defied synthesis for several years. Four low yield syntheses (Borden et al., 1979; Mori, 1979) held little promise for the practical use of lineatin, but a more productive synthesis now yields gram quantities and also can be adapted to produce optical isomers (Slessor et al., 1980).

Only (±)-sulcatol is produced commercially (Aldrich Chemical Co.). If pheromone-based management of ambrosia beetles passes from research and development (R&D) to an industrial stage, a commercial source will be essential.

Pheromone Compatability

Because *G. sulcatus* responds only when both enantiomers of sulcatol are present, and *G. retusus* responds optimally to pure *S*-(+)-sulcatol (Table 1), there is no hope of practically trapping the 2 species together. However, either species is often found in the same host with *T. lineatum*, and our preliminary experiments disclose no mutual inhibition of response between lineatin and either *S*-(+)- or (±)-sulcatol. Therefore, 2 sets of paired pheromones could potentially be used together. This tactic would reduce by 1/3 the number of bait stations, traps, or trap logs needed to reach all 3 species.

Role of Synergists and Trap Form

The activity of chemical synergists for the 3 pheromones (Table 1) must be defined and regulated before compatible pheromones can be combined. Ethanol is an effective synergist for the response of *G. retusus* to (*S*-(+)-sulcatol, while both ethanol and α-pinene synergize the response of *G. sulcatus* males to (±)-sulcatol. An increase in the catch of male *Gnathotrichus* spp. is important, because any uncaptured males are free to set up competing pheromone sources, thus reducing the efficiency of any program aimed at intercepting host-selecting beetles (Borden et al., 1980c).

A critical problem arises when the prospect of combining compatible pheromones and synergists are considered. The inactivity of α-pinene and the inhibitory effect of ethanol when combined with lineatin in wire mesh traps (Table 1) is surprising, because T. lineatum responds to either ethanol alone or synergized by α-pinene (Bauer and Vité, 1975; Nijholt and Schönherr, 1976). Moreover, ethanol and α-pinene are excellent synergists for lineatin in Scandinavian drain pipe traps, which provide a prominent vertical silhouette (Vité and Bakke, 1979). Therefore, additional research is necessary to find a type of trap which is effective with synergized pheromones for both Trypodendron and Gnathotrichus spp. and to standardize pheromone and synergist release rates for use in that trap.

STRATEGY, TACTICS, AND LOGISTICS

Numerous strategies and tactics for ambrosia beetle management have been considered (Table 2). Some of these are in practice; all could potentially be integrated.

Table 1.

Pheromone or response characteristic	Species		
	Gnathotrichus sulcatus	Gnathotrichus retusus	Trypodendron lineatum
Pheromone produced	65:35 Mixture of S-(+)- and R-(-)-6 methyl-5-hepten-2-01 (sulcatol) (Byrne, et al., 1974).	S-(+)-Sulcatol (Borden, et al., 1980a).	3,3,7-Trimethyl-2,9-dioxa-tricyclo-[3.3.1.0^{4,7}] nonane (lineatin) (MacConnell, et al., 1977; Borden, et al., 1979).
Role of optical isomers	Both enantiomers necessary for response (Borden, et al., 1976), (±)-sulcatol as effective as natural pheromone (Borden, et al., (1980a).	Response inhibited by large, but not small, amounts of R-(-)-sulcatol (Borden et al., 1980a).	Chirality of natural pheromone unknown. Both (±)- and (+)-lineatin active in field, (-)-lineatin inactive (Borden, et al., 1980b).
Compatability with other pheromones.	Can be used with lineatin (Borden, et al., unpubl.).	Can be used with lineatin (Borden, et al., unpubl.).	Can be used with either, S-(+)- or (±)-sulcatol (Borden, et al., unpubl.).
Role of ethanol and α-pinene as synergists in wire mesh traps.	Significant increase in response of males when both ethanol and α-pinene deployed with (±) sulcatol (Borden, et al., 1980c).	Significant increase in response of both sexes when ethanol deployed with S-(+)-sulcatol. (Borden, et al., 1980c).	Significant reduction in response when ethanol deployed with lineatin. No effect of α-pinene (Borden, et al., unpubl.).
Effect of trap form and position on response to pheromone and synergists.	Vertical wire mesh cylinder better than horizontal (McLean, 1976). Dark, vertical silhouette, Scandinavian drain pipe traps do not influence activity of synergists (Borden, et al., 1980c).	Not investigated	Ethanol and α-pinene effective synergists only when deployed in dark, vertical silhouette traps (Vité and Bakke, 1979).

Table 2. Practiced or potentially feasible strategies and tactics for ambrosia beetle management.

Strategy	Tactic
Habitat management	Eliminate potential or occupied habitat by:
	a. removing vulnerable logs from the forest before they are attacked during the beetle flight season, i.e., "hot logging,"
	b. minimizing storage time in dry-land sorts and sawmill yards,
	c. disposal or burning of logging slash.
Protection of product	Deter ambrosia beetle attack by use of:
	a. insecticides
	b. water misting
	c. repellents
Suppression of beetle populations	Intercept and kill host seeking beetles by use of:
	a. unbaited trap logs or lumber piles,
	b. pheromone-baited trap logs or piles,
	c. pheromone-baited, insecticide-treated trap logs or piles,
	d. pheromone-baited traps.

Habitat Management

Forest companies are generally responsive to the habitat management tactic of hot logging (Richmond, 1968; 1969). By doing so, they minimize the risk of ambrosia beetle attack, but lose the luxury of a large reserve of logs "cold decked" in the forest. However, catastrophic weather, equipment breakdowns, strikes, depressed markets, and even the demands of normal operation, often result in vulnerable logs or lumber being exposed to beetle attack.

Elimination of slash burning could intensify the problem by conserving Gnathotrichus spp. populations, which overwinter in the slash. However, this danger may be reduced by new utilization standards which result in far less potential host material being left in the forest.

The need to preserve water quality and fish habitats has encouraged conversion from sorting logs to species and grade in water to processing them in dryland sorts. This move has resulted in more log surface being exposed for a longer period of time. Moreover, the submersion-induced mortality of beetles in infested logs in the water (McLean and Borden, 1977a) is curtailed. Therefore, large resident populations of ambrosia beetles, particularly T. lineatum, build up around such processing areas.

Protection of Product

Aerial applications of chlorinated hydrocarbon insecticides were routinely and effectively used in the past to protect water-stored logs from ambrosia beetle attack (Richmond, 1961; Lejeune and Richmond, 1975). However, spraying logs in Cowichan Lake with benzene hexachloride led to the reported discovery of the pesticide in oysters at the mouth of the Cowichan River (Finegan, 1967). This event led to the eventual elimination of all insecticide use on stored logs (Lejeune and Richmond, 1975).

One alternative protective measure is water misting. On high value logs in a small area with a good water supply and no environmental problems from excessive runoff, this technique has proven economically effective (Richmond and Nijholt, 1972; Nijholt, 1978b).

Repellents based on tree extractives can protect logs from attack (Nijholt, 1973a, 1980a,b), but are still in the basic research phase. The possibility of using a male-produced, anti-aggregation pheromone for T. lineatum (Nijholt, 1973b; Borden, 1974) has not been investigated.

Suppression of Beetle Populations

An attractive, alternative strategy for ambrosia beetle management is to suppress beetle populations by intercepting them in flight before they reach their hosts. This strategy may be of debatable use in a vast forest, but is applicable to the relatively confined areas covered by dryland sorts and sawmills.

Some companies routinely place trap bundles of logs to absorb ambrosia beetles around the peripheries of dryland sorts. This tactic relies on the beetles finding the trap logs on their own, chosing them in preference to the valuable product to be protected, and setting up a pheromone source to induce mass attack. Prior to brood emergence, the trap logs are processed through a chipper to kill the beetles and are used for pulp. The logs in these piles are often culls, and consequently may be unattractive or unacceptable to the beetles. Therefore, the tactic of attracting beetles to trap logs or lumber piles by supplementing the log odor with synthetic pheromones may greatly increase the efficiency of the procedure. An application of residual pesticide to the trap host, as used for bark beetles (Dyer, 1973; Smith, 1976), could make the tactic even more effective, by protecting baited logs from deterioration by attack, while killing every attracted beetle.

Pheromone-baited traps may be easier to deploy and service, and may be less disruptive to industrial operations than trap logs or lumber piles. However, to be competitive with them, traps must incorporate all the essential features of form and odor found in a normal host.

Logistics

The logistics of a pheromone-based management operation are critical to its success. Off season planning should lead to the following: a supply of pheromones of known autheticity and purity; confirmed financing; provision for hiring, training and accommodating personnel; vehicles; adequate numbers of functional traps; a supply of suitable trap logs; clearance for insecticide use, if necessary; and briefing of all relevant agencies and administrators.

Assuming that all requirements are met and contingencies anticipated, the most crucial concern is timing. If personnel are hired on the first of April, in anticipation of a first *T. lineatum* flight in late April, and the first flight occurs in mid-March (as in 1979), an otherwise well planned program may fail. It is vitally important to intercept the first beetles that fly and to continue to intercept the majority of beetles. The longer a breakthrough can be prevented, so that the beetles are not permitted to attack the product and set up competing pheromone sources, the more effective the operation will be.

Results

Ultimately, the nature of the timber processing operation will render all strategies and tactics partially ineffective, regardless of perfect logistics. This failure is because infested and susceptible logs will inevitably arrive in the dryland sort and depart for the mill. In the spring, some of these logs will attract resident beetles away from synthetic pheromone sources. From some, *Gnathotrichus* spp. will emerge to attack adjacent logs or lumber, without ever encountering a distant pheromone source. From midsummer onward, brood *T. lineatum* will emerge, and fly to overwinter in the margins of the processing area, thus ensuring a new pest population for next year.

The primary objectives of a suppression strategy should be to intercept as many as possible of the emergent, resident T. lineatum before they fly into the processing area, and to capture as many as possible of the Gnathotrichus spp. as they fly from old to new hosts. The degree of success of the operation will depend on the extent to which a company pursues the habitat management and product protection strategies (Table 2) to create a truly integrated pest management program.

OPERATIONS IN THE CHEMAINUS SAWMILL

An early opportunity to test (±)-sulcatol for suppression of G. sulcatus arose in the MacMillan Bloedel, Ltd., Chemainus Division Sawmill, Vancouver Island, B.C. (Fig. 2). Much of the lumber produced by this mill is exported abroad. It is sawn from logs trucked to the mill and dumped into the adjacent bay, as well as from log booms towed into the bay from other coastal logging divisions. Coincident with this operation, the Chemainus Woodlands Division water sorts and booms logs for towing to other mills (Fig. 2).

G. sulcatus is a persistent problem in the sawmill. The beetles are brought to the mill in infested logs from which they emerge and fly into the mill site to attack freshly sawn, unseasoned lumber (McLean and Borden, 1977a). Since G. sulcatus can survive for many weeks, and even complete development in this lumber (McLean and Borden, 1975a), it is a severe threat to be exported. The confined area of the mill, surrounded mainly by the town of Chemainus, the single source of beetles from logs in the water, the relatively few beetles in the mill site compared to the forest, increasing problems associated with treating lumber piles with insecticides, and the fact that any attack is a major problem in export-bound lumber made the Chemainus sawmill an ideal location for a trial mass-trapping operation.

Pre-operational Research

In 1974, 0.3 m^2, cylindrical, wire mesh sticky traps (20-cm diam. x 30-cm long), coated with Stikem Special® (Michel and Pelton Co., Emeryville, CA) were placed at 10 locations throughout the mill site (McLean and Borden, 1975b). These "survey" traps were baited with (±)-sulcatol released at an estimated 0.10 mg/h, and were in operation for the last week of the months April-October.

The trapping program yielded 3,098 G. sulcatus. There was a distinct, bimodal seasonal pattern, with a minor peak in April-May and a major peak in August-September, consistent with data from studies in the forest (Daterman et al., 1965; Cade, 1970; Byrne et al., 1974). The large numbers in late April and the surprising catch at the end of October indicated that an effective suppression program should operate from at least the first of April through October.

Unexpectedly few beetles were caught on the shores of the bay. The majority were captured in 4 locations (nos. 4, 6, 1 and 3, Fig. 2), which accounted for 57.0, 13.0, 10.2 and 5.6% of the total, respectively. All 4 locations are major storage yards for vulnerable, newly sawn lumber.

In 1975, experiments were designed to determine if suppression of G. sulcatus populations could be achieved in a mass trapping program (McLean and Borden, 1977a).

Immediately, the problem of population assessment arose. Unlike a field or orchard crop, in which insect populations can be routinely sampled, the "crop" i.e., logs and lumber, in a sawmill continually arrives, is modified and departs. Moreover, there are no proven, practical sampling methods that can indicate ambrosia beetle populations in infested hosts. Our conclusions were that beetles that reattacked logs in the water could not be assessed, and any that flew out of the mill site were of no consequence. The only G. sulcatus of concern to the operation either were caught in traps, or attacked the lumber at risk. Therefore, a suppression ratio (SR) was developed as an index of effectiveness of the trapping program as follows:

$$SR = \frac{\text{No. beetles caught on traps}}{\text{No. beetles caught on traps + no. beetles attacking sampled lumber.}}$$

The relative attack on lumber within a storage yard was sampled weekly by checking at least 25 loads of stacked lumber around the yard perimeter. G. sulcatus entrance holes were marked so that they would not be recounted if the load remained in the yard. In dissections of freshly attacked boards, 35 of 50 males were paired with females. Therefore, the number of attacking males was equal to the number of active, frass-producing entrance holes, and the number of females was calculated as 0.7 x the number of males.

The 4 yards in which G. sulcatus populations were highest in 1974 were used in 1975. One control yard (no. 3, Fig. 2) contained 4 unbaited traps. Two yards (nos. 1 and 6, Fig. 2) were serviced by 4 sulcatol-baited traps as in 1974, but for alternate, 2-week periods. A 4th yard (no. 4, Fig. 2) contained 4 continuously-baited traps. The traps were in operation from April 24 to the end of September.

Of the 5,798 G. sulcatus captured, 85.6% were in the continuously-baited location. The SR's in this yard were 0.70 and 0.87 for males and females, respectively, as opposed to 0.06 and 0.06 in the unbaited yard and means of 0.34 and 0.65 in the 1/2-time baited yards. When the trapping regime was suspended for 2 weeks the beetles apparently attacked lumber and set up competing pheromone sources, which probably contributed to the lower overall SR.

A strike closed the mill from July 3 through October. Progressively fewer beetles were caught as the summer progressed, and the expected late summer peak failed to materialize, probably because no new beetles entered the mill site in infested logs.

In the baited yards, 55% of the G. sulcatus attacks on lumber were in loads adjacent to traps or on the next 3 nearest loads. This result suggested that trap loads of cull lumber might be incorporated into a management program.

Test loads were put out weekly by the mill so that attack rate could be monitored. The plan had been to remove each load after 4 weeks, but the mill closure led to loads of increasing age in the mill from July on. There was some attack on these loads in the first week, the greatest number of attacks were on loads 2-4 weeks old, and the attack rate declined in weeks 6-8. Apparently, the lumber became more attractive as it aged while exposed to air, a hypothesis supported by other studies (McLean, 1976), suggesting that unidentified host compounds exist which could be used with (±)-sulcatol. The need for rapid inventory turnover before lumber reaches maximum susceptibility is obvious.

Operational Research

The encouraging results in 1975 led to a full scale, operational suppression program from April 28 - October 26, 1976 (McLean and Borden, 1979). The suppression trap comprised 2, 70 x 95-cm grey fiberglass screen vanes (1.7 x 1.4-mm mesh) coated with Stickem Special and suspended at right angles to a 1.3 m high central support post. The pheromone release rate was increased to an optimal 4 mg/h (McLean, 1976). Six yards were serviced, including the 5 highest catch sites in 1974. Because of consistently high catches in the past, 2 traps were set out at each end of the eastern boundary of the planer mill yard and these were supplemented by a survey trap on the western edge of the yard (no. 4, Fig. 2). One trap was placed at each of the other locations, with location 6 an unbaited control. Beside each trap a trap-load of 5 x 10 cm sapwood lumber was placed every 4 weeks to absorb beetles not caught on the traps.

Sampling for attacks in the yards was increased to encompass the sides of at least 100 loads each week. The estimated total number of attacks was extrapolated each week according to the total number of loads in the yard, either counted or determined from mill inventory records. The results of additional sampling revised the number of attacking females to 0.8 x the number of attacks. The effectiveness of the trapping program was calculated as percent suppression (PS) as follows:

$$PS = \frac{\text{No. beetles caught on traps and trap loads} \times 100}{\text{No. beetles caught on traps and trap loads} + \text{estimated no. beetles attacking lumber}}$$

The seasonal catch and attack patterns were bimodal (Fig. 3A, B). The highest attack and capture rates coincided in the spring, but peak attack rate was offset by one week in late summer. The most effective PS occurred when beetle flight activity was low in June, July and August (Fig. 3C). Quite probably some of the numerous males at peak flight times attacked lumber, and set up competing pheromone sources, lowering the suppression rate. This hypothesis is supported by a significant negative correlation ($r = -0.73$) between PS and the number of attacks in the previous 3 weeks. When beetle numbers were lower, the few males which initiated attacks were evidently unable to compete with the copious amounts of pheromone emanating from the baited traps, and the PS rose. Therefore, the numbers of traps should probably be increased during flight peaks. Inclusion of ethanol and α-pinene with (±)-sulcatol (Table 1) should also attract more males, thereby reducing the potential for establishment of competing pheromone sources.

Of the 42,907 *G. sulcatus* captured, 84.3% were in locations 3 and 4, the 2 yards near the planer mill (Fig. 2). Only 47% of the total attacks occurred in these locations. The overall suppresion rates for all baited yards were 58 and 71% for males and females, respectively, and 17 and 19% in the unbaited yard.

Although complete suppression was not achieved, the mill personnel evaluated the attack rate on lumber as exceedingly low based on past experience. The suppression traps required less than one man-day/month to maintain, and offered an effective alternative to conventional pesticides. Therefore, commencing in 1977, the mill incorporated the *G. sulcatus* trapping program into its normal quality control operations.

Fig. 3. Weekly 1976 results in the Chemainus sawmill, summed for all baited locations (1-5, Fig. 2). A) Numbers of beetles captured on sulcatol-baited sticky traps. B) Estimated total numbers of attacks on lumber. C) Percentage suppression of *G. sulcatus*. (From McLean and Borden, 1979).

→

NEW TOOLS, MORE RESEARCH

The availability of *S*-(+)-sulcatol and lineatin (Table 1) has stimulated renewed activity. Accordingly, 1979 was the first year of a large R&D project designed to develop integrated management programs for all 3 ambrosia beetle species in 3 dryland sorts and the Chemainus sawmill.

Resolution of Nagging Problems

Two persistent challenges to pest management researchers are economic justification of the research, and determining what impact the management program has on the target population.

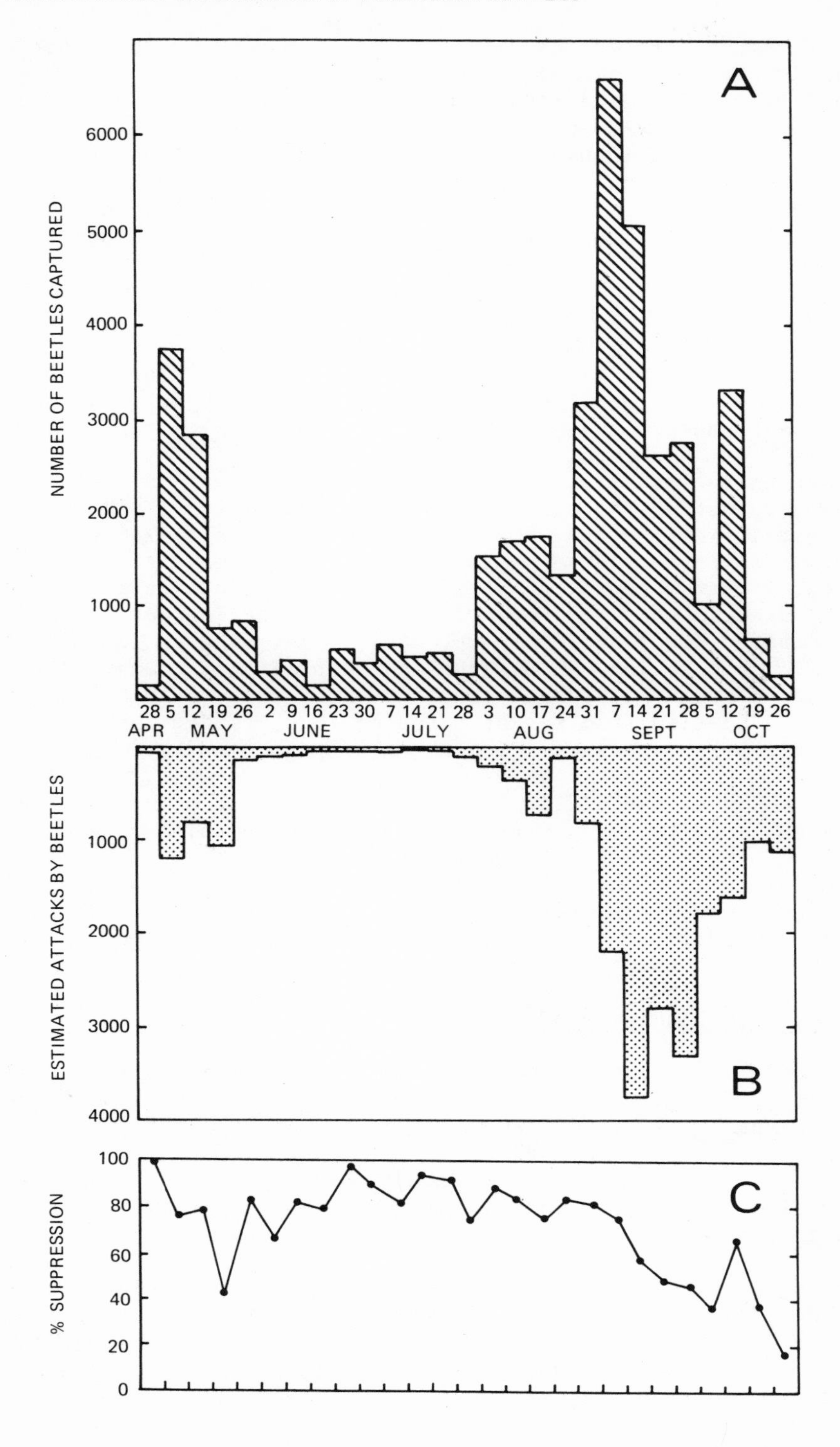
A
NUMBER OF BEETLES CAPTURED
6000
5000
4000
3000
2000
1000
28 5 12 19 26 2 9 16 23 30 7 14 21 28 3 10 17 24 31 7 14 21 28 5 12 19 26
APR MAY JUNE JULY AUG SEPT OCT
ESTIMATED ATTACKS BY BEETLES
1000
2000
3000
4000
B
% SUPPRESSION
100
80
60
40
20
0
C

In timber processing areas, those on the job "know" that ambrosia beetles cause severe degrade. Yet this loss is not nearly so apparent to an executive on the 25th floor of a city office building. Moreover, studies that determine what the degrade value can be on a per unit basis (Dobie, 1978) never indicate the total loss to an operation. And rough estimates like $7 million for 1975-76 (Nijholt, 1978a) tend to become fixed in the mind, even though the actual loss may be much greater. Therefore, current research is focused on sampling for attack severity throughout the milling process, and correlating this estimate with the actual values of the product with and without degrade (Shore and McLean, 1979). This study should allow more precise loss figures to be obtained for an entire mill operation. It should also facilitate the development of realistic cost-benefit ratios for pheromone-based pest management.

As in the sawmill, population assessment again becomes the key to evaluating a program in dryland sorts. For *T. lineatum*, the most logical approach is to modify Chapman's (1974) volumetric sampling approach to one that samples a unit area of forest floor. The total overwintering population can then be estimated for the entire area in which the majority of overwintering beetles would be concentrated (Dyer and Kinghorn, 1961). Any changes in the overwintering population can be assessed each winter. Consistently low level populations would probably represent the yearly influx of immigrant beetles from forest-infested logs, rather than a failure to eradicate the resident population.

Because of the difficulty in sampling attacks on logs, and the formidable and rapid operation of log moving with heavy equipment, it is not as feasible to assess attacks within a dryland sort as it is on lumber in a sawmill. Therefore, an attempt is planned to use a capture, mark and recapture system to determine the percent of the flying population of beetles, particularly *Gnathotrichus* spp., that is captured on traps or trap logs.

Pre-operational Research

While it is premature to describe or evaluate current research in detail, the results of the first year of pre-operational research in the expanded program can be summarized as follows:

1. Conrel® fibers are excellent devices for the long term, controlled release of lineatin, which is optimally active at a release rate of 40 µg/day.

2. A catch of 108,000 *T. lineatum* indicates a high overwintering population in the Chemainus sawmill. *T. lineatum* attacks lumber infrequently (McLean and Borden, 1977a) and the more significant threat is to boomed logs in the water. There is a good probability of carrying out an effective trapping program.

3. A season's total catch of only 2,361 G. retusus in the Chemainus sawmill suggests that this species is not a significant problem in the mill site.

4. The 1978-79 overwintering population of T. lineatum in the margin of the MacMillan Bloedel Ltd., Shawnigan Division dryland sort was estimated to be between 0.3-0.5 million, and at Pacific Logging's Sooke Division dryland sort, 1.4-2.6 million.

5. Catches of T. lineatum in pheromone-baited survey traps were 66,468, 66,718 and 47,070 in the Shawnigan dryland sort, the Sooke dryland sort and B.C. Forest Product's Renfrew Division dryland sort, respectively. Respective catches of G. sulcatus were 33,678, 46,455 and 35,219 and of G. retusus were 11,157, 6,139 and 4,743. These results suggest that all 3 species could be mass trapped in these sorts.

6. Higher early spring catches of T. lineatum inside the forest margin than in the open suggest that beetles could readily be intercepted before they attack logs within a sort.

7. Experimental, pheromone-baited trap logs at the Shawnigan sort absorbed ca. 80,000 ambrosia beetles, indicating that trap logs might be used effectively.

8. Problems encountered and their solutions in dryland sorts include: a) Dust from logging trucks and heavy equipment accumulating on sticky traps, demands judicious trap placement and maintenance, or adoption of a non-sticky trap, b) vespulid wasp populations so heavy in 1979 that protective clothing required, c) destruction of traps by equipment and errant logs, demands judicious trap placement and care by workers, and d) bears, Ursus americanus Pillas, around a garbage dump at the Renfrew sort precluded operations in much of the critical area of forest margin. So far, the bears are winning!

Operational Research

The second year of R&D will include:

1. Development of mass trapping procedures for T. lineatum in the Chemainus sawmill.

2. Development of mass trapping procedures for all 3 species at the Sooke sort, with Renfrew as a control area.

3. Further development and testing of trap log tactics at Shawnigan.

Research in the third year will comprise the intergration and final evaluation of strategies and tactics for all 3 species in both industrial settings prior to takeover of operations by industry.

INDUSTRIAL TAKEOVER

It is uncertain whether industrial takeover of ambrosia beetle management will mean that each company directly will operate its own program, or whether a separate company will contract its management services to the forest industry. The latter option is more attractive from a scientific point of view. A consulting company would most probably be formed by trained pest management personnel who would be more likely to conduct and evaluate their operations in a uniform, quantitative manner, to be more attuned to the progress of research, and thus to be more likely to incorporate new research results widely and rapidly into their operations.

ACKNOWLEDGMENTS

We thank MacMillan Bloedel, Ltd., British Columbia Forest Products, Ltd. and Pacific Logging, Ltd. for welcoming research on their premises, and participating actively in it. B. S. Lindgren and T. L. Shore provided much of the recent data. For their advice and assistance in various phases of the research, we acknowledge in particular N. Barnett, L. Chong, L. Cobb, W. Coombs, W. J. B. Devitt, P. Dobson, D. R. Gray, H. Hagg, J. Holman, T. Huff, J. Lavis, W. W. Nijholt, A. C. Oehlschlager, D. A. Palmer, H. A. Richmond, K. N. Slessor, E. Stokkink, D. Tuckey and G. von Westarp. The research was supported by the Natural Sciences and Engineering Research Council, Canada, the National Science Foundation, U. S. A., the Council of Forest Industries of B. C., and The Science Council of B. C.

REFERENCES

Atkins, M. D., 1966, Behavioral variation among scolytids in relation to their habitat, Can. Entomol., 98: 285.

Bain, J., 1974, Overseas wood- and bark-boring insects intercepted at New Zealand ports, N. Z. For. Serv. Tech. Paper No. 61.

Bauer, J., and Vité, J. P., 1975, Host selection by Trypodendron lineatum, Naturwiss., 62: 539.

Borden, J. H., 1974, Pheromone mask produced by male Trypodendron lineatum (Coleoptera: Scolytidae), Can. J. Zool., 52: 533.

Borden, J. H., and McLean, J. A., 1979, Secondary attraction in Gnathotrichus retusus and cross attraction of G. sulcatus (Coleoptera: Scolytidae), J. Chem. Ecol., 5: 79.

Borden, J. H., and Stokkink, E., 1973, Laboratory investigation of secondary attraction in Gnathotrichus sulcatus (Coleoptera: Scolytidae), Can. J. Zool., 51: 469.

Borden, J. H., Chong, L., McLean, J. A., Slessor, K. N., and Mori, K., 1976, Gnathotrichus sulcatus: synergistic response to enantiomers of the aggregation pheromone sulcatol, Science, 192: 894.

Borden, J. H., Hadley, J. R., Johnston, B. D., MacConnell, J. G., Silverstein, R. M., Slessor, K. N., Swigar, A. A., and Wong, D. T. W., 1979, Synthesis and field testing of lineatin, the aggregation pheromone of Trypodendron lineatum (Coleoptera: Scolytidae), J. Chem. Ecol. , 5: 681.

Borden, J. H., Handley, J. R., McLean, J. A., Silverstein, R. M., Chong, L., Slessor, K. N., Johnston, B. D., and Schuler, H. R., 1980a, Enantiomer-based specificity in pheromone communication by two sympatric Gnathotrichus species, J. Chem. Ecol., 6: 445.

Borden, J. H., Oehlschlager, A. C., Slessor, K. N., Chong, L., and Pierce, H. D., Jr., 1980b, field tests of isomers of lineatin, the aggregation pheromone of Trypodendron lineatum (Coleoptera: Scolytidae), Can. Entomol., 112: 107.

Borden, J. H., Lindgren, B. S., and Chong, L., 1980c. Ethanol and α-pinene as synergists for the aggregation pheromones of two Gnathotrichus species. Can. J. For. Res., (In press).

Byrne, K. J., Swigar, A. A., Silverstein, R. M., Borden, J. H., and Stokkink, E., 1974, Sulcatol: population aggregation pheromone in the scolytid beetle, Gnathotrichus sulcatus, J. Insect Physiol., 20: 1895.

Cade, S. C., 1970, The host selection behavior of Gnathotrichus sulcatus LeConte (Coleoptera: Scolytidae), Ph. D. Dissertation, Univ. Washington, Seattle.

Cade, S. C., Hrutfiord, B. F., and Gara, R. I., 1970, Identification of a primary attractant for Gnathotrichus sulcatus isolated from western hemlock logs, J. Econ. Entomol., 63: 1014.

Chapman, J. A., 1966, The effect of attack by the ambrosia beetle Trypodendron lineatum (Olivier) on log attractiveness, Can. Entomol., 98: 50.

Chapman, J. A., 1974, Ambrosia beetle, guidelines to population estimates near dryland log-storage areas and damage hazard assessment, Can. For. Serv., Pac. For. Res. Cent. Inf. Rep. BC-X-103.

Daterman, G. E., Rudinsky, J. A., and Nagel, W. P., 1965, Flight patterns of bark and timber beetles associated with coniferous forests of Western Oregon. Oreg. State Univ. Tech. Bull. No. 87.

Dobie, J., 1978, Ambrosia beetles have expensive tastes, Can. For. Serv., Pac. For. Res. Cent. Rep. BC-P-24.

Dyer, E. D. A., 1973, Spruce beetle aggregated by the synthetic pheromone frontalin. Can. J. For. Res., 3: 486.

Dyer, E. D. A., and Chapman, J. A., 1965, Flight and attack of the ambrosia beetle Trypodendron lineatum (Oliv.) in relation to felling date of logs, Can. Entomol., 97: 42.

Dyer, E. D. A., and Kinghorn, J. M., 1961, Factors influencing the distribution of overwintering ambrosia beetles, Trypodendron lineatum (Oliv.), Can. Entomol., 93: 746.
Finegan, R. P., 1967, Environmental contamination by benzene hexachloride used for control of ambrosia beetles, B. C. Fish Wildl. Rep. (Unnumbered).
Francke-Grossman, H., 1963, Some new aspects in forest entomology, Ann. Rev. Entomol., 8: 415.
Funk, A., 1965, The symbiotic fungi of certain ambrosia beetles in British Columbia, Can. J. Bot., 43: 929.
Funk, A., 1970, Fungal symbionts of the ambrosia beetle Gnathotrichus sulcatus, Can. J. Bot., 48: 1445.
Graham, K., 1968, Anaerobic induction of primary chemical attractancy for ambrosia beetles, Can. J. Zool., 46: 905.
Graham, K., Kinghorn, J. M., and Webb, W. E., 1950, Measurement of a damage index in logs infested by ambrosia beetles, B. C. Lumberman 35(8): 43, 45, 98, 100, 102, 104.
Johnston, B. D., and Slessor, K. N., 1979, Facile synthesis of the enantiomers of sulcatol, Can. J. Chem., 57: 233.
Kinghorn, J. M., 1957, Two practical methods of identifying types of ambrosia beetle damage, J. Econ. Entomol., 50: 213.
Kinghorn, J. M., and Chapman, J. A., 1959, The overwintering of the ambrosia beetle Trypodendron lineatum (Oliv.), For. Sci., 5: 81.
Lejeune, R. R., and Richmond, H. A., 1975, Striped ambrosia beetle Trypodendron lineatum (Oliv.), in "Aerial Control of Forest Insects in Canada," M. L. Prebble, ed., pp. 246-249, Can. Dept. Environ., Ottawa.
MacConnell, J. G., Borden, J. H., Silverstein, R. M., and Stokkink, E., 1977, Isolation and tentative identification of lineatin, a pheromone from the frass of Trypodendron lineatum (Coleoptera: Scolytidae), J. Chem. Ecol., 5: 549.
McLean, J. A., 1976, Primary and secondary attraction in Gnathotrichus sulcatus (LeConte) (Coleoptera: Scolytidae) and their application in pest management, Ph. D. Dissertation, Simon Fraser University, Burnaby, B. C., Canada.
McLean, J. A., and Borden, J. H., 1975a, Gnathotrichus sulcatus attack and breeding in freshly sawn lumber, J. Econ. Entomol., 68: 605.
McLean, J. A., and Borden, J. H., 1975b, Survey for Gnathotrichus sulcatus (Coleoptera: Scolytidae) in a commercial sawmill with the pheromone, sulcatol, Can. J. For. Res., 5: 586.
McLean, J. A., and Borden, J. H., 1977a, Suppression of Gnathotrichus sulcatus with sulcatol-baited traps in a commercial sawmill and notes on the occurrence of G. retusus and Trypodendron lineatum, Can. J. For. Res., 7: 348.
McLean, J. A., and Borden, J. H., 1977b, Attack by Gnathotrichus sulcatus (Coleoptera: Scolytidae) on stumps and felled trees baited with sulcatol and ethanol, Can. Entomol., 109: 675.
McLean, J. A., and Borden, J. H., 1979, An operational pheromone-

based suppression program for an ambrosia beetle, Gnathotrichus sulcatus, in a commercial sawmill, J. Econ. Entomol., 72: 165.

Mathers, W. G., 1935, Time of felling in relation to injury from Ambrosia beetles, or pinworms, B. C. Lumberman, 19(8): 14.

Milligan, R. H., 1970, Overseas wood- and bark-boring insects intercepted at New Zealand ports, N. Z. For. Serv. Tech. Paper No. 57.

Moeck, H. A., 1970, Ethanol as the primary attractant for the ambrosia beetle Trypodendron lineatum (Coleoptera: Scolytidae), Can. Entomol., 102: 985.

Moeck, H. A., 1971, Field test of ethanol as a scolytid attractant, Can. Dept. Fish. and For., Bi-mo. Res. Notes 27: 11.

Mori, K., 1979, Synthesis of (+)-lineatin, the unique tricyclic pheromone of Trypodendron lineatum (Olivier), Tetrahedron Lett., 15: 1329.

Nijholt, W. W., 1973a, Ambrosia beetle attacks delayed by turpentine oil, Can. Dept. Fish. and For., Bi-mo. Res Notes, 29: 36.

Nijholt, W. W., 1973b, The effect of male Trypodendron lineatum (Coleoptera: Scolytidae) on the response of field populations to secondary attraction, Can. Entomol., 105: 583.

Nijholt, W. W., 1978a, Ambrosia beetle, a menace to the forest industry, Can. For. Serv., Pac. For. Res. Cent. Rep. BC-P-25.

Nijholt, W. W., 1978b, Evaluation of operational watermisting for log protection from ambrosia beetle damage, Can. For. Serv., Pac. For. Res. Cent., Rep. BC-P-22-78.

Nijholt, W. W., 1980a, Pine oil delays attack of ambrosia beetles on piled log sections. Environ. Can, Bi-mo. Res. Notes, 35: 22.

Nijholt, W. W., 1980b, Pine oil and oleic acid reduce attacks on logs by ambrosia beetles (Coleoptera: Scolytidae), Can. Entomol., (In press).

Nijholt, W. W., and Schohherr, J., 1976, Chemical response behavior of scolytids in West Germany and western Canada, Environ. Can., Bi-mo. Res. Notes, 32: 31.

Prebble, M. L., and Graham, K., 1957, Studies of attack by ambrosia beetles in softwood logs on Vancouver Island, British Columbia. For. Sci., 3: 90.

Richmond, H. A., 1961, Helicopters protect log booms in B.C. Can. Lumberman, 81(12): 40.

Richmond, H. A., 1968, The ambrosia beetle on the British Columbia coast, B.C. Loggers Div., Coun. For. Ind. of B.C., Mimeo Rep.

Richmond, H. A., 1969, Appetite for wood, B. C. Lumberman, 53(8): 34.

Richmond, H. A., and Nijholt, W. W., 1972, Water misting for log protection from ambrosia beetles in B.C. Can. For. Serv., Pac. For. Res. Cent., Rep. BC-P-4-72.

Rudinsky, J. A., and Daterman, G. E., 1964a, Response of the ambrosia beetle Trypodendron lineatum (Oliv.) to a female-produced pheromone, Z. Angew. Entomol., 54: 300.

Rudinsky, J. A., and Daterman, G. E., 1964b, Field studies on flight patterns and olfactory responses of ambrosia beetles in Douglas-fir forests of western Oregon, Can. Entomol., 96: 1339.

Schuler, H. R., and Slessor, K. N., 1977, Synthesis of enantiomers of sulcatol, Can. J. Chem., 55: 3280.

Shore, T. L., and McLean, J. A., 1979, Ambrosia beetle research in the Chemainus sawmill, 1979, including proposals for 1980. Unpublished report, Univ. of B. C. Faculty of Forrestry.

Slessor, K. N., Oehlschlager, A. C., Johnston, B. D., Pierce, H. D., Jr., Grewal, S. K., and Wickremsinghe, L. K. G., 1980, A regio-selective synthesis and resolution leading to the chiral pheromone of Trypodendron lineatum. J. Org. Chem., (In press).

Smith, R. H., 1976, Low concentration of lindane plus induced attraction traps mountain pine beetle, USDA For. Serv. Res. Note, PSW-316.

Vité, J. P., and Bakke, A., 1979, Synergism between chemical and physical stimuli in host colonization by an ambrosia beetle, Naturwiss., 66: 528.

DEPLOYMENT OF TRAPS IN A BARRIER STRATEGY TO REDUCE POPULATIONS OF THE EUROPEAN ELM BARK BEETLE, AND THE INCIDENCE OF DUTCH ELM DISEASE

John W. Peacock

Roy A. Cuthbert

USDA Forest Service
Delaware, OH 43015 USA

Gerald N. Lanier

College of Environmental
Science and Forestry
State University of New York
Syracuse, NY 13210 USA

INTRODUCTION

Dutch elm disease (DED) is caused by a fungus, *Ceratocystis ulmi* (Buisman) C. Moreau, that infests the vascular systems of elms and forms sticky masses of conidial spores (coremia) in cavities under the bark of the diseased portions of the tree. These spores have been detected on many arthropods that are found beneath the bark of diseased elms, but only bark beetles are an efficient vector because only they frequently visit or inflict wounds in healthy elms; deposition of spores in wounds is necessary for infection.

In North America, the 2 important vectors of DED are the native elm bark beetle, *Hylurgopinus rufipes* (Eichhoff), and the European elm bark beetle, *Scolytus multistriatus* (Marsham). The European beetle is more aggressive and dominates in urban areas where the climate is permissive (winter low temperature higher than -20°F) (Lanier, 1978a). The native beetles often predominate in wood lots and green spaces where dense shade inhibits attack by *S. multistriatus*, and they may be the only vector in the colder regions of the Lake States, New England and Canada.

Both insects transmit DED while feeding on the bark of healthy elms. Injury by S. multistriatus occurs in twig crotches following the initial dispersal flight from the brood wood. H. rufipes can transmit the DED fungus in the fall when it forms feeding and overwintering tunnels in the lower bole, or in the spring when the overwintered adults move up the tree and feed on the thin bark of branches.

Contrary to most accounts, "maturation feeding" in twig crotches is not a prerequisite to breeding by S. multistriatus in diseased or moribund elm wood. H. rufipes also does not require a separate feeding attack in the spring before it can breed, but it probably does require a fall feeding period for the adult to overwinter.

The flight capacity of S. multistriatus considerably exceeds the 300 m limit frequently stated. We have found that the normal dispersal flight is 400 m or more and flights exceeding 15 km have been observed (M. C. Birch, personal communication). There is little documentation of the flight capacity of Hylurgopinus, but circumstantial evidence indicates that it is capable of directed flights of several hundred meters.

Current (1980) integrated programs for the control of DED include: (1) Sanitation (the removal of diseased, dead, or dying elm material that serves as a breeding site for the beetles and fungus); (2) insecticide spraying (to prevent beetle feeding on healthy elms); (3) systemic fungicide injection (as a prophylactic treatment to protect healthy trees or as a therapeutic treatment to cure diseased trees); and (4) chemical treatment to prevent root graft transmission of the fungus between adjacent elms (Schreiber and Peacock, 1979). Additional gains in DED control might be made by mass trapping beetles using pheromone-baited traps.

MASS TRAPPING ELM BARK BEETLES

Aspects of urban areas that favor mass trapping as a strategy to control elm bark beetles are as follows:

a. High value of amenity elms justifies high expenditures for their protection. Individual ornamental trees are usually valued at $100 to tens of thousands of dollars, depending on size, placement, species and condition. Costs of removal and replacement must also be considered; in Syracuse, NY, the current average cost of removing a mature tree plus planting a sapling is about $300.

b. Mass trapping is an attractive alternative to insecticides. Spraying of chemicals in urban environments is increasingly difficult due to regulations and public opposition. The cost is unusually high because the types of equipment that can be used are limited and extensive precautionary actions (publicity, routing traffic) must be taken.

c. Available alternative methods may not adequately control the pest.

d. Access for survey, trap placement, and servicing is excellent.

Several attributes of elm bark beetles that make them amenable to mass trapping were outlined by Lanier (1978b):

a. The action is directed against the damaging stage. The adult bark beetle kills trees and disseminates disease. Interfering with or killing adults will have immediate impact on the level of damage. This is in contrast to Lepidoptera for which the strategy is to reduce mating by adults, but it is the late instar larva that does most of the damage. The numbers of breeding adults and the eggs may be only loosely correlated with the numbers of larvae surviving to late instars.

b. Both sexes are affected; thus, the reproductive potential of the population is directly reduced. If only males were affected, survivors could compensate by mating more frequently.

c. Successful breeding depends upon mustering enough individuals over a short period to overcome the resistance of the host tree. Destroying part of the population by trapping or disrupting its aggregation will result in many beetles lost in flight or absorbed in fruitless scattered attacks on trees that cannot be killed.

In addition to mass trapping, behavior-modifying chemicals might be employed as disruptants or repellents of elm bark beetles in urban areas. However, these applications are beyond the scope of this paper.

THE MULTILURE TRAPPING SYSTEM

Developmental Research

After it was shown conclusively that virgin female *S. multistriatus* produce an aggregation pheromone (Peacock et al., 1971), studies were begun to isolate and identify the pheromone of the beetle. The objective was to determine how pheromones could be used in the integrated program for the control of DED. After 5 years of study, the pheromone complex of *S. multistriatus* was isolated, identified, and synthesized, and field tests in 1974 showed conclusively that mass attraction of elm bark beetles to suitable breeding material is the result of their response to a host-produced chemical, α-cubebene, and to 2 beetle-produced pheromones, α-multistriatin and 4-methyl-3-heptanol (Pearce et al., 1975; Gore et al., 1977). Since 1975, studies have been aimed toward developing an efficient trapping system and determining the efficacy of mass trapping or other pheromone use strategies for beetle suppression and the reduction of beetle-vectored incidence of DED.

Components of the Multilure Blend

The principal aggregation attractants are (-)-4-methyl-3-heptanol (H), (-)-α-multistriatin (M), (-)-α-cubebene (C) (Pearce et al., 1975; Gore et al., 1975). Extensive laboratory and field trials with synthetic H and M, and C from cubeb oil have demonstrated that the aggregation attractant for this species is a synergistic mixture of H, M and C (Lanier et al., 1976; Lanier et al., 1977; Cuthbert and Peacock, 1978).

Bioassays and chemical analyses have shown that H and M are produced by virgin female beetles and that both chemicals are present in the female abdomen; compound H is concentrated in the upper abdominal area and M in the abdomen tips (Gore et al., 1977). The production of H ceases after females are mated and it is the lack of H that apparently accounts for diminished attraction to mated beetles (Elliott et al., 1975). The release of compound C from elm wood is enhanced by beetle boring activity, and the presence of *C. ulmi* (the DED pathogen) (Gore et al., 1977).

Pearce et al. (1975) reported that both the α- and β-isomers of multistriatin are associated with beetle-infested logs, but that only the α-isomer was active in the laboratory. In the field, Lanier et al. (1977) confirmed that only the α-isomer is attractive. Since synthetic racemic α-multistriatin was as attractive as natural α-multistriatin, Lanier et al. (1977) proposed that both the (-) and (+) enantiomers are attractive. However, Elliott et al. (1978) showed that only (-)-α-multistriatin was active. Gerken et al. (1978) reported that *S. multistriatus* in the Upper Rhine Valley of Germany did not respond to the tripartite attractant containing (-)-α-multistriatin, but did respond to (-)-δ-multistriatin combined with 4-methyl-3-heptanol and cubeb oil. One explanation for this apparent difference is that North American and European populations of *S. multistriatus* are genetically distinct and behaviorally different. Another possibility is that the concentrations of attractant released in the Rhine Valley experiments far exceeded the 400:100:800 μg/day (H:M:C) release rates used in North American studies. Such excessive release rates could deter response to mixtures containing α-multistriatin (Cuthbert and Peacock, 1978) and possibly elicit response to δ-multistriatin, even though the δ-isomer has no activity at low release rates (Lanier et al., 1977).

4-Methyl-3-heptanol consists of a pair of diastereomers, each with 2 enantiomers. Only the (-)-threo enantiomer is produced by virgin female beetles (Pearce et al., 1975). Data from field tests reported by Lanier et al. (1977) suggest that of the 4 enantiomers of synthetic H, only the naturally occurring enantiomer is active.

Studies on various doses and mixtures of the pheromone components indicate that the ratio of H to M is clearly the most important factor in the attractiveness of the pheromone blend. Within a wide range, the total dose of these 2 components is relatively insignificant, as long as H exceeds M. The dose of C, on the other hand, is an important factor -- attractiveness of the pheromone mixture increases with increasing doses of C (Cuthbert and Peacock, 1978).

The Pheromone Dispensing Systems

Several systems have been evaluated for dispensing the multilure components. Glass capillary tubes (Peacock et al., 1975) or polyethylene snap cap vials (Pearce et al., 1975; Cuthbert et al., 1977) were used in initial field tests. For various reasons (cost, poor delivery rates, etc.), neither were as suitable as Hercon® laminated plastic or Conrel® hollow fiber dispensers that have been used in field trials since 1975 (Lanier et al., 1976). Although the nominal release rates of H, M and C have been 400:100:800 μg/day, respectively, in actual practice the release rates vary greatly from those specified, regardless of the type of dispenser.

Cuthbert (unpublished) has shown that initial release rates from Hercon and Conrel baits far exceed those specified for the 1st week or more after deployment in the field. However, after an initial 2 weeks of weathering, both types of dispenser degrade at acceptable rates, following a typical decay model, and release the attractant components at ratios suitable for beetle attraction for an additional 4 to 6 weeks.

Traps and Trap Deployment

Presently, the traps used for beetle suppression are 45 x 66 cm sheets of white, double poly-coated cardboard coated with Tangle-Trap®. This trap design is the product of the evaluation of a number of systems, including Sectar-I® traps, Dixie® cups coated with Stikem®, Stikem-coated hardware cloth, and different sizes, shapes, and colors of flat, Stikem-coated cardboard (Lanier et al., 1976). None was effective as the present system, probably because most were too small to effectively capture incoming beetles.

Trap visibility is important in trap deployment (Lanier et al., 1976). Hardware cloth traps compared poorly with paper traps of equal size, apparently because the former are less visible. Black traps, or the black areas of traps with black and white stripes, capture more beetles than white traps when the traps are hung in the open. But when the traps are attached to the bole of elm trees or to utility poles, white traps capture more beetles than black ones. Apparently, contrast is critical, so white traps are now placed on utility poles.

The influence of trap height on beetle capture was reported by Cuthbert and Peacock (1975). When placed in elm trees, traps at the 3 m level captured significantly more beetles than traps at higher levels. Wollerman (unpublished) found that beetle catches on a vertical line of traps on an 18 m high metal tower in an open field were largest at 1 m. Over 90% of captured beetles were caught on traps below 4.6 m and only 4 beetles were captured at the 18 m level. No studies have been conducted with traps placed at different heights on utility poles in less open areas, but it is possible that the optimal placement height is below the 3 m level at which traps were deployed in mass-trapping studies.

APPLICATION OF MASS TRAPPING FOR BEETLE SUPPRESSION AND DED CONTROL

Beetle suppression studies with multilure-baited traps have used 2 basic trap deployment strategies: (1) Deployment in a grid within areas to be protected; or (2) deployment in one or more "barrier" rows around areas to be protected. Both systems were evaluated extensively between 1975 and 1979. The following is a discussion of the results and implications from these studies.

Mass Trapping -- Grid Deployment of Traps

Grid trapping for beetle suppression has been studied in Detroit, MI (1975), Independence, CA (1976-79), and Aberdeen, SD. The study in Aberdeen was initiated in 1979, so there are no data to report at this time.

In the Detroit study -- the first attempt at mass trapping with multilure -- more than 400 Stikem-coated, multilure-baited, hardware cloth traps were deployed 50 m apart on the boles of healthy elms in a 1 km^2 plot. Beetles on traps were counted after the 1st flight period and again at the end of the season to determine total catch for the 2nd flight. The number of beetles emerging from diseased trees within the plot was also estimated to compare total catch and total emergence. In 1975, the 1 million beetles captured on traps represented only 20% of the beetles that we estimated to have emerged within the plot. In light of this low catch of emerging beetles, it was concluded that trapping had little appreciable effect in suppressing the beetle population (Cuthbert et al., 1977). More important is that DED surveys conducted before, during, and after the experiment indicated that DED levels in the plot actually increased during the trapping period (Cuthbert and Peacock, 1979). Several possible reasons can be given for failure to suppress beetles or DED in this experiment. First, beetle populations were extremely high because of numerous diseased trees in the plot that served as suitable beetle sources, and beetle immigration from surrounding areas added significantly to the beetles already in the plot. Second, the baits were later found to be attractive for only 2 of the 6 weeks they were in the field, and the traps captured only a portion of the incoming beetles (as indicated by higher catches on traps of different design). Finally, the traps were placed on the bole of elms, where incoming beetles could feed and inoculate trees with the DED pathogen -- if they missed the traps. This experiment demonstrated the need to use improved baits and traps, to place traps on some standard other than the bole of elms (e.g. utility poles), and to conduct the studies in an area where DED was less rampant (Lanier et al., 1977). Grid-trapping studies in Independence, CA and Aberdeen, SD, take into account these considerations.

In Independence, a grid of traps (placed on utility poles) has been used in an attempt to suppress a population of *S. multistriatus* developing in a population of about 500 elms in an area of about 135 ha. In 1976, 1977 and 1978, 52 traps with a spacing of 50-100 m were used. In 1979, the number of traps was increased to 200, and the spacing between traps was reduced to about 10-15 m.

Not only does Independence have a relatively small population of beetles in a relatively small area, but it also has no DED. These factors would seem to favor a significant suppression of the beetle population, one that could be recognized by a reduction in 2nd generation catches, and a downward trend in the total catches from year to year.

Catches on 52 traps totaled about 224,000 beetles in 1976 and 71,000 in 1977 (Birch, 1979) (Table 1). Suprisingly, the number of *S. multistriatus* trapped in 1978 increased above 1976 levels, indicating that the drop in catches in 1977 was only coincidental with trapping and resulted from normal fluctuations in the beetle population. In 1979, when the number of traps nearly quadrupled, total trap catches were about equal to those recorded in 1978.

An estimated 857,000 beetles were captured in Independence over a 4-year period. It is clear that even at this experimental site where the beetle and elm population are completely isolated from beetle pressure outside of the trapping area, trapping could not affect a significant year-to-year reduction in the beetle population. Rather, year-to-year fluctuations in beetle catches were probably controlled mainly by the amount of weakened elm wood available for colonization which, in turn, is affected by the weather (drought).

A given bark surface area produces about the same number and brood over a wide range of attack densities. Thus, trapping beetles will affect the size of the next generation only if it actually prevents some of the suitable host material from being colonized.

Table 1. Yearly catches of _Scolytus multistriatus_ at multilure-baited traps deployed in grid in Independence, CA, 1976-79.

Year*	Trap spacing (m)	Total beetles captured (1000s)**
1976	50-100	224
1977	50-100	71
1978	50-100	239
1979	10-15	323

*52 Traps used in 1976, 1977, and 1978; 200 traps used in 1979.

**Total trap catches rounded to the nearest 1000 beetles.

Results from the Independence study seem to support the conclusion drawn from the earlier study in Detroit: trapping does not directly affect size of generations within or between years. However, if DED is reduced by trapping, the subsequent beetle populations can be indirectly affected because there will be less weakened elm wood for colonization. It remains to be seen whether the Aberdeen study in which 500 traps are deployed in a grid with 50-100 m spacing, will change the basic conclusion about grid trapping.

Mass Trapping -- Barrier Deployment of Traps

By far, the most intensive effort to evaluate the multilure-trapping system for beetle suppression has involved barrier deployment of traps. These studies have been conducted in a variety of elm, beetle, and DED situations: (1) Isolated groves on private estates, university campuses, etc., with relatively few elms, little DED and relatively small areas; and (2) large area, city-wide conditions where there are large numbers of elms, beetles and DED-infested trees (Evanston, IL, Fort Collins, CO and Minneapolis, MN).

a. _Protection of elms in groves_. During 1975-79, multilure-baited sticky traps were positioned around groves of 26-250 healthy elms at 12 locations in 8 states in the East and Midwest (Table 2). In 1975, traps were deployed (1) in a plot in East Syracuse, NY, (2) at Hinnerwadel's Grove, North Syracuse, NY; and (3) on the campus of the University of Delaware (Newark). Traps were deployed for the

1st time in 1976 at 9 other locations: (4) Dewitt, NY; (5) Phillips Academy, Andover, MA; (6) Brown University, Providence, RI; (7) State House, Providence, RI; (8) Carcraft, Wilmington, DE; (9) Chevy Chase, MD; (10) Barton Heights, Richmond, VA; (11) NC State University, Raleigh; and (12) Bradley Estate, River Hills, WI. Trapping continued through 1978 in all areas, but was discontinued in 4 areas (7, 10, 11, 12) in 1979. Trapping continued in the other areas through 1979.

Table 2. Yearly catches of *Scolytus multistriatus* captured at multilure traps deployed as barrier around isolated groves of elms, 1976-79.

	$\overline{X}$ Trap catches (1000s)			
Location*	1976	1977	1978	1979
1 (430/104)	5.1	8.2	9.7	4.5
2 (4/7)	7.1	14.3	8.7	7.0
3 (37/20)	11.8	12.4	10.7	6.6
4 (303/32)	5.3	12.0	I.D.	4.3
5 (124/12)	.9	3.4	2.4	2.8
6 (26/12)	I.D.	13.5	18.1	No traps
7 (8/4)	I.D.	7.5	4.3	No traps
8 (40/12)	30.6	12.5	16.9	12.5
9 (5/19)	22.0	28.1	24.4	17.5
10 (32/12)	22.4	37.7	35.6	No traps
11 (36/12)	8.5	33.3	25.4	No traps
12 (16/9)	8.5	4.2	7.0	No traps

*See text for explanation of locations. Figures in ( ) = area in ha/number of traps in area. I.D. = Incomplete data.

Criteria for selecting the "grove" trapping sites (Lanier, 1979) were as follows (several of the sites were at variance with 1 or 2 of the criteria):

1. Fifty or more elms constituted a rather insular grove; i.e., the site was not adjacent to extensive areas of elms yielding beetles.

2. An accurate history of the DED rate since 1970 was available.

3. DED rate during 1974 and 1975 was 2-15%.

4. Permission for testing and placement of traps could be obtained from the local authorities.

5. A qualified individual would oversee the local operation including: (1) Careful diagnosis of every elm within the area; (2) trap placement and maintenance; (3) Return of all traps to Syracuse.

6. The sites would represent regional differences in climate, tree population, and beetle density.

7. There would be no changes in the existing DED control program during the period of experimentation (other than fungicide therapy of diseased elms).

Rates of DED (Table 3) within the area were assessed by thoroughly inspecting each tree from the ground and sampling branches or boles when appropriate during early (May-June) and late (August-September) summer. Results for any year were not final until the early summer inspection of the following year. Each case of DED was listed by probable cause (beetle vectored or root graft) and year. Case histories were constructed for each area from historical data on DED rates. A system of experimental "check" areas was not established because of difficulties in finding comparable groves in the area near the treatment plots. Beetle catches on traps were estimated by placing templets with circular holes over the trap and then using a regression equation to project the estimated number of beetles (Table 2) on the trap from the number in the sample holes.

In all areas, multilure-baited sticky traps were deployed at a height of 3 m on utility poles, with a usual trap/area relationship of about 1 trap/4 ha. Baited traps were established in mid-May, replaced with new traps and baits in mid-July, and terminated in early October.

The general trend in disease rates, except in 2 areas, indicates that a decrease in DED incidence coincided with the advent of beetle trapping (Table 3, Fig. 1). Only in area 9 (Chevy Chase) and area 11 (NC State University, Raleigh) did DED incidence increase (in 1978) despite a trapping program. In other areas, there were some year-to-year fluctuations in disease rates. In several areas, DED rates of zero were reached and maintained (Table 3). In areas 4-12, comparisons were made between beetle-vectored and total (beetle-vectored + root graft cases) DED for the period 1976-79. In these 9 areas, beetle-vectored cases of the disease decreased even more dramatically than the rates for all cases of DED (Fig. 1).

Trapping was discontinued in 4 areas (7, 10, 11, 12) in 1979. DED data are not available for area 12, but it is important to note that in 2 areas (#10 - Barton Heights, Richmond, VA and #11 - NC State Univ.), DED levels rose significantly with the absence of

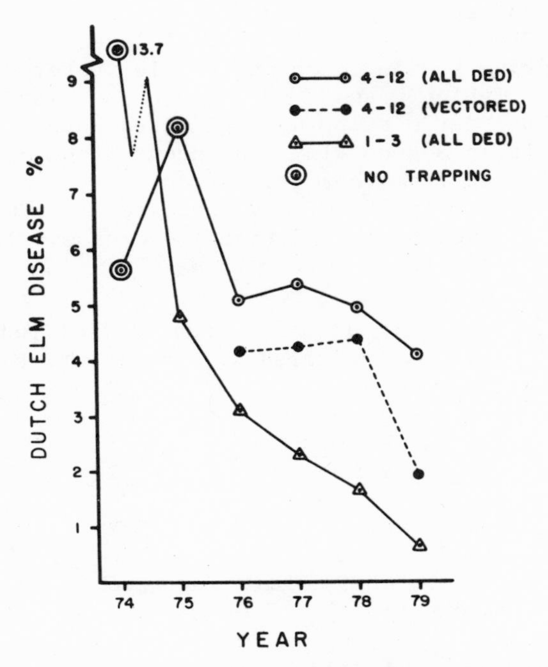

Fig. 1. Trends in Dutch elm disease incidence in 12 elm groves and scattered elm stands surrounded with multilure-baited traps. Trapping in groves 1-3 (see Table 3) began in 1975; trapping in areas 4-12 began in 1976. Large circles indicate disease rates before trapping.

traps in 1979 (Table 3). Figure 2 shows the steady decline in DED rates during the trapping period in areas 7, 10, 11 and the immediate increase in DED incidence coincidental with the cessation of trapping in areas 10 and 11.

Trap catches were extremely variable between areas and there is no clear correlation between numbers of beetles trapped and DED rate (Tables 2 and 3). However, it is generally true that within areas the number of beetles/trap in the spring moves in the same direction as DED. Exceptions are in areas 10 and 11 (Richmond, VA, and Raleigh, NC) where in 1978, spring catches increased while DED decreased. One variable that may be operative in these southern localities is that substantial beetle flight often occurs before elms are in leaf; perhaps traps are more effective when elm foliage is not present.

The effect of spring-emerging beetles on DED rate is illustrated in Chevy Chase when, due to a lapse in brood tree removal, the spring beetle catch and beetle-vectored DED increased sharply in 1978 (Tables 2 and 3). Chevy Chase also illustrates that control of DED by beetle trapping is limited when the trees to be protected are within a generally continuous stand and/or are not adequately protected from root graft transmission of the DED fungus. It is probable that trap efficiency increases as DED and the competing sources of attractant (brood trees) decrease. Trap density may also affect trapping efficiency.

Table 3. Dutch elm disease rates in isolated groves of elms encircled with multilure-baited traps.*

Location*	Dutch elm disease rate (%)					
	1974	1975	1976	1977	1978	1979
1	(26.7)**	9.6	6.3	5.2	2.6	1.7
2	(7.7)	0	0	0	4.2	0
3	(5.0)	2.9	1.8	0.6	0.6	0
4	Unknown	(18.5)	5.6	5.6	3.3	0
5	(4.8)	(3.0)	0	2.0	1.1	1.6
6	(3.3)	(5.2)	0	0	0	1.8
7	(18.5)	(9.1)	0	0	0	(0)
8	(14.8)	(40.1)	36.6	33.3	23.8	12.5
9	(1.1)	(5.1)	6.5	7.7	10.2	6.8
10	Unknown	(3.8)	2.3	2.9	1.1	(4.3)
11	(3.4)	(5.9)	6.3	6.6	3.0	(6.3)
12	(7.7)	(8.9)	5.5	4.5	7.6	(blank)

*See text for explanation of locations. Trapping began at locations 1, 2 and 3 in 1975; trapping in areas 4-12 commenced in 1976.

**( ) = DED rates when no traps were in place.

Any beetle control procedure that significantly reduces twig crotch feeding by S. multistriatus should have an effect on DED incidence. A direct measure of the beetle population is the number of injuries made by beetles feeding in the twig crotches of healthy elms. In Syracuse, where twig feeding was monitored during mass-trapping experiments, twig crotch feeding was reduced by 70% within the trapping area as compared to areas outside of the Syracuse plot (see paper by Lanier, this symposium).

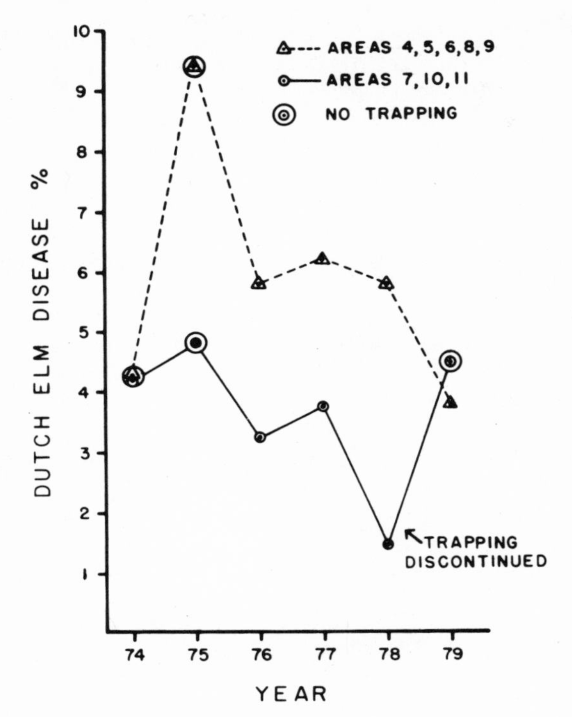

Fig. 2. Comparison of Dutch elm disease incidence in areas 4-9 with trapping from 1976 through 1979, and in areas 7, 10, 11 where trapping was discontinued in 1979. Large circles indicate disease rates before or after (areas 7, 10, and 11 in 1979) trapping.

Analyzing the effectiveness of multilure-baited traps by DED rates is complicated by the effects of weather on infection, root graft inoculation, etc. But results from these grove-trapping studies indicate that mass trapping of European elm bark beetles on pheromone-baited traps can contribute to the control of DED where conditions are appropriate. Multilure-baited traps will be most effective where elms within groves to be protected are free from diseased or weakened wood that may serve as competitive pheromone sources if infested by even a few beetles. The effect of multilure-baited traps on the in-flight beetle population should increase as the number of potentially competing sources of attractant (diseased, cut or injured elm wood) decreases. An intensive sanitation program (the prompt removal of infested elm wood and breeding material) is therefore essential to the success of grove trapping or, for that matter, any mass trapping efforts.

b. Protection of elms in city-wide, large-area situations. Suppression studies in Fort Collins, CO were carried out from May 1976 to October 1979. Fort Collins was chosen because the elm and beetle populations in the city were relatively isolated (elms are not native to this region), and because the DED rate among the amenity elms in the city was low. The experimental area was approximately 4,300 ha and contained about 4,000 American elms (*Ulmus americana*) and 4,000 Siberian elms (*U. pumila*). Some 650 suppression traps were deployed around the city in a single barrier row (50 m spacing between traps) in 1976 and 1977. In 1978 and 1979, the experimental area was encircled with 2 barrier rows (which included one of the rows used in 1976 and 1977) with a total of about 900 traps. Each year, a 5 x 6 grid of 30 "core" traps (200-300 m apart) was deployed in a 1 x 1.2 km area within the experimental area of Fort Collins, and 2 similar grids were deployed in nearby "check" cities, Greeley and Loveland, CO. These core traps were used to monitor changes in the beetle population in the trapping city (Fort Collins) vs. the check cities. Beetles captured on selected suppression traps and on all core traps were counted at the end of the 1st and 2nd beetle flight periods. Catches were compared between the cities and between generations and years within the cities to determine the impact of trapping on beetle abundance. DED surveys were conducted each year in all 3 cities to determine the effect of trapping on DED incidence.

Cuthbert and Peacock (1979) reported the results of trapping in 1976 and 1977. Trapping appeared to have little influence on either elm bark beetle populations within and between years or on DED incidence. There was a 5- to 10-fold increase in beetle catches between generations within years and little change in trap catches between years. Although average catches at core traps in Greeley and Loveland greatly exceeded those in Fort Collins each year, these differences could not be directly attributed to beetle suppression due to trapping in Fort Collins. Most important, despite the annihilation of more than 20 million beetles a year in Fort Collins, the DED rate seemed little, if at all, affected. Although there was a general decreasing trend in DED rates in 1976 and 1977, this same trend occurred in the check cities.

Beetle suppression and DED rates did not change significantly in 1978 and 1979 despite the addition of another barrier row of about 400 traps. The relative number of beetles captured between cities remained virtually unchanged, but catches, in general, were higher in all 3 cities (Table 4). DED rose in 1979 in all 3 cities despite a downward trend in 1976-78. There was a surprising increase in Fort Collins where the disease rate rose from 70 cases in 1978 to 111 cases in 1979 (Fig. 3). Most of the 1979 cases were attributed (by symptom development) to beetle transmission of the pathogen.

Table 4. Yearly catches of Scolytus multistriatus at multilure-baited traps in Fort Collins, Loveland and Greeley, CO, 1976-1979.

City and number of traps	Total beetles captured (1000s)*			
	1976	1977	1978	1979
Fort Collins				
685 (1976, 1977-barrier)	1,973	2,250	-	-
900 (1978-barrier)	-	-	4,410	-
670 (1979-barrier)	-	-	-	2,211
30 (core)	35	49	111	99
Loveland				
30 (core)	133	156	261	222
Greeley				
30 (core)	179	156	261	135

*Total trap catches rounded to nearest 1,000 beetles.

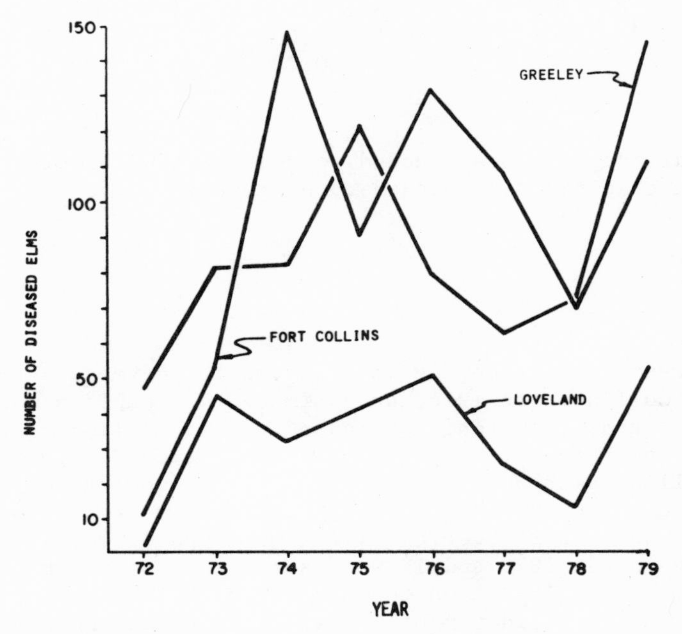

Fig. 3. Dutch elm disease trends in Fort Collins, CO (barrier traps installed in 1976), and in Greeley and Loveland, 2 nearby communities without mass trapping.

Barrier trapping for beetle and DED suppression was evaluated in Evanston, IL from May 1976 to October 1979. Suppression traps (800 in 1976, 1977 and 1979; 1,000 in 1978) were deployed in a 3-row barrier (30-50 m spacing within rows, 50-100 m between rows) around the 20-km perimeter of the city. These rows of traps encircled an area of approximately 18,000 ha containing an estimated 20,000 American elms.

Table 5. Yearly catches of Scolytus multistriatus at multilure-baited traps deployed as a barrier in Evanston, IL, 1976-1979.

Year	No. of traps	Total beetles captured*	Mean trap catch
1976	800	4,500,000	5600
1977	800	3,500,000	4400
1978	1000	8,800,000	8800
1979	900	8,500,000	9400

*Total trap catches rounded to nearest 1000 beetles; mean trap catches rounded to nearest 100.

The results of trapping in Evanston for 4 years parallel those from Fort Collins. In 1976, an estimated 4.5 million beetles were captured on traps. Catches decreased to 3.5 million beetles in 1977 (Cuthbert and Peacock, 1979). But in 1978 and 1979, catches jumped to 8-9 million each year (Table 5). Between-generation capture ratios were about the same for each of the 4 years. Despite the tremendous numbers of beetles killed on traps in each year, especially in the 1978 and 1979 seasons, this beetle mortality apparently had no impact on the incidence of beetle-vectored DED. In comparing the incidence of DED in Evanston and nearby cities where there was no trapping, it is clear that in all cities there was a steady increase in DED rates from 1975-78. There was a decline in incidence in 1979, but this decline also was recorded for surrounding cities (Fig. 4).

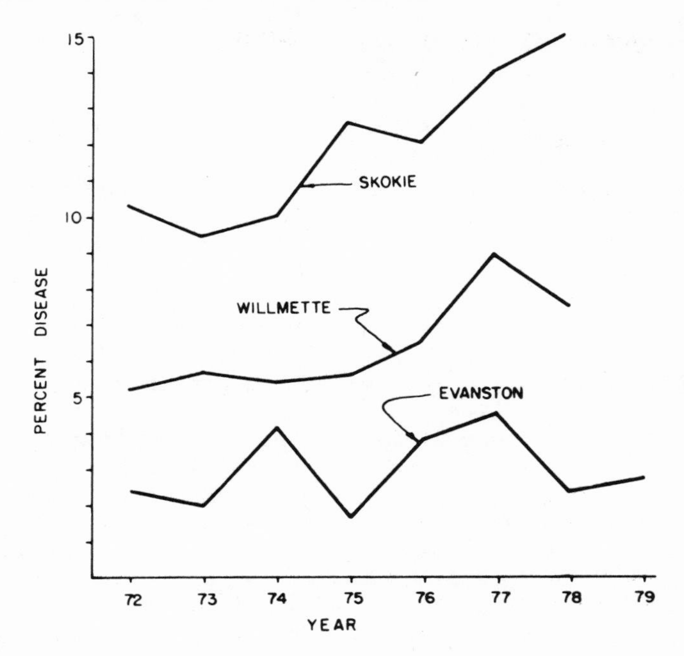

Fig. 4 Dutch elm disease trends in Evanston, IL (barrier traps installed in 1976), and in Skokie and Willmette, 2 nearby communities without mass trapping.

We have concluded after several years of trapping in Fort Collins and Evanston that changes in the DED rate in both of the treated cities and in untreated surrounding cities are a reflection of random fluctuations in the detection and expression of disease symptoms, in environmental conditions, in beetle feeding behavior, and/or in the amount and virulence of the fungus. Trapping has had no significant effect on the DED rate.

SUMMARY

We have attempted to describe the research program on the pheromone of the European elm bark beetle, *Scolytus multistriatus*. Emphasis has been placed on the current status of studies to determine the efficacy of applying the pheromone-trapping system to manipulate elm bark beetle populations and reduce DED incidence in an integrated disease control program.

It is clear that there is a potent, attractant-trap system for killing the beetle. Unfortunately, there are conflicting results on the effect of this system in reducing the incidence of DED. There is the apparently successful use of trapping to protect small groves or scattered stands of elms to reduce disease incidence in these situations. Proceedings to register the attractant-trapping system are in progress on the basis of the results from these "grove" tests. On the other hand, the application of the existing multilure-trapping system for reducing DED in large-area, city-wide barrier trapping efforts has so far been unsuccessful or inconclusive -- despite the annihilation of millions of beetles each year in these city-wide programs.

The trapping program obviously has had a significant impact on flying beetle populations wherever it has been used. It seems logical that DED rates should also have been affected as a result of this impact on the beetles, but DED rates have not been significantly altered when the system was applied over large areas. One reasonable explanation for the discrepancy in results from "grove" and "city-wide" studies is that elms in the groves or scattered stands are free of diseased elm wood and beetles, and thus are free of sources of attractant that compete with traps. But in the city-wide efforts, diseased elm wood suitable for beetle attack -- and, in some cases, producing beetles -- provides sources of attractant that effectively compete with the barrier traps deployed around the experimental area. Thus, in one case (groves or scattered stands), the traps need only to intercept incoming beetles that could inoculate the experimental trees with the pathogen, and they apparently do this successfully. In the other case (city-wide studies), the traps must intercept incoming beetles from surrounding areas; but more important, they must also compete with attractive brood wood in luring beetles from within the experimental site. They may be ineffective in doing the latter, with the result being continued inoculation of elms with the pathogen, and disease rates that are unaffected by trapping.

We are investigating refinements in trap deployment strategy, in the chemical makeup and formulation of the lure as well as more intensively studying beetle behavior so that the pheromone-trapping system might be used successfully in city-wide programs.

REFERENCES

Birch, M. C., 1979, Use of pheromone traps to suppress populations of _Scolytus multistriatus_ in small, isolated Californian communities, _Bull. Entomol. Soc. Am._, 25: 112.

Cuthbert, R. A., and Peacock, J. W., 1975, Attraction of _Scolytus multistriatus_ to pheromone-baited traps at different heights, _Environ. Entomol._, 4: 889.

Cuthbert, R. A., and Peacock, J. W., 1978, Response of the elm bark beetle, _Scolytus multistriatus_ (Coleoptera: Scolytidae), to component mixtures and doses of the pheromone, multilure, _J. Chem. Ecol._, 4: 363.

Cuthbert, R. A., and Peacock, J. W., 1979, The Forest Service program for mass-trapping Scolytus multistriatus, Bull. Entomol. Soc. Am., 25: 105.

Cuthbert, R. A., Peacock, J. W., and Cannon, W. N., Jr., 1977, An estimate of the effectiveness of pheromone-baited traps for the suppression of Scolytus multistriatus (Coleoptera: Scolytidae), J. Chem. Ecol., 3: 527.

Elliott, E. W., Lanier, G. N., and Simeone, J. B., 1975, Termination of aggregation by the European elm bark beetle, Scolytus multistriatus, J. Chem. Ecol., 1: 283.

Elliott, W. J., Hromnak, G., Fried, J., and Lanier, G. N., 1979, Synthesis of multistriatin enantiomers and their action on Scolytus multistriatus (Coleoptera: Scolytidae), J. Chem. Ecol., 5: 279.

Gerken, B., Grüne, S., and Vité, J. P., 1978, Response of European populations of Scolytus multistriatus to isomers of multistriatin, Naturwissenschaften, 65: 110.

Gore, W. E., Pearce, G. T., and Silverstein, R. M., 1975, Relative stereochemistry of multistriatin (2,4-dimethyl-5-ethyl-6,8-dioxabicyclo [3.2.1] octane), J. Org. Chem., 40: 1705.

Gore, W. E., Pearce, G. T., Lanier, G. N., Simeone, J. B., Silverstein, R. M., Peacock, J. W., and Cuthbert, R. A., 1977, Aggregation attractant of the European elm bark beetle, Scolytus multistriatus. Production of individual components and related aggregation behavior, J. Chem. Ecol., 3: 429.

Lanier, G. N., 1978a, Vectors, in: "Dutch Elm Disease, Perspectives After 60 Years," W. A. Sinclair and R. J. Campana, eds., Search (Agric.), 8(5): 1.

Lanier, G. N., 1978b, Behavior-modifying chemicals as a basis for managing bark beetles of urban importance, in: "Perspectives In Urban Entomology," G. W. Frankie and C. S. Koehler, eds., Academic Press, New York.

Lanier, G. N., 1979, Protection of elm groves by surrounding them with multilure-baited sticky traps, Bull. Entomol. Soc. Am., 25: 109.

Lanier, G. N., Silverstein, R. M., and Peacock, J. W., 1976, Attractant pheromone of the European elm bark beetle (Scolytus multistriatus): Isolation, identification, synthesis and utilization studies, in: "Perspectives in Forest Entomology," J. E. Anderson and H. K. Kaya, eds., Academic Press, New York.

Lanier, G. N., Gore, W. E., Pearce, G. T., Peacock, J. W., and Silverstein, R. M., 1977, Response of the European elm bark beetle, Scolytus multistriatus (Coleoptera: Scolytidae), to isomers and components of its pheromone, J. Chem. Ecol., 3: 1.

Peacock, J. W., Lincoln, A. C., Simeone, J. B., and Silverstein, R. M., 1971, Attraction of Scolytus multistriatus (Coleoptera: Scolytidae) to a virgin female-produced pheromone in the field, Ann. Entomol. Soc. Am., 64: 1143.

Peacock, J. W., Cuthbert, R. A., Gore, W. E., Lanier, G. N., Pearce, G. T., and Silverstein, R. M., 1975, Collection on Porapak Q of the aggregation pheromone of *Scolytus multistriatus* (Coleoptera: Scolytidae), *J. Chem. Ecol.*, 1: 149.

Pearce, G. T., Gore, W. E., Silverstein, R. M., Peacock, J. W., Cuthbert, R. A., Lanier, G. N., and Simeone, J. B., 1975, Chemical attractants for the smaller European elm bark beetle *Scolytus multistriatus* (Coleoptera: Scolytidae), *J. Chem. Ecol.*, 1: 115.

Schreiber, L. R., and Peacock, J. W., 1979, Dutch elm disease and its control, U. S. Dept. Agric., *Agric. Info. Bull.*, 193 (revised).

PRACTICAL RESULTS FROM THE MASS TRAPPING OF *IPS TYPOGRAPHUS* IN SCANDINAVIA

Reidar Lie

Borregaard Industries, Ltd.
N-1701 Sarpsborg
Norway

Alf Bakke

Norwegian Forest Research
Institute, N-1432 AS-NLH
Norway

INTRODUCTION

The spruce bark beetle, *Ips typographus* (L.), is one of the most serious pests for Eurasain spruce. Norway spruce, *Picea abies* (L.) Karst, is the usual host tree. Under normal conditions, the beetles breed in the subcortical tissue of very weak trees or trees broken or overthrown by wind. Thus, such breeding accelerates the natural decay of such trees. Healthy trees are killed only when the beetle population is high and many beetles attack simultaneously. During the last 10 years there has been a rapid growth in the population of *I. typographus* in Scandinavia. The bark beetles have become a real threat to spruce trees in southeast Norway and west-central Sweden. Several factors have caused this situation to occur. Since 1969 rainfall has been lower than average over large areas of Scandinavia.

In addition to drought, changes in logging and transportation methods during the 1960s provided excellent conditions for beetle reproduction. Likewise, storage of unbarked logs and slash left in the forest created favorable breeding places. In southern Norway more than 25% of the spruce stands are mature or overmature. Therefore, the reasons for the present bark beetle situation in Scandinavia is a combination of drought, a high initial beetle population and a large proportion of overmature trees.

THE PHEROMONE OF *IPS TYPOGRAPHUS*

In the early 1970s field experiments demonstrated that boring males of *I. typographus* produced components that attract both male and female beetles (Bakke, 1970; Rudinsky et al., 1971). *cis*- and *trans*-Verbenol (I, II); ipsdienol (2-methyl-6-methylene-2,7-octadiene-4-ol) (III); and ipsenol (2-methyl-6-methylene-7-octene-4-ol) (IV) have been identified (Silverstein et al., 1966; Vité et al., 1972) from other *Ips* species. *cis*-Verbenol (I), ipsdienol (III), ipsenol (IV), and 2-methyl-3-butene-2-ol (V) were identified as substances made by the male *I. typographus* (Bakke, 1976; Bakke et al., 1977). Field experiments showed that both sexes will respond to a mixture of *cis*-verbenol (I), ipsdienol (III), and methylbutenol (V) (Bakke et al., 1977). The behavioral function of ipsenol (IV) is still under investigation by Norwegian Forest Research Institute.

THE INTEGRATED CONTROL PROGRAM

In October of 1978 the Norwegian government formed a committee to evaluate the bark beetle problem and offer remedial advice. Forest owner organizations, forest product industries, the Department of Agriculture, the Civil Forest Administration, and the Norwegian Forest Research Institute made up the committee and recommendations were issued in February, 1979 (NORGES OFFENTLIGE UTREDNINGER 1979). The committee suggested short- and long-term remedies to reduce the beetles: the major long-term remedy was to stimulate harvesting of over-mature forest stands; the major short-term remedy was a recommendation to mass-trap *I. typographus*. In 1979 the integrated control program was estimated to cost $20 million. Total expenses for mass-trapping *I. typographus* was $7.5 million. The economic argument in favor of the integrated control program was that a possible loss of up to 57% of the spruce stands in south Norway could be caused by *I. typographus* in future years.

THE MASS TRAPPING PROGRAM

Pheromone

Based on field experiments, a mixture of methylbutenol (V), *cis*-verbenol (I), and ipsdienol (III) was used in the traps (Bakke and Saether, 1978). Methylbutenol (V) is commercially available and can be bought from several producers while *cis*-verbenol (I) and ipsdienol (III) are produced by Borregaard Industries, Ltd. Starting from α-piene, *cis*-verbenol (I) was synthesized in 5 steps (Whitham, 1961); the product was 95% pure and had a stereoselectivity of 85%. Ipsdienol (III), 90% pure, was produced as a racemate in 8 steps beginning with 1,4-butyndiol and *tert*. butylalcohol (Karlsen et al., 1976).

Dispenser

Large-scale use of pheromones requires a practical dispenser system which must provide continuous release of the pheromone over the prescribed period while, at the same time, being able to withstand the forest environment. The 3 pheromone components of *I. typographus* have very different volatilities: methylbutenol's (V) b.p. is 97°C at normal pressure, *cis*-verbenol (I) melts at 64°C, and ipsdienol's (III) b.p. is 55°C at 0.15 torr. During the summers of 1977 and 1978, 4 dispenser systems were tested in the field: (1) Hollow fiber produced by Conrel (Norwood, MA, USA); (2) polyethylene bag produced by Celamerck (Ingelheim, West Germany); (3) paste produced by Celamerck (Ingelheim, West Germany); and (4) multilayer laminate produced by Hercon (New York, NY, USA). Evaporation rates for different dispensers were measured both in the laboratory and under field conditions. The polyethylene bag and the laminated strip proved to be the most suitable dispensers for trapping *I. typographus*. We found that even though 3 pheromone components can be mixed in one dispenser, (there was a great difference in volability). The evaporation rate of these 2 dispenser systems is very temperature-dependent. Dispensers were exposed to field conditions in traps and then analyzed for pheromone residues. While trapping, trap temperature varied from +5 to +35°C, and on a sunny day trap temperature was 5 to 10°C above the air temperature. Evaporation rate was high when first exposed but stabilized shortly thereafter. Next, we studied dispenser efficiency during the summer. Findings showed a reduced attractiveness of about 45% after 2-4 weeks and about 60% after 6-8 weeks. Initially, each dispenser contained the pheromone methylbutenol (V) 1500 mg, *cis*-verbenol (I) 70 mg and ipsdienol (III) 10 mg.

Traps

I. typographus can be caught with pheromone-baited traps (Bakke, 1979). Pheromone-baited bark beetle traps were used in the past, i.e., cardboard cylinders (Pitman, 1971), sticky paper, fiberglass screen vanes (Borden, 1978; McLean and Borden, 1979), perforated PVC cylinders (Bakke and Saether, 1978). During summer of 1978 we tested several cylindrical traps the best being a polyethylene drainpipe (Fig. 1). The main body of the trap is a black, ridged, cylindrical drainpipe of polyethylene (12.5 x 135 cm). The pipe has 900 holes (diam 3.5 mm); a lid covers its top while its bottom has a funnel and a collecting bottle. Traps should be placed in open areas near trees killed previously while minimum distance to healthy trees should be at least 10 m. One trap for every 3 infested trees was recommended based on infested trees in 1978, 600,000 traps were used in Norway in 1979. The total area covered by the Norwegian campaign was 4 million ha (Table 1).

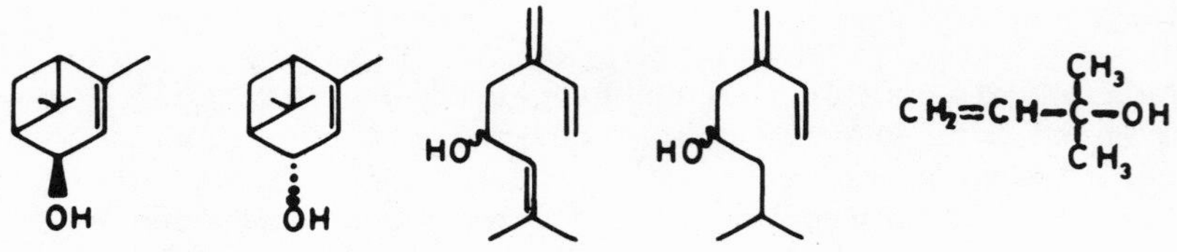

I II III IV V

Fig. 1. Compounds used as attractants by various *Ips* species.

Table 1. Bark beetle-killed spruce volume in Norway.

County	Productive forest area (ha X 1000)*	Beetle-killed spruce (1000 m^3) 1978	1979	$\overline{X}$ No. beetles/trap
Østfold	217 (20)	19	11	3,630
Akershus	318 (60)	60	74	4,340
Hedmark	1,246 (60)	42	75	3,170
Oppland	649 (75)	71	44	2,810
Buskerud	555 (100)	176	198	4,680
Vestfold	117 (100)	283	201	6,920
Telemark	530 (150)	410	400	5,650
Aust-Agder	324 (40)	47	45	4,660
Total	3,956 (605)	1,108	1,048	4,850

*Number of traps deployed in 1979 (X 1000) shown in ( ).

FINDINGS OF THE MASS-TRAPPING PROGRAM

Trapping of *Ips typographus*

Traps were subsidized by the government who paid 1/3rd of their cost, while dispensers were free for forest owners. The Forest Service Administration organized the campaign, and supplied information to forest owners. By late April, all 600,000 traps had been delivered to forest owners and baited traps were ready before May 15.

To evaluate the efficiency of the mass-trapping campaign, the Norwegian Forest Research Institute randomly selected about 6,000 traps (1% of total) and dispersed them throughout the entire trapping program area.

Trap location and beetle density caused trap catches to vary considerably. The average catch for each trap was 4,850 beetles while the total beetle catch in Norway was about 2.9 billion.

Trapping of Non-target Insects

More than 99% of trapped insects were *Ips* beetles, the majority of which were *I. typographus* with a few *I. duplicatus* (Sahlberg) and most traps captured the predators, *Thanasimus formicarius* (L.) and *Thanasimus femoralis* (Zett.). Bakke and Kvamme (unpublished data: Kairomone response in *Thanasimus* predators to provide components of *I. typographus*) discovered ipsdienol and *cis*-verbenol to be a kairomone for these beetles. The average catch of predators was only 1.6/1,000 spruce bark beetles trapped.

Tree Mortality

During the fall of 1978 Norway's spruce stand was examined. The Civil Forest Administration calculated that bark beetles had killed 1.1 million m^3 of spruce that year (about 3 million trees). Moreover, during that same year very few infested trees were felled before the new beetle generation emerged. A large increase in killed trees was predicted for 1978-79 if no campaign was organized to destroy bark beetles. In 1979 (Table 1) bark beetles killed 1.05 million m^3 of spruce (less than 3 million trees) with results varying among districts.

More than 20,000 persons participated in the campaign and generally followed its guidelines. However, many traps were erroneously placed too near the forest's edge or were placed in forest stands. Apparently, the 10 m distance was too close between traps and healthy trees were in heavily-populated beetle areas.

FINDINGS IN SWEDEN

Guidelines for the campaign against *I. typographus* in Sweden during 1979 were the same as Norway's. Sweden used 320,000 pheromone-baited traps which captured an estimated 1.6 billion beetles. Sweden also reported a great reduction in trees killed by beetles as a result of the mass-trapping program (Regnander, personal communication).

FUTURE TRAPPING PROGRAM

Civil Forest Administrations of Norway and Sweden have decided to launch a new campaign against *I. typographus* in 1980. Mass trapping will follow 1979 guidelines with the following modifications: ipsdienol in the dispensers will be increased from 10 to 15 mg, and distance between traps and healthy trees will be 30 m. Trap design also will be changed (Fig. 2). Improvements in the trap include increased pipe diameter, better landing platforms, and wider funnels for beetle collection, (formerly, beetles would fall off when they tried to land) and perforated collecting bottles. The odor of decomposing beetles was found to reduce the catch by about 50% (Bakke, 1979). Three additional improvements would be to place the traps at ground-level (in 1979 traps were placed 1.5 m aboveground), to place the dispenser in the lower part of the pipe (in 1979 the dispenser was placed in the middle of the pipe), and to place traps in groups instead of singly as in 1979.

Fig. 2. Pheromone traps used in 1979 (left) and 1980 (right) for capturing *Ips typographus*.

We feel that the integrated control program will be a major aid in suppressing the *I. typographus* population and that the mass-trapping campaign will suppress spruce damage, thus producing a healthier spruce stand.

REFERENCES

Bakke, A., 1979, Evidence of a population aggregating pheromone in *I. typographus* (Coleoptera: Scolytidae), *Conbrib. Boyce Thompson Inst.*, 24: 309.

Bakke, A., 1976, Spruce bark beetle, *I. typographus*: Pheromone production and field response to synthetic pheromones,

Naturwissenschaften, 63: 92.
Bakke, A., 1979, Foreløpige forskningsresultater gir godt grunnlag for billeaksjonen 1980, Norsk Skogbr., 25: 12.
Bakke, A. and Saether, T., 1978, Granbarkbillen kan fanges i rørfeller (The spruce bark beetle can be trapped in drainage pipes), Skogeieren, 65: 100.
Bakke, A., Frøyen, P. and Skattebøl, L., 1977, Field response to a new pheromonal compound isolated from I. typographus, Naturwissenschften, 64: 98.
Bedard, W. S., Tilden, P. E., Wood, D. L., Silverstein, R. M., Brownlee, R. G. and Rodin, J. O., 1969, Western pine beetle: Field response to its sex pheromone and a synergistic host terpene, myrcene, Science, 164: 1284.
Browne, J. E., 1978, a trapping system for the western pine beetle using attractive pheromones, J. Chem. Ecol., 4: 261.
Karlson, S., Frøyen, P. and Skattebøl, L., 1976, New syntheses of the bark beetle pheromones 2-methyl-6-methylene-7-octen-4-ol (Ipsenol) and 2-methyl-6-methylene-2, 7-octadien-4-ol (Ipsdienol), Acta Chem. Scand., B 30: 664.
McLean, J. A. and Borden, J. H., 1979, An operational pheromone-based suppression program for an ambrosia beetle, Gnathotrichus sulcatus, in a commercial sawmill. J. Econ., Entomol., 72: 165.
Norges offentlige utredninger, 1979, Granbarkbillen. Granbarkbille- og grantørkesituasjonen i Sør-Norge, Norges offentlige utredninger 1979:22, Universitetsforlaget, Oslo-Bergen-Tromsø, 71 pp.
Pitman, G. B., 1971, trans-Verbenol and alpha-pinene; their utility in manipulation of the mountain pine beetle, J. Econ. Entomol., 64: 426.
Rudinsky, J. A., Novak, V., Svihra, P., 1971, Attraction of the bark beetle, I. typographus L., to terpenes and a male-produced pheromone, Z. ang. Entomol., 67: 179.
Silverstein, R. M., Rodin, J. O., and Wood, D. L., 1966, Sex attractants in frass produced by male I. confusus in ponderosa pine, Science, 154: 509.
Vité, J. P., Bakke, A., Renwiek, J. A. A., 1972, Pheromones in Ips (Coleoptera: Scolytidae) occurrence and production, Can. Entomol., 104: 1967.
Whitman, C. H., 1961, The reaction of α-pinene with lead tetra-acetate, J. Chem. Soc., 2232.

MASS TRAPPING FOR SUPPRESSION OF JAPANESE BEETLES

Michael G. Klein

Japanese Beetle and Horticultural
Insect Pests Research Laboratory
AR-SEA, USDA
Wooster, OH 44691 USA

INTRODUCTION

Little was known on the biology of the Japanese beetle, *Popillia japonica* Newman, upon its discovery in a New Jersey nursery in 1916. Apparently, it was introduced with some Japanese iris bulbs before the Plant Pest Act of 1912 which required inspection of incoming plant materials. In the United States the beetle found large areas of turf essential to development during its immature stages, a favorable climate with sufficient rainfall to prevent soil-inhabitants from desiccating and at least 350 plant species to act as hosts for adult beetles. These favorable conditions and an absence of natural enemies contributed to beetles inhabiting 26 states in the midwestern, eastern and southeastern United States. Both larvae and adults are severe crop pests and cause vast damage annually.

Japanese beetle larvae (grubs) eat roots of various grasses, ornamental plants, garden or truck crops, and often cause severe damage to golf courses and lawns; they can also reduce pasture yields. Adult Japanese beetles eat foliage, flowers or fruit of such diverse plants as deciduous and citrus fruits, berry crops, garden vegetables, farm crops (corn, soybeans, clover and alfalfa), ornamental and floral plants, and trees. Odors are highly important for attracting the beetle to a certain plant species (Fleming, 1972), and it has been used in developing semiochemicals to check the Japanese beetle.

DEVELOPMENT OF SEMIOCHEMICALS

Food-Type Lures

Fleming (1969) reviewed the extensive literature on beetle attractants generated between 1919 and 1964. Within 3 years from the discovery of the Japanese beetle, scientists noticed that some essential oils and fruity odors attracted the beetle. Their investigations resulted in patenting the attractant, geraniol, (Fig. 1) in 1926. Studies since 1964 led to discovering another highly attractive material, a mixture of phenethyl propionate (PEP) and eugenol (Fig. 1) in a ratio of 3:7 (Ladd et al., 1976), and its adoption as the standard survey lure for state and federal agencies. In addition, Ladd et al. (1975, and unpublished data) found that a 3-part lure of (3:7:3) PEP:eugenol:geraniol attracted more beetles than the 3:7 mixture alone.

Fleming et al. (1940) found that while traps may not protect plants from Japanese beetles, removing beetles will reduce plant damage and removing females also will reduce the next generation. Thus, they predicted that tangible results would occur if many traps were used throughout a community for several years. Cory and Langford (1955) similarly reported that during the height of an extensive trapping campaign in Maryland, over 369 tons of beetles were captured in one season and that trapping-isolated infestations prevented establishing beetle populations for several years.

Hamilton et al. (1971) too found that mass trapping at Nantucket, MA reduced beetles by at least 50% over 3 years. Although a general drought during this experiment may have contributed, large areas were irrigated so traps appeared to be primarily responsible for beetle reduction.

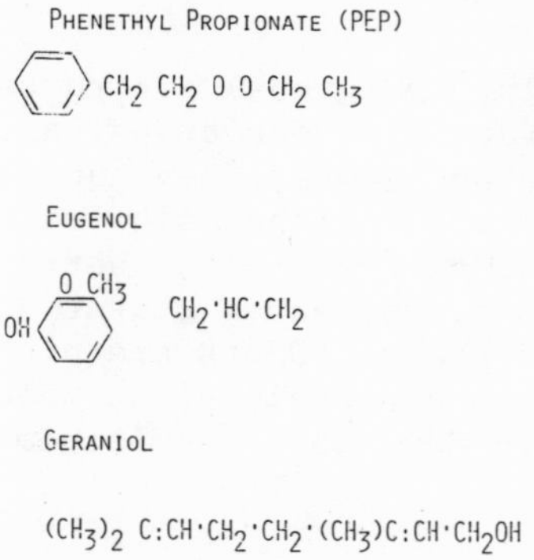

PHENETHYL PROPIONATE (PEP)

$C_6H_5\ CH_2\ CH_2\ O\ O\ CH_2\ CH_3$

EUGENOL

$OH,\ O\ CH_3\ \ CH_2{\cdot}HC{\cdot}CH_2$

GERANIOL

$(CH_3)_2\ C{:}CH{\cdot}CH_2{\cdot}CH_2{\cdot}(CH_3)C{:}CH{\cdot}CH_2OH$

Fig. 1. Structures of important food-type lures for the Japanese beetle.

Japonilure

Chemistry and biological activity--Smith and Hadley (1926) observed that male Japanese beetles approaching a female against the wind apparently are attracted by an odor. Balls of beetles composed of 1 female and up to 200 males would then form. Fleming (1969) reported that traps baited with virgin females captured no more beetles than unbaited traps. However, Ladd (1970) found that females collected from balls of beetles had a powerful volatile sex pheromone, and that many males were captured by traps baited with these unmated females. Consequently, Japanese Beetle Laboratory personnel and the USDA Insect Attractants, Behavior and Basic Biology Laboratory in Gainesville, FL cooperated in an effort to isolate and identify the pheromone and produce its biologically active synthetic counterpart (Tumlinson et al., 1977, 1979; Doolittle, 1980).

The pheromone was gathered by collecting benzene washes of glass vessels used to hold virgin females. These washes were filtered, concentrated, fractionated by gel-permeating liquid chromatography and purified by gas chromatography; each step was monitored by a field bioassay to ensure continued presence of the active material. Although a compound, (Z)-5-(1-deceny)dihydro-2(3H)-furanone, proved identical to the natural pheromone on all chromatographic and spectral runs and seemed to cause activity in the male beetles, it nevertheless prevented males from finding virgin females in the field. However, the Z furanone identified as the pheromone had an optically active carbon. Since not enough natural pheromone was available to determine enantiomeric composition, a stereospecific synthesis of the R and S forms of the pheromone was undertaken so that both enantiomers might be tested. The (R,Z)-5-(1-decenyl) dihydro-2 (3H)-furanone (Fig. 2) had a biological activity like the natural pheromone; subsequently, it was named Japonilure. Adding as little as 1% of the S,Z enantiomer to the R,Z enantiomer would significantly reduce male attraction and adding 20% or more of the S,Z enantiomer would completely inactivate the R,Z enantiomer.

Formulating Japonilure--Early tests showed that ca. 5 μg of Japonilure on a stainless steel planchet was just as attractive to Japanese beetle males as 4 virgin females were for 1 day (Tumlinson et al., 1977). Subsequent unpublished tests at this laboratory indicated that 500 μg of Japonilure on a planchet would be attractive ca. 1 week and that a similar amount dispensed with a rubber septum or polyethylene cap would be effective for 1 month. Subsequently, Hercon Luretape® pheromone dispensers containing as little as 1 mg of Japonilure/6.5 cm^2 retained activity for at least 100 days; similar laminated plastic tape dispensers containing 0.5 mg/6.5 cm^2 were never as effective. Also, Conrel® fibers having 5 mg/unit were active at least 100 days. However, additional tests are needed since pheromone was left in all dispensers at the end of the test.

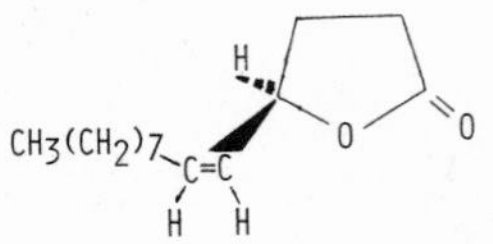

$CH_3(CH_2)_7CH=CH$ (lactone ring structure)

(R,Z)-5-(1-DECENYL)DIHYDRO-2(3H)-FURANONE

$[\alpha]_D^{25}$ -69.6° AT CONC OF 5G/100ML IN $CHCl_3$

Fig. 2. Structure of Japonilure, synthetic Japanese beetle sex pheromone.

A dual-lure system--Anticipating the availability of synthetic sex pheromone, Klein et al. (1973) tested simultaneous exposure of PEP:eugenol and virgin female beetles. Unexpectedly, although females only attract males, female captures using the chemical lure plus females were almost twice as great as female captures using PEP:eugenol alone. Since Adler and Jacobson (1971) concluded from electroantennogram studies that males make a pheromone attractive to both sexes, mated females and males were tested in combination with PEP + eugenol. Results (Table 1) showed that the synergism produced by combining virgin females and PEP:eugenol was lacking when mated females or males were used with PEP:eugenol.

Table 1. Captures of Japanese Beetles by Simultaneous Exposure of PEP:Eugenol and Adult Japanese Beetles.

	Mean Captures Per Trap		
Lure	Males	Females	Total
PEP:Eugenol + Virgin Females	1268 a	727 a	1996 a
Virgin Females	933 b	14 b	947 b
PEP:Eugenol + Mated Females	330 c	496 c	826 b
PEP:Eugenol + Males	289 c	446 c	735 b
PEP:Eugenol	256 c	407 c	663 b

[a]Means in the same columns followed by the same letter are not significantly different at the 5% level, (Duncan's new multiple range test).

Subsequently, Klein et al (1980) tested the effects from simultaneously exposing Japonilure and PEP:eugenol (7:3) during the adult flight season. The attractants combined would always capture significantly more beetles than either Japonilure or the food-type lure alone, but during peak emergence Japonilure captured significantly more males than PEP:eugenol. Then, ca. 2 weeks later (and thereafter), these 2 lures were not significantly different when comparing attractancy to males.

Japonilure's enantiomeric purity is not as important when the synthetic pheromone is combined with food-type lures (Klein, unpublished data). For example, a 95:5 ratio of R and S enantiomers is only ca. 1/3rd as attractive to male beetles as the pure R enantiomer. Early in the season, however, traps baited with the 95:5 mixture or with the pure R enantiomer and PEP:eugenol would capture 3.5 and 3.7X as many beetles, respectively, as the food-type lure alone. Since this difference was not significant, possibly alternate synthesis routes [such as that suggested by Sato et al. (1979) resulting in an optical purity of ca. 90%] may have a place in producing and utilizing Japanese beetle lure.

Factors Influencing Effectiveness of Semiochemicals

Fleming (1969) summarized work on the importance of trap color and luster to capture beetles; a yellow trap with a high degree of luster proved more effective for his food-type lures. Likewise, Ladd et al. (1973) found that yellow, red, and black traps were not significantly different when using virgin females as lure. Our recent unpublished data indicates that traps of several shades of yellow and green as well as white and blue traps are highly attractive to the beetles.

Fleming (1969) noted the importance of trap placement--3-8 m from favored trees, shrubs and vines, and not too close to plants. Although he found that traps exposed ca. 1 m above ground to be the best, Ladd and Jurimas (1972) found no difference between 0.5 and 1 m, the type of lure involved may be of importance. Klein et al. (1972) reported that when traps were baited with virgin females, traps ca. 30 cm above ground were most attractive; Ladd and Klein (unpublished data) recently noted that traps baited with Japonilure and PEP:eugenol:geraniol (3:7:3) captured more when placed at 56 cm above the surface. They also found that traps in direct sunlight were 2X as effective as those in the shade and that maximum captures occurred between 10:00 AM and 1:00 PM (EST).

FUTURE PROSPECTS FOR UTILIZING SEMIOCHEMICALS

A positive aspect in trap use against Japanese beetles is availability of trapping systems from 3 commercial companies (Fig. 3). Both components of the dual-lure system have patents assigned to the Secretary of Agriculture and may be licensed (McGovern, et al., 1973; Tumlinson, et al., 1979). Two companies market lures composed of PEP:eugenol dispensed from a bran mixture. The 1st company's trap is all-metal while the other's trap has plastic fins and funnel with a plastic bag to hold beetles. The 3rd company is the only one now using the dual-lure system of Japonilure and PEP:eugenol; their trap consists of plastic fins and a plastic bag that serves as a combination funnel and beetle holder.

Fig. 3. Commercially available traps for the Japanese beetle.

The USDA did not previously recommend using traps to suppress Japanese beetles; however, Schwartz (1975) listed using these traps as a noninsecticidal method for reducing adult damage and destroying numerous beetles. Also, Maryland strongly endorsed trapping (Cory and Langford, 1955) but more information is needed on how traps can best be used. Although a popular belief is that beetles are drawn into areas where traps are located, Fleming et al. (1940), Cory and Langford (1955), and myself noticed that presence (or absence) of favored host plants is more important than presence of traps in concentrating beetles. Also, more research on mass trapping large areas are needed. In 1979, the Animal and Plant Health Inspection Service of the USDA deployed 1400 traps at Dulles

International Airport; for the first time since 1975, airplanes leaving that facility did not have to be treated for beetles (W. McLane, personal communication). More data is needed concerning the best type and color of trap, dispensing systems, and optimum lure, in addition to demonstrations of their practical utility.

Besides needing data on effectiveness of mass trapping isolated infestations, airports, or neighborhoods, we must also explore other uses of semiochemicals. Trap cropping seems feasible but Schwartz et al. (1969) and Ladd (unpublished data) noted that the most effective insecticides seem to repel beetles. Disruption or confusion of male searching behavior is a possibility, but males find females quite easily in a heavy population even after females have been mated and are not emitting the sex attractant. However, given the synthetic sex pheromone, a good food-type lure for both sexes, and the unique synergism between the 2 semiochemicals seem destined to have a place in suppressing Japanese beetles.

REFERENCES

Adler, V. E., and Jacobson, M., 1971, Electroantennogram response of adult Male and Female Japanese beetles to their extracts, J. Econ. Entomol., 64: 1561.

Cory, E. N., and Langford, G. S., 1955, The Japanese beetle retardation program in Maryland, University of Marland, Extension Bulletin 156, 20 pp.

Doolittle, R. E., Tumlinson, J. H., Proveaux, A. T., and Heath, R. R., 1980, Synthesis of the sex pheromone of the Japanese beetle, J. Chem. Ecol., 6: 473.

Fleming, W. E., 1969, Attractants for the Japanese beetle, U. S. Department of Agriculture Tech. Bull. 1399, 87 pp.

Fleming, W. E., 1972, Biology of the Japanese beetle, U. S. Department of Agriculture Tech. Bull. 1449, 129 pp.

Fleming, W. E., Burgess, E. D., and Maines, W. W., 1940, The use of traps against the Japanese beetle, U. S. Department of Agriculture Circ. 594, 12 pp.

Hamilton, D. W., Schwartz, P. H., Townshend, B. G., and Jester, C. W., 1971, Traps reduce an isolated infestation of Japanese beetle, J. Econ. Entomol., 64: 150.

Klein, M. G., Ladd, T. L., Jr., and Lawrence, K. O., 1972, The influence of height of exposure of virgin female Japanese beetles on captured males, Environ. Entomol., 1: 600.

Klein, M. G., Ladd, T. L., Jr., and Lawrence, K. O., 1973, Simultaneous exposure of phenethyl propionate-eugenol (7:3) and virgin female Japanese beetles as a lure, J. Econ. Entomol., 66: 373.

Klein, M. G., Tumlinson, J. H., Ladd, T. L., Jr., and Doolittle, R. E., 1980, Japanese beetle (Coleoptera: Scarabaeidae): Response to synthetic sex attractant plus phenethyl propionate: eugenol, J. Chem. Ecol., (In press).

Ladd, T. L., Jr., 1970, Sex attraction in the Japanese beetle, J. Econ. Entomol., 63: 905.
Ladd, T. L., Jr., and Jurimas, J. P., 1972, Influence of trap height on captures of Japanese beetles, J. Econ. Entomol., 65: 407.
Ladd, T. L., Jr., Klein, M. G., and Lawrence, K. O., 1973, Japanese beetles: Attractive females as lures in colored traps, J. Econ. Entomol., 66: 1236.
Ladd, T. L., Jr., Buriff, C. R., Beroza, M., and McGovern, T. P., 1975, Japanese beetles: Attractancy of mixtures of lure containing phenethyl propionate and eugenol, J. Econ. Entomol., 68: 819.
Ladd, T. L., Jr., McGovern, T. P., Beroza, M., Buriff, C. R., and Klein, M. G., 1976, Japanese beetles: Attractancy of phenethyl propionate + eugenol (3:7) and synthetic eugenol, J. Econ. Entomol., 69: 468.
McGovern, T. P., Beroza, M., and Ladd, T. L., 1973, Phenethyl propionate and eugenol, a potent attractant for the Japanese beetle (Popillia japonica Newman), U. S. Patent 3,761,584, Sept. 25, 1973.
Sato, K., Nakayama, T., and Mori, K., 1979, New synthesis of both enantiomers of (Z)-5-(1-decenyl-oxacyclopentan-2-one, the pheromone of the Japanese beetle, Agric. Biol. Chem., 43: 1571.
Schwartz, P. H., Jr., 1975, Control of insects on deciduous fruits and tree nuts in the home orchard-without insecticides, U. S. Department of Agriculture, Home & Garden Bulletin 211, 36 pp.
Schwartz, P. H., Hickey, L. A., Hamilton, D. W., and Ladd, T. L., Jr., 1969, Combinations of insecticides and baits for the Japanese beetle, J. Econ. Entomol., 62: 738.
Smith, L. B., and Hadley, C. H., 1926, The Japanese beetle, U. S. Department of Agriculture USDA Circular 363, 67 pp.
Tumlinson, J. H., Klein, M. G., Doolittle, R. E., Ladd, T. L., and Proveaux, A. T., 1977, Identification of the female Japanese beetle sex pheromone: Inhibition of male response by an enantiomer, Science, 197: 789.
Tumlinson, J. H., Klein, M. G., Doolittle, R. E., and Ladd, T. L., Jr., 1979, Sex pheromone produced by the female Japanese beetle: Specificity of male response to enantiomers, U. S. Patent 4,179,446, Dec. 18, 1979.

MASS TRAPPING FOR DETECTION, SUPPRESSION AND INTEGRATION WITH OTHER SUPPRESSION MEASURES AGAINST THE BOLL WEEVIL

E.P. Lloyd
G.H. McKibben

Boll Weevil Eradication
Research Laboratory
AR-SEA, USDA
Raleigh, NC 27607 USA

J.A. Witz
A.W. Hartstack

Cotton Pest Control
Equipment Research
AR-SEA, USDA
College Station, TX. 77843
USA

D.F. Lockwood

NC Agric. Research Service
NC State University
Raleigh, NC 27650
USA

E.F. Knipling

Pest Management
AR-SEA, USDA
Beltsville, MD 20705
USA

J.E. Leggett

Cotton Insects Research
Laboratory
AR-SEA, USDA
Florence, SC 29502
USA

INTRODUCTION

The male boll weevil, _Anthonomus grandis_ Boheman, after feeding on cotton plants, emits an airborne pheromone which attracts both sexes of emerged overwintered adults prior to squaring of cotton plants. Cross et al. (1969) reported that male-baited sticky wing traps attracted female boll weevils. Hardee et al. (1969), in concurrent tests, observed that both male and female boll weevils responded to male-baited traps in the spring and in the fall. However, relatively few boll weevils responded to traps around cotton fields during the fruiting period from late June until late August. Mitchell and Hardee (1974) found that pheromone-baited traps placed in cotton fields during mid-season captured females primarily. Thus, the pheromone appears to serve as an aggregating attractant for both sexes of overwintered boll weevils and as a mating attractant for females seeking mates in cotton fields.

Tumlinson et al. (1969) reported the isolation, identification and synthesis of the pheromone of male boll weevils. Hardee et al. (1972) and McKibben et al. (1974) reported that traps baited with synthetic pheromone (grandlure) were at least as attractive as those baited with live males. Subsequently, the synthetic lure replaced the live males in pheromone traps.

Initially, the sticky wing traps developed by Cross et al. (1969) were painted John Deere green, 10-45®, so they were similar in color to healthy plant tissue. Subsequent tests by Cross et al. (1976) showed that traps painted a fluorescent (Caution®) yellow were more attractive than traps painted certain other colors. However, other fluorescent yellow paints, Solar® and Saturn®, were almost as attractive.

The sticky wing trap developed by Cross et al. (1969) was later replaced with a floral liner, painted Saturn yellow, that had a capture device above the liner. This trap, developed by Leggett and Cross (1971), is referred to as the Leggett trap. Mitchell and Hardee (1974) later developed an infield trap (Mitchell trap) constructed with a smaller base for use inside cotton fields. This trap was constructed with a 1-liter inverted soda cup painted Saturn yellow and topped with a capture device. Dickerson et al. (unpublished data) modified the Mitchell trap so it could be mass-produced inexpensively for use in the Boll Weevil Eradication and Optimum Pest Management Trials.

DETECTION AND SUPPRESSION

Detecting the presence of low density boll weevil populations is one of the major problems in evaluating the effectiveness of boll weevil containment and eradication programs. Visual examination of a large number of plants for live insects or damage is laborious and costly, and it is also a relatively insensitive procedure when populations are low, particularly when progeny of reproducing females are closely clumped. However, the availability of the boll weevil pheromone, grandlure, for use in infield traps provided an opportunity for the development of a more sensitive detection method. The difficulty was that it is not possible to determine precisely how sensitive trap survey methods will be in detecting natural low level populations because densities of natural populations vary and are difficult to measure. Also, boll weevil infiltration into experimental fields confounds the results. Therefore, in 1978 Leggett et al. (unpublished data) undertook special investigations in isolated fields. The experiments were designed to determine the sensitivity of different densities of traps for the detection of the F_1 progeny that would be expected from a single overwintered female weevil reproducing in a cotton field. The authors who used 2.3 or 10 traps/ha reported the detection of 4 of 7 and 7 of 7 reproducing clumps, respectively, in the F_1 generation. Such a high probability of detecting very low populations represents a major advance for boll weevil detection.

In order to more fully understand the influence of trap densities and spacing on the probability of detecting and controlling small isolated reproducing clumps of boll weevils, we conducted a series of experiments in 1979 in a non-cotton area (Surry County, NC) more than 80 miles from the nearest commercial cotton plantings. By using such a site and by creating known infestations, we could eliminate competition from unknown numbers of competing males. Twenty-four cotton plots averaging 1.28 ha, therefore, were planted 2 or more miles apart. These plots were then infested as follows: Beginning July 9 and for 9 subsequent days, overwintered female weevils were allowed to deposit eggs in cotton squares at the Cotton Insects Research Laboratory (USDA), Florence, SC. Squares were held on moist sand in an outdoor insectary to simulate natural conditions as closely as possible. Approximately 10-12 days later, they were transported to Surry County by automobile. Infested squares were placed near the center of the test cotton fields. Sites were selected to simulate the usual egg deposition and subsequent emergence pattern of progeny. Treatments included were as follows: 20 Infested squares/clump with traps (1) 210, (2) 149, (3) 121, and (4) 105 ft apart; and 10 infested squares/clump with traps (5) 149 and (6) 105 ft apart.

The traps (infield--modified Mitchell) had been placed in the cotton fields as indicated approximately 1 week prior to installing infested squares. They were baited initially and weekly thereafter with 3 mg of grandlure impregnated in polyester-wrapped cigarette dispensers. The squares were infested so that emergence from them would occur uniformly over a 10-day period. To simulate spatial deposition of eggs, for the 1st 4 days, the 1 or 2 infested squares were spaced 4.6 m apart on a row and 3 m apart for the remaining 6 days. The emergence area was approximately 33 m long. Theoretically, 2 weevils would emerge each day in plots with 20 infested squares and 1 weevil/day would emerge in plots with 10 infested squares. The emergence rate averaged about 75%. On the basis of equal sex distribution, emerging females averaged about 8 and 4/field plot at the 2 density levels, respectively.

RESULTS AND DISCUSSION

Captures of emerged F_1 generation weevils with infield traps are summarized in Table 1. The spatial arrangements of traps 210,

149, 121, and 105 ft apart represent slightly more than 1, 2, 3 and 4 traps/0.4 ha, respectively, for large cotton fields. However, in the smaller specially planted cotton fields used in this study, the initial trap was placed about 15 ft from a field corner, and the other traps were arranged for the desired spacings.

Table 1. Summary of capture of F_1 and F_2 generation boll weevils with infield traps at indicated spacings (ft) in test fields in boll weevil detection experiment. Surry County, NC, 1979.

		20 Infested squares/clump			10 Infested squares/clump	
Average	210'	149'	121'	105'	149'	105'
F_1 captured	2 ♀	3 ♀ 0.2 ♂	5.5 ♀ 1 ♂	6.7 ♀ 0.7 ♂	2.7 ♀ 0.2 ♂	3 ♀ 0.5 ♂
F_2 captured	17.5	15.2	9.5	5.7	5	2
% Emergence F_1 from squares	77	78	73	63	77	76
Traps/0.4 ha	1.9	3.6	4.2	5.7	3.0	6.6

Knipling (1976, 1979) and Knipling and McGuire (1966) advanced the theory that the degree of control of insects subjected to pheromone-baited traps will be governed by the relative attractiveness of the pheromone produced by calling insects. On the basis of the competitive attraction theory, theoretical models were set up to estimate the potential value of traps for detection and for control. The simulation models and assumed parameters have proven useful in the experimental design and interpretation of data obtained in these investigations.

The more important variables that govern efficiency of traps for boll weevil detection and control include the: (1) Relative attractiveness of the pheromone emanating from the traps vs. the pheromone disseminated by the males, (2) number of traps and distribution pattern of traps in relation to the number of distribution pattern of the pheromone-producing males and the mate-seeking females, (3) size of the area in which a trap and competing males can elicit a response by mate-seeking females, and (4) actual efficiency of the traps in capturing the responding females (and males, in the case of overwintered populations).

If all traps used in each field are as attractive as males at any given distance and are in competitive range of mate-seeking females, we can calculate the proportion of the available females that will be captured by assuming that all females responding to traps are captured. Such calculations, when related to actual capture rates, offer clues to the type of action and the effectiveness of different densities of infield traps relative to the distribution pattern of sexually active boll weevils. The distribution pattern of a given number of boll weevils entering a cotton field from hibernation will likely differ greatly from the distribution pattern of the same number of F_1 progeny emerging in a restricted area from one or more egg-laying females. If traps are uniformly distributed in a cotton field, it is unlikely that overwintered males and females will be capable of aggregating because of the confusion (interference) produced by the synthetic pheromone that is liberated by the traps. Therefore, males would have maximum competition from the pheromone emanating from traps for the attraction of mate-seeking females. In contrast, F_1 males emerging in a greatly restricted area (clumps) would be expected to be more attractive than the synthetic pheromone for females emerging in close proximity, since the traps usually would be positioned at a greater distance from the emerging aggregations.

In order to determine the theoretical capture rates for traps based on the assumption that the traps and males are uniformly distributed and that traps are fully competitive in all respects with males for the attraction of mate-seeking females, we made the calculations presented in Table 2. These estimates can then be compared with actual capture rates at the various trap spacings and densities in order to arrive at the relative effectiveness of traps and males for the attraction of females. A critical comparison of such data should help in identifying and evaluating factors governing the action and effectiveness of pheromone traps for the detection and control of low level boll weevil populations.

The data in Table 2 show the expected daily and cumulative female capture rates until male and female emergence is complete for each of the trap spacings and densities. The area of competition in this case is assumed to be 0.4 ha. A brief description of the method of calculations based on the equal competitiveness assumption may be helpful. The data for days 1 and 2 at the 1.95 trap/0.4 ha density will be cited as examples. On day 1, the numerical ratio of traps to 1 competing male is 1.95:1 for the attraction of 1 mate-seeking female. The probability of female capture would be 0.66 and the probability of the female mating, therefore would be 0.34. On day 2 the traps would be in competition with 2 males for the attraction of the 1 female seeking a mate on that day. The ratio of traps to males would be 1.95:2, and the probability of capturing the female would be 0.49. All other calculations were made in the same manner.

Table 2. Daily probability of capturing F_1 female boll weevils emerging on 0.4 ha with indicated no. of traps/0.4 ha assuming males and traps equally attractive - 1♀, 1♂, daily.

Day of emerg.	Distance between traps (ft)*			
	210 (1.95)	149 (3.59)	121 (4.23)	105 (5.67)
1	0.66	0.78	0.81	0.85
2	0.49	0.64	0.68	0.74
3	0.39	0.55	0.59	0.65
4	0.33	0.47	0.51	0.58
5	0.28	0.42	0.46	0.53
6	0.25	0.38	0.42	0.48
7	0.22	0.34	0.38	0.44
8	0.19	0.31	0.35	0.41
Total	2.81	3.89	4.20	4.68

*Number traps/0.4 ha shown in ( ).

By the time the emergence of the approximately 8 males and 8 females in each field is complete, the traps were more competitive than the males in capturing females. The 4.23 traps/0.4 ha, spaced 121 ft apart, should theoretically capture 4.2 females/replicate, but actual capture was 5.5 females/replicate. The 5.67 traps/0.4 ha, spaced 105 ft apart, should have captured 4.66 females/replicate, but acutally captured 6.75 females/replicate. These results suggest that the higher trap densities and closer proximity of the traps to the clumps of F_1 progeny resulted in a sufficiently high concentration of pheromone to effectively compete with the pheromone produced by the males. Thus, most of the females were eventually drawn to the traps rather than to the competing males.

These results are of great practical significance. They suggest that high density trapping may be highly effective for the suppression and management of low level boll weevil populations. Not only is the ratio of traps to competing males a major factor in efficiency, but so is the greater volume of pheromone due to the shorter distance from the traps to females. Because of the importance of the distribution factor, the method will likely be most effective against scattered overwintered boll weevils entering trapped fields. It is particularly encouraging that a high proportion of the females was captured despite the closely clumped emergence pattern. If the hypothesis that the traps will be even more effective against the scattered boll weevils is correct, this ecologically desirable system of suppression deserves high research priority. Such research should include investigations to reduce the cost of the traps and the cost of the pheromone, prolong the duration of effectiveness, and increase the actual capture or killing efficiency at the pheromone sources.

Field observations made by Cross (personal communications) indicate that only about 50% of females responding to traps are actually captured. Even at the 2 higher trap densities and closer trap spacing, the traps on the average were not as close to the emergence area of the females as were the competing males. When we consider these variables, it seems probable that the attractiveness of a trap is several times that of a single competing male. Therefore, our original hypothesis that a trap and a male will have equal attraction for mate-seeking females provides too conservative an estimate of the potential of mass trapping to support either eradication or management programs for the boll weevil.

In an analysis of the results discussed thus far, the area of influence of traps was assumed to be 0.4 ha. It seems desirable to also consider the actual versus the theoretical results that could be expected on the basis of the total present in the field replicates. Therefore, all data from the 4 replicates of each spacing and trap density were pooled and are summarized in Table 3.

In this analysis, 8 of 30 females emerging in the field having a trap spacing of 210 ft were captured. This represents 27% control. With 26 traps in all replicates and 30 females emerging, the theoretical capture would be 19.13 females if traps and males are equally competitive. Therefore, the actual capture was 42% of the theoretical capture. The results for the 149-ft trap spacing were somewhat more favorable. The traps captured 40% of the females or 53% of the theoretical capture. The capture rate of females when trap spacing was 121 ft was even more favorable since 22 of 28 females were captured; this is 79% effectiveness, and it equals the theoretical (21.72 females). In fields having a trap spacing of 105 ft, 27 females were captured. This exceeded the total number estimated to have emerged in the 4 fields. The capture of more females than were estimated to have emerged cannot be fully explained. However, it is possible that there was considerable chance bias in emergence of females versus males. In any event, the higher trap density (14.1/ha) and the closer spacing of traps must have resulted in the capture of more than 90% of the available females. As will be discussed elsewhere, the number of females captured during the F_2 emergence period in the fields where trap spacing was 105 ft was very low. This strongly supports the idea that a high percentage of the females was captured in these fields during the F_1 generation.

Table 3. An analysis of pooled trap numbers/field at indicated trap spacings comparing theoretical and actual captures of boll weevils. Surry County, NC, 1979.

Total no. traps*	Estimated no. of emerging weevils		Theoretical capture of weevils in 10 days	Actual capture	% Actual captures vs. theoretical
			20 Infested squares/clump		
26 (210)	30	30	19.13	8	42
46 (149)	30	30	22.50	12	53
50 (121)	28	28	21.72	22	101
67 (105)	23	23	19.44	27	139
			10 Infested squares/clump		
52 (149)	15	15	12.97	11	85
52 (105)	16	16	14.06	12	86

*Number in ( ) = distance between traps (ft).

Additional Detection and Suppression Analysis

When infield traps were spaced 210 ft apart, they detected 4 of 4 clumps in these experiments. However, the traps were spaced in such a manner that the entire field was covered with a reasonably uniform array of traps. The first trap was installed approximately 15 ft from one field corner and the remainder of the traps were 210 ft apart. The data indicate that 94% of the time this spatial arrangement will detect F_1 progeny of a single-mated overwintered female. These results fully support the conclusions reached from prior studies (Leggett et al., 1980) that infield traps provide a highly sensitive method of detecting the existence of boll weevil infestations resulting from a single reproducing female in a cotton field of any size.

A simplified computer simulation routine was written to place confidence limits on the probability of detection estimates. Multiple regression analyses were run to relate capture of female weevils to various factors to provide parameters for the model. A significant relationship was found to exist between capture and trap density and the trap-to-male ratio. Trap density values used were based on a field of infinite size since effectiveness of a given absolute trap density varies with size of the field. Capture probabilities were not expressed in the model as a function of distance from the trap since this is a difficult relationship to define in the cotton field. Capture probability was thus expressed as a function of log trap density and trap-to-male ratio.

One mated overwintered female weevil was assumed to enter the field and deposit her eggs in a normal clumped arrangement. The number of progeny was determined by Monte Carlo simulation probability distribution. A 1:1 sex ratio of F_1 progeny was assumed. Daily mortality was randomly assumed to be 5%. Probability of capture was determined for each female weevil each day by the Monte Carlo simulation by using the log trap density and trap-to-male ratio function. Five hundred simulations were made for a 14-day trapping period with a trap density of 1 trap/0.4 ha (Table 4). In 5.5% of the simulation runs, no boll weevils were detected. Thus, under these conditions, ca. 95% of the F_1 weevil clumps would be detected. The size of the reproducing clump (number of weevils) can also be estimated by these procedures; however, in an eradication program, detection of the clump is of greater importance.

As already discussed, the results of this experiment indicate that significant suppression of the F_1 generation was achieved when traps were spaced 121 ft and 105 ft apart. This conclusion is strongly supported by the rate of female capture of F_2 progeny at the different trap densities (Table 1). Since infested squares were used in the natural simulation rather than adult weevils (previous experience had shown that a high proportion of the adult weevils placed in fields leaves immediately and cannot be accounted for), the sex ratio of the boll weevils that emerged in the various fields is not known. Therefore, the exact degree of suppression cannot be calculated. However, the data in Table 1 are indicative of the degree of suppression of the boll weevils that emerged in the fields. If reproduction had been normal in the absence of control, the average F_2 populations should have been similar in the 4 replicates for each trap spacing and density. Under such circumstances, the female capture rate during the F_2 emergence period should have been low at the lowest trap density and high at the highest trap density. However, the actual capture rate in fields having the 4 different trap densities was completely in reverse of such expectation. The highest number was captured in the low trap density fields and the lowest number was captured in the highest trap density fields. While precise calculations may not be possible, we estimate that 80% or more of the F_1 females were captured when traps were spaced 121 ft apart and probably more than 90% of the F_1 females were captured when traps were 105 ft apart. Such levels of control of F_1 progeny from the clumped cotton were higher than anticipated. Therefore, the use of high density traps may be highly effective against both low level populations of overwintered boll weevils and low level populations of F_1 and F_2 progeny.

Table 4. Estimated number of boll weevils in cotton field based on the number of female weevils captured with 1 trap/0.4 ha (500 computer simulations with the % capture of a function of log trap density and trap/♂ ratio during a 14-day trapping period).

% Weevils captured*	Estimated no. weevils Mean	Conf. limits**
5.5 (0)	6	4-8
10.7 (1)	13	11-15
21.5 (2)	18	16-19
20.8 (3)	20	19-21
21.7 (4)	21	20-22
9.3 (5)	23	22-25
7.3 (6)	24	22-25
2.2 (7)	29	25-33
0.6 (8)	35	31-38
0.0 (9)	-	- -
0.0 (10)	-	- -
2.0 (11)	37	35-39

*Weevils captured by simulation in ( ).
**95% Confidence level.

Possible Use of Grandlure for Boll Weevil Management

The experiments with grandlure-baited traps were designed primarily as a means of developing a reliable detection system for possible use in a boll weevil eradication program. However, critical analysis of the data obtained indicates the potential value of this behavioral chemical both for detection and control in eradication efforts. The results also suggest the possibility of employing high density trapping to manage boll weevil populations that originate from normally low overwintered populations or populations reduced to low levels by diapause and cultural controls undertaken the previous fall.

Results obtained in the 1979 Surry County, NC experiments indicate that infield traps used at high densities are sufficiently competitive with males to achieve a high degree of suppression of low level populations. Traps properly spaced may be several times as effective as males for the attraction of mate-seeking females. Pheromone sources other than traps may also achieve additional control through the disruption of mating by confusion. If the observed and estimated effects are confirmed by suitable field trial programs, the use of low cost traps and long lasting pheromone formulations could provide an effective, economical, and ecologically desirable system for managing overwintered or low density F_1 boll weevil populations. As an alternative or supplement to traps, suitable devices that would result in the death of boll weevils attracted to pheromone sources may be more effective and economical than traps alone. The ultimate goal in boll weevil management should be to achieve adequate seasonal control while minimizing the need for insecticides. In view of the encouraging results obtained in the preliminary investigations herein reported, there is strong justification for research to fully develop ways of using the pheromone as a continuing management system. The use of grandlure offers prospects of a management procedure with a built-in monitoring system to serve as a guide to the type and intensity of supplemental suppressive measures that may be required. Trap capture data would serve to identify any cotton fields that may need supplemental suppression in order to maintain sub-economic populations throughout the management area.

CONCLUSIONS

As a result of the research efforts by many cooperating scientists during the past 2 years, the potential of grandlure-baited infield traps for detection and population assessment has become apparent. This advance has been possible largely because of the use of artificially created boll weevil infestations in isolated cotton fields. Also, the procedures used and results obtained in investigations give some insight into the mechanism of action and efficiency of the traps for the attraction and possible confusion of mate-seeking females. The attractiveness of the traps and formulations employed seem high enough to effectively manage low level overwintered and F_1 populations with little or no need for chemical treatments during the growing season. Using infield traps as a survey and monitoring tool also can be an important aid to other management systems. Before the full benefits of grandlure can be realized, additional innovative research remains to be done. A variety of trapping and pheromone release systems must be investigated and tested in field experiments to determine the optimum use of the pheromone in boll weevil suppression and management systems.

ACKNOWLEDGMENTS

Thanks are due to Jack Waddell, Danny Key, Mary Beth Southern and Charles Jackson, Surry County, members of North Carolina Agricultural Extension Service, for invaluable assistance in gaining grower cooperation to culture special cotton fields in their county, and to participating farmers as well. Miss Barbara Leatherman is also due thanks for her tireless efforts in data collection and many other duties assigned to her during this experiment.

REFERENCES

Cross, W. H., Hardee, D. D., Nichols, F., Mitchell, H. C., Mitchell, E. B., Huddleston, P. M., and Tumlinson, J. H., 1969, Attraction of female boll weevils to traps baited with males or extracts of males, J. Econ. Entomol., 62: 154.

Cross, W. H., Mitchell, H. C., and Hardee, D. D., 1976, Boll weevils: Response to light sources and colors on traps, Environ. Entomol., 3: 565.

Hardee, D. D., Cross, W. H., Mitchell, E. B., Huddleston, P. M., Mitchell, H. C., Merkl, M. E., and Davich, T. B., 1969, Biological factors influencing responses of the female boll weevil to the male sex pheromone in field and large-cage tests, J. Econ. Entomol., 62: 161.

Hardee, D. D., McKibben, G. H., Gueldner, R. C., Mitchell, E. B., Tumlinson, J. H., and Cross, W. H., 1972, Boll weevils in nature respond to grandlure, a synthetic pheromone, J. Econ. Entomol., 65: 97.

Knipling, E. F., 1976, Biomathematical basis for suppression and elimination of boll weevil populations, Proceedings of a Conference on Boll Weevil Suppression, Management and Elimination Technology, Feb. 13-15, Memphis, TN, USDA Tech. Bull. ARS-S-71.

Knipling, E. F., 1979, The basic principles of insect population suppression and management, USDA Handbook No. 512, 623 pp.

Knipling, E. F., and McGuire, J. U., Jr., 1966, Population models to test theoretical effects of sex attractants used for insect control, USDA Inform. Bull. 308, 20 pp.

Leggett, J. E., and Cross, W. H., 1971, A new trap for recapturing boll weevils, USDA Coop. Econ. Insect Rept., 21: 773.

McKibben, G. H., Davich, T. B., Gueldner, R. C., Hardee, D. D., and Huddleston, P. M., 1974, Polymeric compositions for attracting cotton boll weevils, U. S. Patent No. 3,803,303.

Mitchell, E. B., and Hardee, D. D., 1974, Infield traps: A new concept in survey and suppression of low populations of boll weevils, *J. Econ. Entomol.*, 67: 506.

Tumlinson, J. H., Hardee, D. D., Gueldner, R. C., Thompson, A. C., Hedin, P. A., and Minyard, J. P., 1969, Sex pheromones produced by the male boll weevil: Isolation, identification, and synthesis, *Science*, 166: 1010.

FUTURE THRUSTS FOR DEVELOPING AND UTILIZING MASS-TRAPPING TECHNOLOGY

The members of the mass-trapping panel recognized the potential role of mass-trapping technology in suppression of populations of insects which damage forests, forest products, shade trees, ornamentals, and cultivated crops. While specific thrusts are related to specific insect problems and commodities, the following were identified as needs common to most of the suppression systems now under development.

Research and Development

Judicious selection of appropriate pests and commodities which are amenable to the mass-trapping strategy is essential. Resources should thus be provided to intensify research on mass-trapping concepts on these pests and commodities. Research areas should include basic behavioral studies of specific insect pests; isolation, identification, synthesis, and formulation of attractants and pheromones; optimization of attractant/pheromone systems with suitable dispensers and traps; and quantitative population research where needed to define population densities at which mass-trapping can be effectively employed, and to provide efficient sampling systems by which the impact of a mass-trapping program can be measured. Open communications between scientists developing mass-trapping procedures and others researching different suppression systems are imperative for the rapid development of integrated systems of suppression where mass-trapping may represent but one component of the system. We must encourage entomologists, extension personnel, professional foresters, growers, and others to consider the use of mass-trapping procedures as an alternative to current control procedures which require extensive use of insecticides or where substantial economic losses are occurring in the absence of an acceptable system of suppression or control.

Assessment of Results and Potentials

Techniques for measuring the input of mass-trapping on the population and the determination of injury levels at varying population levels must be developed. These techniques will identify the potential biological, economic, and environmental benefits which will accrue from employment of mass-trapping technology. Effective communications between scientists conducting mass-trapping research with systems economists, environmental systems modelers, and potential users of new technology are imperative before estimates of potential economic benefits of new systems of suppression can be realistically assessed.

Technology Transfer

Technology transfer is an important step in bringing these programs to fruition. The scientist must identify appropriate methods of transferring technology from scientists to potential users. Large-scale field trials involving scientists, the extension services, growers, and others are proposed as the primary method for accomplishing this technology transfer. Open and continuing communication between the originating scientists and the operational personnel, extension specialists, private consultants, industry, and growers are essential in effective transfer of technology from scientists to potential users. We must also continue to offer technical assistance to programs in the operational stage to assure that misapplication will not result in failure and setback in the use of this promising technique.

PANEL: E. P. Lloyd, USDA; W. D. Bedard, USFS; J. H. Borden, Canada; R. T. Cunningham, USDA; R. T. Huber, University of Arizona; G. H. Lanier, State University Environmental Services and Forestry, NY; R. Lie, Norway; M. Klein, USDA; J. W. Peacock, USFS.

MATING DISRUPTION OF LEPIDOPTEROUS PESTS: CURRENT STATUS AND FUTURE PROSPECTS*

G. H. L. Rothschild

Division of Entomology
C.S.I.R.O., P. O. Box 1700
Canberra, A.C.T. 2601
Australia

INTRODUCTION

The possibility of using synthetic sex pheromones and other behavior-modifying compounds for interfering with sexual communication between male and female moths was first proposed about 20 years ago, and was followed soon after by a large-scale field trial to disrupt mating of the gypsy moth, *Lymantria dispar* (L.). The results were exceedingly disappointing but it soon became apparent that the compound used in the trial, "gyplure" was inactive. This material was a homologue of "gyptol" which had been incorrectly identified as the major component of the pheromone. This failure highlighted the need for accurate identification of sex pheromones and led to greatly increased research activity in this area. Indeed, the major emphasis in most centers for the remainder of the 1960s was on pheromone identification and synthesis. Very few studies on adult behavior relevant to mating disruption were undertaken during this period. Notable exceptions included the work of Shorey and colleagues on sexual communication of noctuids, particularly the cabbage looper, *Trichoplusia ni* (Hübner) (Shorey et al., 1968).

The identification of the active component of the female cabbage looper pheromone enabled Shorey et al. (1967) to undertake the first meaningful field trial on mating disruption, albeit on a very small scale. The level of work on disruption after this preliminary trial, based on the number of publications, is shown in Table 1. There was a steady rise in the number of published papers dealing with field trials on disruption until 1976, after which

*This review was based on 119 references, but for space reasons it has been possible to cite only those listed in the references.

remained constant and even declined somewhat. This decline possibly resulted from problems encountered in mating disruption trials, which necessitated a return to more basic studies of adult behavior and pheromone perception, as well as increased emphasis on more practical aspects such as the release characteristics and persistency of formulations.

Table 1. Number of papers describing mating disruption trials published between 1967 and 1979*.

Year	Number of papers	Year	Number of papers
1967	1	1976	22
1972	8	1977	18
1973	2	1978	18
1974	14	1979	11
1975	19		

*Papers seen by author, not based on exhaustive search.

THE CONCEPT OF MATING DISRUPTION

The behavioral mechanisms whereby synthetic sex pheromones or other behavior-modifiers prevent or reduce mating in the field are only partially understood and form part of the somewhat large "black box" component of most disruption studies in the field. Possible mechanisms include:

1) Constant exposure to a relatively high level of synthetic sex pheromone leading to adaptation of the antennal receptors and habituation of the central nervous system, following which males fail to respond to pheromone signals from the calling females or can only perceive and respond to stimuli greater than those produced by the latter. Laboratory studies of Shorey et al. (1967), Traynier (1970), Bartell and Lawrence (1973) and others have confirmed that pre-exposure to synthetic or natural pheromone components can reduce or suppress the response of male moths to sex pheromone. These experiments were designed to answer specific questions relating to pre-exposure and were conducted under conditions very different from those prevailing in the field; females were, for example, not confined together with males in any of the habituation studies, and pre-exposure was not extended over the entire diel period of male responsiveness. Work of Sower et al. (1975) has shown that the "spontaneous" or random activity of apparently habituated males of the Indian meal moth, Plodia interpunctella (Hübner), in the presence of females can lead to chance encounters that in turn result in copulation; initial physical contact during these counters could perhaps have dishabituated the males (Barrer and Hill, 1980).

Shorey et al. (1967), on the other hand, have demonstrated that mating of Trichoplusia adults held in sealed containers can be totally suppressed when subjected to sufficiently high concentrations of synthetic pheromone even when under very crowded conditions. The effects of continuous exposure to pheromone components on male responsiveness probably vary from species to species. Additional pheromone components, other non-olfactory cues, and factors leading to dishabituation may contribute to this variability.

It seems unlikely that aerial pheromone concentrations presently used in field disruption trials can be maintained at sufficiently high levels, both in space and time, to habituate males to the extent that they are no longer sexually responsive. Variations in the aerial distribution of pheromone probably enable flying males to disadapt or dishabituate, at least partially. It is, however, possible that the treatments raise the perception threshold of sexually responsive males to a level where they can no longer detect the pheromone signal from the calling females (Marks, 1976a).

2) A second mechanism may be actual "confusion" where the environment is permeated with synthetic pheromone components in the correct ratio, and sexually responsive males flying within the treated area are diverted from the calling females by the large number of "competing" synthetic pheromone sources (false trails). If confusion is occurring, there should be a greater probability of males locating females when adult population densities are high. A number of disruption studies with moth pests include reports of such density effects.

3) Another possible mechanism for disruption may arise where the female sex pheromone consists of a particular blend of components. Sexual communication may be suppressed when the ratio of components perceived by the responding male is drastically altered by treatments with one compound only. When this component mediates "long range" behavior, males may be sexually aroused and fly upward but are unable to locate the calling female [the somewhat subjective terms "close" and "long" range to describe male behavior are used in the sense of Kennedy (1977) accepting that these are open to differing interpretations]. The observations of Marks (1976b) on the behavior of red bollworm, Diparopsis castanea Hampson, males near virgin females in sites permeated with one of the 4 components of the female pheromone support this possibility. Roelofs (1978) has also observed that upwind flight behavior of male redbanded leafrollers, Argyrotaenia velutinana (Walker), toward a source containing the normal blend of pheromone components in a wind tunnel is greatly altered when air containing only one component is circulated through the apparatus.

Research work to establish which one or more of these behavioral mechanisms contributes to decreased mating still remains to be done. This requires the development of field bioassay methods to compare arousal, orientation, landing, and the precopulatory behavior of males in treated and untreated areas. Currently, all that can be stated is that there is evidence that confusion is commonly involved in disruption, with some additional support for the third mechanism, namely, distortion of the ratio of pheromone components when only one compound is used. So far, there is little evidence that habituation occurs in the field; this may be due to inadequate experimentation. These comments on mechanisms refer to disruption control based on synthetic female sex pheromones, but may also apply to parapheromones -- compounds that mimic pheromone activity and presumably stimulate the same antennal receptors. An example of a parapheromone is "hexalure" [(Z)-7-hexadecenyl acetate], which was used in experiments for mating control of the pink bollworm, *Pectinophora gossypiella* (Saunders), before the female sex pheromone [(Z,Z)- and (Z,E)-isomers of 7,11-hexadecadienyl acetate -- "gossyplure"] was characterized. Shorey (1977) considered that a parapheromone would be less suitable than a true pheromone for disruption because far greater quantities of the former would be required to give the same degree of control; this implies that parapheromones are perhaps less stimulatory to antennal receptors. Significant disruptive effects, however, have been recorded when hexalure was released at the relatively low mean rate of 2 mg/ha/h (McLaughlin et al., 1972a), given that gossyplure produced a comparable effect at the even lower rate of 0.5 mg/ha/h (Shorey et al., 1976). The available information from field trials with parapheromones is still so inadequate that it is probably unwise to speculate in general terms about the potential application of these compounds for mating disruption.

Similar caution is necessary in assessing the likely role of antipheromones -- compounds that interfere with pheromone perception in male moths. Conflicting results have been obtained when these materials were tested in close proximity to pheromone sources or were tested as background treatments (McLaughlin et al., 1972b). Sanders (this volume) considers that this may be due to the effects of differences in the aerial distribution of the pheromone and inhibitory compounds. When both materials are released together from one point, the male receives synchronous stimuli and is unable to distinguish the pheromone signal; however, there is no such synchrony when the compounds are released individually from different sites and males may continue to locate pheromone sources. Shorey (1977) suggested that if there were specific receptors for inhibiting compounds, males could become habituated to these materials but still remain responsive to pheromone stimuli. There is certainly strong evidence for the existence of specific receptors for inhibitory compounds in some species, but Minks et al. (1976) and others have shown that antipheromones can be used to disrupt sexual communication in a number of moth pests.

Little is also known about the impact of behavior-modifiers on the process of mating. Such treatments must almost certainly delay mating, even in situations where the final level of mating suppression does not appear to be high. Barrer (1976) has demonstrated that mating delays in the almond moth, Ephestia cautella (Walker), can result in a significant reduction of both fecundity and fertility of the female. From work on the codling moth, Laspeyresia pomonella (L.) it has been suggested that some process associated with aging occurs during such delays, which lowers the probability of mating (Nowosielski and Suski, 1977). This process could perhaps be related to senescence of the sensory receptor system (Seabrook et al., 1979). Very little is known about the mating behavior of most common moth pests in the field. This is remarkable when it is considered that the objective of disruption is to interfere with mating. Notable exceptions include the studies of Marks (1976a,b) and Richerson (1977), who have made careful observations of adult behavior both in pheromone-treated and check areas.

The lack of basic knowledge on mechanisms, however, should not be regarded as a deterrent to further attempts at disruption in the field on a small or large scale. It can be argued that difficulties in interpreting the results of disruption trials can be attributed in part to a lack of knowledge of mechanisms involved, but it is equally apparent that there is an even greater ignorance of ecological factors relating to the outcome of mating disruption. These include the quantitative effects of reduced mating on population trends and the role of dispersal.

A general impression is gained that there are few "universal truths" that can be applied to mating disruption, and the likely outcome of treatments with pheromones, parapheromones or inhibitors can be assessed only by careful studies on each species.

PROCEDURES FOR MATING DISRUPTION TRIALS IN THE FIELD

The general procedure in most disruption trials has been to liberate behavior-modifying chemicals from various dispensing systems into a well-defined area for varying periods, and then use a number of criteria for assessing the effect of the treatment, usually in comparison with an untreated check plot.

About 25% of mating disruption trials reviewed (total of 170) extended over a full field-season with the objective of determining whether the treatments resulted in decreased crop infestation. The remaining 75% were essentially exploratory studies with varied objectives including establishment of which behavioral modifiers were most disruptive, formulation characteristics, and (far less frequently) observation of the behavior of adults in treated areas.

Compounds Used for Mating Disruption

Synthesized versions of compounds known to be components of the female sex pheromone have been used in most mating disruption trials. The results of trials with the cabbage looper suggest that treatments with one pheromone component, (Z)-7-dodecenyl acetate, can disrupt all facets of pre-mating communication including arousal, orientation, landing and "close range" precopulatory behavior -- even when adult densities are high (Shorey et al., 1967). In the codling moth, relatively high concentrations of the single known component, (E,E)-8,10-dodecadienol, do not suppress mating when adult densities are high (Cardé et al., 1977), indicating that the compound may not mediate "close range" behavior; recent evidence suggests that the female codling pheromone contains additional components (Bartell and Bellas, unpublished data).

In the eastern spruce budworm, *Choristoneura fumiferana* (Clemens), mating disruption can apparently be achieved only when treatments are based on a combination of the 2 active components (E and Z isomers of 11-tetradecenal) in the same ratio as is found in the female secretion (Sanders, 1976). This suggests that in this species the disruption mechanism may be confusion rather than distortion of the female signal. The female pheromone of the red-banded leafroller consists of a blend of at least 3 components (Baker et al., 1976), of which the first 2, (E)- and (Z)-isomers of 11-tetradecenyl acetate, mediate arousal, upwind orientation and, to a lesser extent, landing. The 3rd component, dodecyl acetate, apparently plays a role in "close range" behavior by increasing the frequency of landings and approaches to an odor source. In field trials, treatments with blends of the geometrical isomers of 11-tetradecenyl acetate, in proportions similar to those in the female secretion, were more disruptive than other ratios. The addition of dodecyl acetate to the other 2 compounds failed to increase the level of disruption obtained (Taschenberg and Roelofs, 1978), supporting suggestions that suppression of both "close" and "long" range behavior may not be essential to achieve mating disruption. In the Oriental fruit moth, *Cydia molesta* (Busck), the female pheromone blend consists of 4 components (Cardé et al., 1979). A combination of 3 of these [Z and E isomers of 8-dodecenyl acetate and (Z)-8-dodecenyl alcohol], in similar proportions to the blend emitted by females, apparently mediates male behavior from arousal and upwind orientation to landing and location of the pheromone source (Baker and Cardé, 1979). It is not possible to assign a particular "close range" role to any one component, although the alcohol does increase landing frequency. Most disruption trials against the oriental fruit moth have involved treatments with (Z)-8-dodecenyl acetate, containing 2-5% of the E isomer (Rothschild, 1975; Gentry et al., 1976). Treatments with the E isomer were far less disruptive than those using the other isomer (Rothschild, 1979), which again suggests that confusion rather than distortion of the female signal is the more effective mechanism for mating disruption in this species.

The identified component of the female gypsy moth pheromone is the (+) optical isomer of (Z)-7,8-epoxy-2-methyloctadecane, disparlure. Synthetic disparlure also contains the (-) isomer, which is inhibitory and modifies the in-flight behavior of males (Roelofs, 1978). All disruption trials have so far been conducted with racemic mixtures and it is not known which of the isomers would be more disruptive if used individually.

The female pheromone blends of a number of other moth species include components which are inhibitory to conspecific (and sometimes congeneric) males. These insects include Spodoptera litura (F.), S. littoralis (Boisd.), Diparopsis castanea Hampson, and Ephestia cautella (Walker). The function of these compounds is not clear, but they are perhaps released after a female has mated to prevent further males arriving, or may play a role in ensuring the reproductive isolation of allied species. In Diparopsis (Marks, 1976a), similar levels of disruption were obtained when the inhibitor and attractant compounds were released at comparable rates [the term "attractant" has been criticized by Kennedy (1972) but remains convenient, providing that its limitations are appreciated]. In S. littoralis, Campion et al. (1976) found that the attractant component, which mediated most facets of behavior leading to location of the female, was more disruptive than the inhibitory material. Treatments with the inhibitory pheromone component of Ephestia had little effect on mating, which Sower et al. (1974) considered to be a result of the male antennae possessing specific receptors for the inhibitory and attractant components.

The varying results of tests with these species probably reflect differences in both the specificity of antennal receptors and processes of integration in the central nervous system.

Almost the only example of disruption with parapheromones is that of pink bollworm with hexalure (referred to earlier). There are also only a few instances of the field use of antipheromones, namely, the control of Adoxophyes orana (Fischer von Röslerstamm) with geometrical isomers of the pheromone components (Minks et al., 1976), small-scale treatments with formate derivatives of the female pheromone components of the Asiatic rice borer, Chilo suppressalis (Walker), (Beevor and Campion, 1979) and Heliothis species (Mitchell et al., 1975), and the use of the acetate derivative of the codling moth pheromone (Hathaway et al., 1979). Comparable levels of disruption were obtained with antipheromones and pheromone components when these were released at similar rates in a number of trials. This perhaps supports suggestions that these compounds are competing for the same receptor sites on the male antennae.

collected by such means as sweep-netting and suction sampling; (4) estimates of number of fertile eggs, larval counts, and the extent of damage to the crop or commodity. The assessments are based on samples from both treated and check plots. Table 3 summarizes the extent to which these various criteria have been utilized in disruption trials based on a survey of publications between 1967 and 1979.

Table 3. Percentage of field trials using various methods of assessing disruption.

Disruption criterion	Kind of Pest			
	Field crop	Orchard	Forest	Stored-product
Pheromone or virgin female trap only	46.3	54.7	39.4	0
Laboratory-reared virgin females	25.9	22.6	48.5	0
Wild females	5.5	15.1	12.1	100.0
Crop damage	14.8	30.2	0	0

A reduction in the number of males trapped at virgin female or synthetic sex pheromone sources was the sole criterion of mating disruption in over 40% of the trials. This method provides some indication of the extent to which the disruption treatment is diverting males from their virgin female "targets" (by whatever mechanism is involved), but fails to demonstrate whether mating is reduced; examples of experiments in which the reduction in mating was equivalent to the decrease in trap catch are rare. In general, fewer males are captured at virgin female sources than at synthetic pheromone baits in pheromone-treated areas. A single male captured at a virgin female trap in a treated area could be regarded as representing a successful "mating," regardless of how many more males were captured at a corresponding trap in the check plot. Work in Australia suggests that a more rigorous and realistic approach to the use of traps as monitors of mating disruption is to merely score the presence or absence of male captures in traps, regardless of the actual catch size. Assessments have frequently been based on recaptures of released laboratory-reared males because there is insufficient wild material in the field or because tests were conducted out of season. Care is needed in interpreting the results of such trials as laboratory-reared males may respond been used commercially (see chapter on mating disruption of pink bollworm). There is also considerable scope for improvement of this formulation. Sower et al. (1979) have demonstrated that up to half of the fibers may not fall in the target area when applied aerially, and fiber losses from foliage over a 2-month period can exceed 75%.

Trials on mating disruption using regularly spaced evaporators have shown that the distribution of these dispensers is less critical than the overall level of the behavior-modifying compound liberated per unit of ground area -- or (more meaningfully, in some situations) per unit volume of air space. There is, of course, a lower limit below which there are insufficient evaporators to prevent the occurrence of "pheromone-free" areas in which mating could continue. Broadcast formulations are essential where very large areas have to be treated, like in field crops and forests, but evaporators may be adequate for small-scale operations -- in orchards and vineyards, for example.

Release Rates

The aerial distribution and concentration of behavior-modifying chemicals is one of the key factors in determining the outcome of mating disruption treatments but has received surprisingly little attention from researchers. Less than 1% of publications on disruption in recent years have dealt with this particular subject (Plimmer et al., 1978), others having been concerned with measuring losses of compounds from various formulations under laboratory conditions. In many disruption studies, release rates are cited in mg/ha/h or /day and are based on estimates obtained under laboratory conditions, often at a single temperature without any indication of the variability of the data. Such figures become somewhat more meaningful if they are modified to take account of temperatures prevailing during the daily period of sexual activity. In other trials, dispensers have been weighed at intervals of several days or even weeks and release rates expressed as an average figure per unit time and area. Such estimates provide a crude indication of the quantities of behavior-modifying compounds required to influence sexual communication, but offer little insight into the aerial concentrations that flying males encounter in a treated area.

Criteria for Assessing the Success of Disruption

The effectiveness of mating disruption treatments has been assessed in various ways, including: (1) Numbers of wild or released laboratory-reared males captured at traps baited with virgin females or synthetic sex pheromone; (2) proportion of "decoy" females (tethered or with clipped wings) that mate with wild or released males; (3) proportion of wild females that are mated --

collected by such means as sweep-netting and suction sampling; (4) estimates of number of fertile eggs, larval counts, and the extent of damage to the crop or commodity. The assessments are based on samples from both treated and check plots. Table 3 summarizes the extent to which these various criteria have been utilized in disruption trials based on a survey of publications between 1967 and 1979.

Table 3. Percentage of field trials using various methods of assessing disruption.

Disruption criterion	Kind of Pest			
	Field crop	Orchard	Forest	Stored-product
Pheromone or Virgin female trap only	46.3	54.7	39.4	0
Laboratory-reared virgin females	25.9	22.6	48.5	0
Wild females	5.5	15.1	12.1	100.0
Crop damage	14.8	30.2	0	0

A reduction in the number of males trapped at virgin female or synthetic sex pheromone sources was the sole criterion of mating disruption in over 40% of the trials. This method provide some indication of the extent to which the disruption treatment is diverting males from their virgin female "targets" (by whateve mechanism is involved), but fails to demonstrate whether mating is reduced; examples of experiments in which the reduction in mating was equivalent to the decrease in trap catch are rare. In general, fewer males are captured at virgin female sources than at synthetic pheromone baits in pheromone-treated areas. A single male captured at a virgin female trap in a treated area could be regarded as representing a successful "mating," regardless of how many more males were captured at a corresponding trap in the check plot. Perhaps a more rigorous and realistic approach to the use of traps as monitors of mating disruption would be to merely score the presence or absence of male captures in traps, regardless of the actual catch size. Assessments have frequently been based on recaptures of released laboratory-reared males because there is insufficient wild material in the field or because tests were conducted out of season. Care is needed in interpreting the results of such trials as laboratory-reared males may respond differently to behavior-modifying chemicals, particularly where the insects have been irradiated (Hutt, 1977).

The next most widely used measure of disruption (Table 3) was the decoy virgin female which was utilized in almost 50% of the trials with forest moth pests (where, in the case of lymantriids, females are often wingless) but in only 20-25% of the experiments with species occurring in field and orchard crops. A comparison of the extent to which wild males are able to locate and mate with decoy virgin females in treated and untreated sites provides a far more meaningful assessment of mating disruption than reduction of captures at monitoring traps. However, it is important that the placement of the decoys should follow the likely spatial distribution of their wild counterparts. Because of sampling problems, little is known about adult distributions in the field, but there is, for example, evidence that male activity and sexual encounters in many species are most frequent in the upper parts of the canopy of field and forest crops (Sower and Daterman, 1977; Kaae and Shorey, 1973). The most meaningful criterion of disruption is the mating status of the wild female population, but this has only rarely been assessed in the field, again, as a result of sampling difficulties. Table 3 shows that only in the well-defined storage environment has the mating condition of wild females been used consistently for evaluating the impact of disruption treatments. Even in these optimum conditions there have been important differences between the mating status of females sampled at rest or in flight (Haines and Read, 1977). In orchard trials, feeding lures (bait pails) have been used to trap females, but the samples appear to be biased toward mated individuals and contain only a small proportion of virgins; there is, however, some evidence to show that there are small but significant increases in the percentage of virgin females trapped with bait pails following pheromone treatments (Rothschild, 1975; Roehrich et al., 1977).

From an economic standpoint, the only relevant criterion of successful disruption is the level of infestation and damage in the crop or commodity. Such information has been obtained in only 15% of disruption trials in field crops and in about 30% of studies in orchards (Table 3). Crop infestation represents the outcome of a series of interacting processes that influence population development and may be far removed in time from the period at which mating disruption occurred. Because of this time factor it is often difficult to attribute differences in crop infestation between treated and untreated plots to the influence of the treatments with any degree of confidence; where possible, it may be desirable to obtain life table information on the target species in both treated and untreated areas. Mating disruption trials are of greatest value when they are assessed in terms of all or most of the criteria discussed above. Among the few studies in which this has been attempted are those of Marks (1976a,b) and Roehrich et al. (1977).

Preliminary attempts have been made to model the influence of mating disruption on the population dynamics of pest species and to predict the outcome of various treatments. Symmons and Rosenberg (1978) constructed a model for S. littoralis based on a reasonable body of field data and found that emigration and immigration of adults (both sexes) was of particular significance. The model indicated that treatments with behavior-modifiers would be effective only if applied very early in the season. Models of the disruption process are likely to become increasingly useful as more basic information necessary for their development is obtained.

FACTORS INFLUENCING OUTCOME OF DISRUPTION TRIALS

Species Characteristics Favoring Mating Disruption

About 40 moth species have been subjected to mating disruption treatments in the field (Table 4). The majority of papers have dealt with Tortricidae (37%), followed by Noctuidae (22%) and Lymantriidae (15%), while the gypsy moth, pink bollworm and codling moth were individually the most-studied species. The main ecological characteristics of a species likely to favor control by mating disruption are that the adults (particularly mated females) are non-migratory and have a narrow host range, both of which would tend to limit invasion of a treated area by mated females. The problem of invasion is discussed in some length here, as this has generally received little attention in disruption studies.

In the more comprehensive disruption trials extending over a full field season, it has often been stated that the results were inconclusive because of probable immigration of mated females from adjacent areas. This appears to have become a major problem in field experiments, and it is indeed difficult (if not impossible) to evaluate the effects of disruption treatments where these are consistently overridden by the influence of immigrating gravid females. The likelihood of immigration is no doubt accentuated by the small scale of most disruption trials (Table 4).

Not a great deal is known about the dispersive behavior of many of the insects listed in Table 4, and the species have been classed as migratory and non-migratory on the limited information available [using Johnson's (1969) definition of migrants and non-migrants]. Of the tortricids, the spruce budworm and pine shoot moths are recognized migrants, and large-scale movements of mated females extending over hundreds of kilometers occur regularly. The scale of disruption experiments on these species (Table 4) would appear to be inadequate, except perhaps for short-term exploratory trials undertaken at times when there were no significant migratory movements. As Sanders (1979) has pointed out, it may be necessary to locate and treat foci of infestation throughout a district to suppress mating within local populations and thereby prevent the production of mated female migrants. The remaining tortricids (Table 4) are mostly orchard and vineyard pests, and are considered to be non-migratory. The olethreutine species (codling moth, grape berry moth and oriental fruit moth) also have a narrow host range and would appear to be particularly amenable to mating disruption. The extent to which these species disperse by "trivial" flights is not altogether clear. Most studies have been concerned with dispersal of males and little is known of the flight behavior of gravid females. The information available for the oriental fruit moth and the codling moth suggests that few mated females disperse further than 300 m. This estimate, however, is still large in relation to the average size of experimental plots (Table 4) and the distance between treated and untreated areas. Disruption treatments have resulted in significant but variable reductions in crop damage in the grapevine moth (Lobesia) and the grape berry moth (Paralobesia), and this variability has been largely attributed to immigration of gravid females (Taschenberg et al., 1974; Roehrich et al., 1977). Among the Tortricinae, only the redbanded leafroller has been the subject of more intensive disruption studies. This species is highly polyphagous, but evidence from mass-trapping experiments suggests that the adults do not disperse widely (Trammel et al., 1974). However, the areas treated in most disruption trials were so small (Table 4) that it seems probable that significant immigration could have occurred.

The migratory behavior of pink bollworm adults is well-known, and there are instances of both sexes being carried great distances on wind fronts (Johnson, 1969; Stern, 1979) as well as records of extensive local movements between cotton fields. What is not apparent from these observations is whether the majority of female migrants are mated. Immigration of unmated females would clearly have less impact on the outcome of disruption treatments than the arrival of mated individuals. In species such as the spruce budworm where mating takes place prior to dispersal, the migrants are largely females, and most males presumably remain near their emergence sites. There is evidence to show that migrating pink bollworm adults include large numbers of males (Stern, 1979); therefore, it is possible that females only mate on arrival in a cotton crop.

Noctuid moth pests have received considerable attention as potential targets for mating disruption, commencing with the pioneering studies of Shorey and colleagues referred to earlier. As indicated in Table 4, most of these species are polyphagous and a high proportion are also known to be migratory. A combination

Table 4. Lepidoptera studies in mating disruption field studies.

Species	Publ.	Mean range†	Migrant status	Host range
Laspeyresia pomonella	9	0.3 (<0.01-1)	-	narrow
Cydia molesta	4	3.7 (0.4-13)	-	narrow
Grapholitha funebrana	3	0.4 (0.1-0.7)	-	narrow
Rhyacionia buoliana	2	0.3 (<0.01-5)	+	narrow
Rhyacionia frustrana	1	0.4	+	narrow
Zeiraphera diniana	1	300	+	narrow
Paralobesia viteana	4	0.3 (0.1-0.4)	-	narrow
Lobesia botrana	3	0.3 (<0.01-1.5)	-	wide
Choristoneura fumiferana	3	47 (<0.01-2.40)	+	wide
Choristoneura rosaceana	1	1.0	-	wide
Argyrotaenia velutinana	7	0.3 (<0.01-1)	-	wide
Adoxophyes orana	1	0.2	-	wide
Archips podana	1	1.0	-	wide
Archips mortuanus	1	1.0	-	wide
Archips argyrospilus	1	1.0	-	wide
Pandemis limitata	1	1.0	-	wide
Platynota idaeusalis	1	1.0	-	wide
Eupoecilia ambiguella	2	0.05 (0.04-0.07)	-	wide
Synanthedon pictipes	3	0.07 (0.01-0.2)	-	narrow
Synanthedon exitiosa	2	0.05 (0.01-0.1)	-	narrow
Pectinophora gossypiella	11	1800 (<0.01-20000)	+	narrow
Sitotroga cerealella	1	30 m^3 (0.2-89 m^3)	-	wide
Ostrinia nubilalis	1	<0.01	-	wide
Chilo suppressalis	2	0.01 (<0.01-0.08)	+	narrow
Ephestia cautella	6	2000 m^3 (0.05-9000 m^3)	-	wide
Plodia interpunctella	3	240 m^3 (0.2-1000 m^3)	-	wide
Lymantria dispar	13	370 (1-6000)	-	wide
Lymantria monacha	2	40 (5-80)	-	wide
Orgyia pseudotsugata	2	2.3 (1-4)	-	wide
Orgia leucostigma	1	0.3	-	wide
Heliothis zea	2	0.06 (0.1-0.2)	+	wide
Heliothis virescens	1	0.02	+	wide
Spodoptera litura	6	0.09 (<0.01-0.5)	?	wide
Spodoptera littoralis	3	6.0 (<0.01-50)	?	wide
Spodoptera exigua	2	0.03 (0.01-0.05)	+	wide
Spodoptera frugiperda	2	0.03 (0.01-0.05)	+	wide
Pseudoplusia includens	1	0.2 (0.09-0.5)	+	wide
Trichoplusia ni	6	57 (0.09-576)	+	wide
Diparopsis castanea	3	0.2 (0.1-0.5)	-	narrow

†Size of trials expressed in cubic meters for grain moths, and hectares for other species.

of both characteristics would appear to limit the prospects for disrupting mating of these insects unless females migrate before they have mated. There is very little information on the mating status of recognized noctuid migrants other than that obtained by Brown et al. (1969) on Spodoptera exempta, which showed that virtually all migrating individuals are unmated. If this also applies to other major noctuid pests, it may well be feasible to attempt mating disruption of even highly mobile species, provided that the areas treated are sufficiently large enough to reduce the impact of trivial flights by mated females. Table 4 shows that most disruption trials with noctuids have been undertaken only on a very small scale.

In the gypsy moth and a number of other Lymantriidae, females are flightless and dispersal arises through the aerial distribution of wind-borne first instar larvae. Recent studies (Cameron et al., 1979) have revealed that most larvae are not carried further than 200 m from their hatching sites, which suggests that dispersal is unlikely to influence the outcome of disruption trials [most of which have been relatively large (Table 4)].

Other Factors Influencing Results of Mating Disruption Trials

Immigration has been selected as one example of a number of possible constraints on mating disruption. Others include: (1) Inadequate release and uneven distribution of behavior-modifying compounds; (2) female pheromone containing components which can be perceived by the male other than those used in the treatments; (3) high adult densities leading to increased opportunities for sexual encounters, particularly in zones where the aerial concentration of behavior-modifying compounds is low -- not only through male activity but as a result of female-initiated sexual encounters (Barrer and Hill, 1980) or females aggregating near sources of synthetic pheromone (Birch, 1977); (4) reductions in the number of fertile eggs resulting from disruption treatments not necessarily leading to equivalent decreases in population density. For example, numbers may be suppressed to a level where density-dependent mortality processes can no longer exert their greatest influence. This could lead to subsequent resurgences stemming from the residual population.

Immigration has been regarded as the most common of these constraints in the field, followed by inadequate release and uneven distribution of behavior modifiers. In grain storage situations, high adult densities and the production of additional olfactory cues (from both males and females) have been suggested as factors limiting the success of attempts at mating disruption (Barrer, personal communication).

PROSPECTS FOR MATING DISRUPTION

Having dwelt on the numerous problems of mating disruption, what are the prospects for practical application of this control technique? Field work to evaluate mating disruption has now been in progress for about 10 years, reaching a maximum level 4 to 5 years ago. As a research field, mating disruption, therefore, is relatively new and it is perhaps not surprising that few programs have progressed beyond the small-scale exploratory phase. As noted earlier, only the pink bollworm project has reached the commercial stage. Despite the misgivings of some authors, the results of this project are encouraging because they demonstrate that mating disruption treatments can be incorporated into a pest management program; they also indicate that it is perhaps possible to disrupt a relatively migratory species if large areas are treated simultaneously. The variability of the results, however, emphasizes the necessity for continuous monitoring in order to determine possible causes of a breakdown in disruption, and to take alternative control measures if necessary. From results so far obtained (Doane and Brooks, 1978), it appears that pheromone treatments against pink bollworm can reduce reliance on pesticides during the earlier part of the season, but cannot provide adequate suppression of numbers later for reasons which are not yet clear. However, the significant reduction in early pesticide treatments against pink bollworm reduces interference with natural enemies of both this and other cotton pests, including *Heliothis*, and thus, there is less likelihood of a resurgence of these insects.

The large gypsy moth project of the 1970s has now been partially wound down. Cameron (1979) considered that useful information had been obtained during this program, but that it had not been conclusively demonstrated that mating disruption treatments had resulted in a significant reduction in the size of ensuing generations. He attributed this situation to an inadequate balance between basic and applied research and the uncoordinated manner in which the overall program operated.

As noted earlier, most other programs have not progressed beyond the exploratory phase, but the results of a number of the orchard trials, in particular, have been sufficiently encouraging to suggest that disruption could be extended to commercial situations. It has been demonstrated that a relatively low rate of release of compounds that mediate only (long-range) behavior can result in a significant reduction of mating. The aerial concentration of behavior-modifying compounds produced in most treatments would probably be far too low to prevent mating under very crowded conditions, but it is most unlikely that adult densities ever reach such levels in the field. A number of trials in orchards and vineyards have confirmed that reductions in mating attributable to the treatments have led to decreases in fruit damage equivalent to those obtained through regular applications of pesticides. However, there have still been no attempts to undertake full-scale disruption trials in these crops for several successive seasons -- which makes discussion of future prospects a difficult task.

At present, there appears to be little likelihood that mating disruption will be adopted in situations where crops are attacked by a complex of pests, particularly where these include insects that are not Lepidoptera and that can currently be controlled only by pesticides. For example, this is the situation in many Asian rice-growing areas where the Asiatic rice borer (*Chilo suppressalis*), one of a complex of borers, causes a consistent low-medium level of damage, the effects of which are masked by the spectacular damage resulting from outbreaks of pests such as plant hoppers. Where the major pests in a complex are Lepidoptera, it may be possible to disrupt mating of several species simultaneously by using a combination of appropriate behavior-modifying compounds. This has been attempted only on a very small scale for orchard tortricids (Roelofs, 1976) and noctuid pests of field crops (Mitchell, 1975). Careful research would be required to determine whether slight variations in the level of disruption of different members of a pest complex would lead to resurgences of those species subjected to the least disruption pressure.

The greatest likelihood of success will be in situations where there is only one important species, and secondary pest problems arise through pesticide treatments directed at the key pest. Codling moth and oriental fruit moth are 2 examples of a key pest, which, at present, can be satisfactorily controlled through the judicious use of pesticides. Mating disruption of these species may only be required as an alternative (or additional) means of control if major problems arise in the existing program. Indeed, this appears to be the present situation for many of the pests that have been the subject of mating disruption studies. Increasing costs of pesticides and pressures from environmental protection agencies and sections of the community, however, are likely to make decreased reliance on pesticides an attractive proposition. Under these circumstances, there may be significantly greater interest in behavior-modifying chemicals, particularly for the control of a number of key pests of horticultural and agricultural crops (the compatibility of mating disruption with other biological control methods as well as with pesticides would be a further advantage in this regard). Pheromone researchers, however, will have to accept that it may be some years before disruption is accepted as a viable control method for any more than 1 or 2 species. Prospects for the control of moths infesting stored grain products through mating disruption would appear to depend on techniques used to control other (often more important) pests including grain beetles. Practices such as thermal disinfestation, controlled atmosphere treatments and fumigation are commonly used to deal with the total pest problem. It is difficult to see how behavior-modifiers could be used in such situations, and their application may only be justified in specific circumstances, for example, in "on farm" storages.

There is a possibility that mating disruption treatments could lead to the development of resistance in target species. Presumably, this could arise where continued disruption with one component of the sex pheromone blend selects for individuals that communicate by means of other components or even non-olfactory cues. There could even be selection for such characteristics as dispersal of mated females or parthenogenesis (known to occur in some Lepidoptera (Wigglesworth, 1967)). However, it would seem to be rather fruitless to speculate on these possibilities until mating disruption has been adopted as a control strategy and used intensively enough for such problems to be investigated.

REFERENCES

Baker, T. C., and Cardé, R. T., 1979, Anlysis of pheromone-mediated behaviors in male Grapholitha molesta, the Oriental fruit moth (Lepidoptera: Tortricidae), Environ. Entomol., 8: 956.

Baker, T. C., Carde, R. T., and Roelofs, W. L., 1976, Behavioral responses of male Argyrotaenia velutinana (Lepidoptera: Tortricidae) to components of its sex pheromone, J. Chem. Ecol., 2: 333.

Barrer, P. M., 1976, The influence of delayed mating on the reproduction of Ephestia cautella (Walker) (Lepidoptera: Phycitidae), J. Stored Prod. Res., 12: 165.

Barrer, P. M., and Hill, R. J., 1980, Insect-oriented locomotor responses by unmated females of Ephestia cautella (Walker) (Lepidoptera: Phycitidae), Int. J. Invert. Reprod., 2: 59.

Bartell, R. J., and Lawrence, L. A., 1973, Reduction in responsiveness of males of Epiphyas postvittana (Lepidoptera) to sex pheromone following previous brief pheromonal exposure, J. Insect Physiol., 19: 845.

Beevor, P. S., and Campion, D. G., 1979, The field use of "inhibitory" components of lepidopterous sex pheromone and pheromone mimics, in: "Chemical Ecology: Odour Communication in Animals," F. J. Ritter, ed., Elsevier/North-Holland Biomedical Press, Amsterdam.

Birch, M. C., 1977, Response of both sexes of Trichoplusia ni (Lepidoptera: Noctuidae) to virgin females and to synthetic pheromone, Ecol. Entomol., 2: 99.

Brown, E. S., Betts, E., and Rainey, R. C., 1969, Seasonal changes in distribution of the African armyworm, Spodoptera exempta (Walk.) (Lep., Noctuidae), with special reference to East Africa, Bull. Entomol. Res., 58: 661.

Cameron, E. A., 1979, Disparlure and its role in gypsy moth population manipulation, Mitt. Schweiz. Entomol. Ges., 52: 333.

Cameron, E. A., McManus, M. L., and Mason, C. J., 1979, Disparlure and its impact on the population dynamics of the gypsy moth in the United States of America, Mitt. Schweiz. Entomol. Ges., 52: 169.

Campion, D. G., 1976, Sex pheromones for the control of lepidopterous pests using microencapsulated and dispenser techniques, Pestic. Science, 71: 636.

Campion, D. G., McVeigh, L. J., Hall, D. R., Lester, R., Nesbitt, B. F., and Marr, G. J., 1976, Communication disruption of adult Egyptian cotton leafworm, Spodoptera littoralis (Boisd.) (Lep., Noctuidae) in Crete using synthetic pheromone applied by microencapsulation and dispenser techniques, Bull. Entomol. Res., 66: 335.

Cardé, A. M., Baker, T. C., and Carde, R. T., 1979, Identification of a four-component sex pheromone of the female Oriental fruit moth, Grapholitha molesta, (Lepidoptera: Tortricidae), J. Chem. Ecol., 5: 423.

Cardé, R. T., Baker, T. C., and Castrovillo, P. J., 1977, Disruption of sexual communication in Laspeyresia pomonella (codling moth), Grapholitha molesta (Oriental fruit moth) and G. prunivora (lesser appleworm) with hollow fiber attractant sources, Entomol. Exp. Appl., 22: 280.

Doane, C. C., and Brooks, T. W., 1978, Suppression of the pink bollworm by disruption of pheromone system with gossyplure H. F., Proc. Int. Cotton Conf., 25 Nov. 1978, San Salvador, pp. 1-14.

Gentry, C. R., Bierl, B. A., and Blythe, J. L., 1976, Air permeation trials with Oriental fruit moth pheromone, in: Proceedings 1976 International Controlled Release Pesticide Symposium, Sept. 13-15, Akron, OH, N. F. Cardarelli, ed., Akron, OH.

Haines, C. P., and Read, J. S., 1977, The effect of synthetic sex pheromones on fertilization in a warehouse population of Ephestia cautella (Walker) (Lepidoptera: Phycitidae), Trop. Prods. Inst. Rept., L. 45.

Hathaway, D. O., McDonough, L. M., George, D. A., and Moffitt, H. R., 1979, Antipheromone of the codling moth: potential for control by air permeation, Environ. Entomol., 8: 318.

Hutt, R. B., 1977, Codling moth: response to synthetic sex pheromone of two laboratory strains released in apples and pears, Environ. Entomol., 6: 666.

Johnson, C. G., 1969, "Migration and Dispersal of Insects by Flight," Methuen & Co., Ltd., London.

Kaae, R. S., and Shorey, H. H., 1973, Sex pheromones of Lepidoptera: 44. Influence of environmental conditions on the location of pheromone communication and mating in Pectinophora gossypiella, Environ. Entomol., 2: 1081.

Kennedy, J. S., 1972, The emergence of behaviour, J. Aust. Entomol. Soc., 11: 168.

Kennedy, J. S., 1977, Olfactory responses to distant plants and other odour sources, in: "Chemical Control of Insect Behavior. Theory and Application," H. H. Shorey and J. J. McKelvey, Jr., eds., John Wiley & Sons, New York.

Klun, J. A., Chapman, O. L., Mattes, K. C., and Beroza, M., 1975, European cornborer and redbanded leafroller: disruption of reproductive behavior, Environ. Entomol., 4: 871.

McLaughlin, J. R., Shorey, H. H., Gaston, L. K., Kaae, R. S., and Stewart, F. D., 1972a, Sex pheromones of Lepidoptera. XXXI. Disruption of sex pheromone communication in Pectinophora gossypiella with hexalure, Environ. Entomol., 1: 645.

McLaughlin, J. R., Gaston, L. K., Shorey, H. H., Hummel, H. E., and Stewart, F. D., 1972b, Sex pheromone of Lepidoptera. XXXIII. Evaluation of the disruptive effect of tetradecyl acetate on sex pheromone communication in Pectinophora gossypiella, J. Econ. Entomol., 65: 1592.

Marks, R. J., 1976a, Field studies with the synthetic sex pheromone and inhibitor of the red bollworm Diparopsis castanea Hmps. (Lepidoptera, Noctuidae) in Malawi, Bull. Entomol. Res., 66: 243.

Marks, R. J., 1976b, The influence of behaviour modifying chemicals on mating success of the red bollworm Diparopsis castanea (Hmps.) (Lepidoptera, Noctuidae) in Malawi, Bull. Entomol. Res., 66: 279.

Minks, A. K., Voerman, S., and Klun, J. A., 1976, Disruption of pheromone communication with micro-encapsulated antipheromones against Adoxophyes orana, Entomol. Exp. Appl., 20: 163.

Mitchell, E. R., 1975, Disruption of pheromonal communication among coexistent pest insects with multichemical formulations, BioScience, 25: 493.

Mitchell, E. R., Jacobson, M., and Baumhover, A. H., 1975, Heliothis spp. disruption of pheromonal communication with (Z)-9-tetradecenyl-ol formate, Environ. Entomol., 4: 577.

Nowosielski, J. W., and Suski, Z. W., 1977, Observations on the mating behaviour of the codling moth Laspeyresia pomonella (L.) II. Temporal pattern of copulatory behaviour in relation to the age of the moths and time of day, Ekol. Pol., 25: 341.

Plimmer, J. R., Caro, J. H., and Freeman, H. P., 1978, Distribution and dissipation of aerially-applied disparlure under a wood-land canopy, J. Econ. Entomol., 71: 155.

Richerson, J. V., 1977, Pheromone-mediated behavior of the gypsy moth, J. Chem. Ecol., 3: 291.

Roehrich, R., Carles, J. P., and Tresor, C., 1977, Essai préliminaiŝe de protection du vignoble contre Lobesian botrana Schiff au moyen de pheromone sexuelle de synthèse (méthode de confusion), Rev. Zool. Agric. Path. Veg., 76: 25.

Roelofs, W. L., 1976, Communication disruption by pheromone components, Proc. Symp. Insect Pheromones and their Application, Nagaoka and Tokyo, Dec. 8-11, 1976, pp. 123-133.

Roelofs, W. L., 1978, Threshold hypothesis for pheromone perception, J. Chem. Ecol., 4: 685.

Rothschild, G. H. L., 1975, Control of Oriental fruit moth (Cydia molesta (Busck) (Lepidoptera: Tortricidae) with synthetic female sex pheromone, Bull. Entomol. Res., 65: 473.

Rothschild, G. H. L., 1979, A comparison of methods of dispensing synthetic sex pheromone for the control of Oriental fruit moth (Cydia molesta (Busck) (Lepidoptera, Tortricidae) in Australia, Bull. Entomol. Res., 69: 115.

Sanders, C. J., 1976, Disruption of sex attraction in the eastern spruce budworm, Environ. Entomol., 5: 868.

Sanders, C. J., 1979, The role of the sex pheromone in the management of spruce budworm, in: "Chemical Ecology: Odour Communication in Animals," F. J. Ritter, ed., Elsevier/North-Holland Biomedical Press, Amsterdam.

Seabrook, W. D., Hirai, K., Shorey, H. H., and Gaston, L. K., 1979, Maturation and senescence of an insect chemosensory response, J. Chem. Ecol., 5: 587.

Shorey, H. H., 1977, Manipulation of insect pests of agricultural crops, in: "Chemical Control of Insect Behavior: Theory and Application," H. H. Shorey and J. J. McKelvey, Jr., eds., John Wiley & Sons, New York.

Shorey, H. H., Gaston, L. K., and Saario, C. A., 1967, Sex pheromones of noctuid moths. XIV. Feasibility of behavioral control by disrupting pheromone communication in cabbage loopers, J. Econ. Entomol., 60: 1541.

Shorey, H. H., McFarland, S. U., and Gaston, L. K., 1968, Sex pheromones of noctuid moths. XIII. Changes in pheromone quantity, as related to reproductive age and mating history, in females of seven species of Noctuidae (Lepidoptera), Ann. Entomol. Soc. Am., 61: 372.

Shorey, H. H., Gaston, L. K., and Kaae, R. S., 1976, Air-permeation with gossyplure for control of the pink bollworm, in: "Pest Management with Insect Attractants," M. Beroza, ed., Am. Chem. Soc. Symp. Ser. 23, Washington, DC.

Sower, L. L., and Daterman, G. E., 1977, Evaluation of synthetic sex pheromone as a control agent for Douglas-fir tussock moths, Environ. Entomol. 6: 889.

Sower, L. L., Vick, K. W., and Ball, K. A., 1974, Perception of olfactory stimuli that inhibit the responses of male phycitid moths to sex pheromones, Environ. Entomol., 3: 277.

Sower, L. L., Turner, W. K., and Fish, J. C., 1975, Population-density dependent mating frequency among Plodia interpunctella (Lepidoptera: Phycitidae) in the presence of synthetic sex pheromone with behavioral observations, J. Chem. Ecol., 1: 335.

Sower, L. L., Daterman, G. E., Orchard, R. D., and Sartwell, C., 1979, Reduction of Douglas-fir tussock moth reproduction with synthetic sex pheromone, J. Econ. Entomol., 72: 739.

Stern, V. M., 1979, Long and short range dispersal of the pink bollworm, Pectinophora gossypiella, over southern California, Environ. Entomol., 8: 524.

Symmons, P. M., and Rosenberg, L. J., 1978, A model stimulating mating behavior of Spodoptera littoralis, J. Appl. Ecol., 15: 423.

Taschenberg, E. F., and Roelofs, W. L., 1978, Male redbanded leafroller orientation disruption in vineyards, Environ. Entomol., 7: 103.

Taschenberg, E. F., Cardé, R. T., and Roelofs, W. L., 1974, Sex pheromone mass trapping and mating disruption for control of redbanded leafrollers and grape berry moths in vineyards, Environ. Entomol., 3: 239.

Trammel, K., Roelofs, W. L., and Glass, E. H., 1974, Sex-pheromone trapping of males for control of the redbanded leafroller in apple orchards, J. Econ. Entomol., 67: 159.

Traynier, R. M. M., 1970, Habituation of the response to sex pheromone in two species of Lepidoptera, with reference to a method of control, Entomol. Exp. Appl., 13: 179.

Wigglesworth, V. B., 1967, The Principles of Insect Physiology, Methuen & Co. Ltd., London.

SMALL-PLOT DISORIENTATION TESTS FOR SCREENING POTENTIAL MATING DISRUPTANTS

W. L. Roelofs

M. A. Novak

Department of Entomology
NY State Agricultural
Experiment Station
Geneva, NY 14456 USA

INTRODUCTION

The manipulation of insect chemical communication systems was promoted to be the savior of insect control programs. Many people have been disappointed because this has not happened. Some reasons for this frustration are that people fail to realize that this is not a technique that is applicable to all insect pest species, that the technique is not applicable to all field or crop situations, and that it is not possible to develop the use of this technique for any species without much knowledge of the insect's biology, behavior, communication system, and its interaction with the environment and economics in any particular control situation. In some cases the cost and effort of registration has been a hindrance to further developmental research, but one of the largest problems from a commercial viewpoint is with the demonstration of efficacy. An AIBS-EPA Task Group reviewed (Roelofs, 1979) various test methods for determining the efficacy of pheromones as insect control agents, but found very few tests that actually proved efficacy. There are many reasons for this, including problems with formulations, finances, finding appropriate isolated test sites and insect populations, as well as with pressures to conduct large field trials with an insufficient knowledge of the insect's behavior. This paper will focus only on one aspect of the problems encountered in developing the mating-disruption technique for insect control. The discussion will center on our efforts to set up small-plot field tests for screening possible disruptants prior to their use in larger field trials.

SMALL-PLOT DISORIENTATION TEST DESIGN

There is a complex of tortricid leafroller moths that includes some potentially serious pest species in New York State apple orchards. The sex pheromones for most of these pest species appear to consist of 2 to 4 components. The principal compounds in these are (Z)- and (E)-11-tetradecenyl acetates (Z and E11-14:Ac) and the corresponding alcohols (Z and E11-14:OH). One aspect of our research program was to develop the use of these 4 components for mating disruption of the whole leafroller complex on apple (Roelofs, 1976; Roelofs et al., 1976). After initiating this project it became obvious that many questions had to be answered before attempting a large field test and that some standard method of testing had to be established in small-plot tests. Although mating disruption could not be tested in these plots, it was assumed that a determination of % male disorientation to a pheromone trap for various materials at a number of release rates would provide data needed to compare potential disruptants at a preliminary level.

We initially conducted tests for disorientation of male redbanded leafroller moths, *Argyrotaenia velutinana* (Walker) in which the pheromone chemicals were evaporated from planchets on an 8 x 13-m spacing in vineyards (Taschenberg et al., 1974) or were sprayed out in vineyards and apple orchards in a microencapsulated formulation (Cardé et al., 1975; Roelofs et al., 1976). Tests with hollow fibers (Conrel Co.) (Taschenberg and Roelofs, 1976), however, indicated that this formulation would be better for setting up small plots at various relative release rates.

Release Rates

The release rate for fiber strips filled with the test material was determined in the laboratory by collection of the volatilized material on Porapak-Q® or provided by the Conrel Co. The release rate measurements were used to provide an estimate of the number of fibers to be used per unit area to obtain a specified release rate/h/ha. The actual release rate in the field would depend on the various environmental influences, but the calculated release rates would at least set some approximate value for comparative purposes.

Plot Size

The optimum size for test plots is determined by the availability of insect populations and finances, as well as by requirements of the experiment. Plots for % disorientation to a pheromone trap must be large enough so that male moths flying into the area are sufficiently exposed to the test material before encountering the pheromone trap in the center of the plot. The plots must be small enough to allow for adequate replication and distribution of a variety of treatments within normal blocks of commercial apple orchards or vineyards.

We chose 25 x 30 m as the size for our plots, with a minimum of 100-150 m between plots.

Release Stations

There are many systems that could be used in distributing the formulated material in a test plot. Mechanisms by which mating disruption is effected are still not understood and could involve confusion by non-biological sources, sensory adaptation and/or central nervous system habituation. Therefore, it also is not known for most species if they are disrupted best by a few high releasing stations or by a large number of point sources, each releasing a small amount of disruptant.

Our standard format was to divide each plot into 15 equal subplots of 50 m^2 and then to release test material from 3 stations in each subplot, except the center one (Fig. 1). The fibers in the center subplot were released from the outside edge of that subplot so that there were no release stations immediately adjacent to the trap. The area within a plot is 3-dimensional, but for the purposes of small-plot tests, the area was visualized as a 2-dimensional plane at a height of ca. 2 m. The fibers were hung at this height and the pheromone trap in the center subplot was hung slightly lower. The Conrel® fibers, which were used in 50-fiber strips with one end cut open, were attached to the branches of the trees by means of bulldog clips (Fig. 2). A pheromone trap was hung from a branch in the calculated center of the plot.

A release rate per fiber as determined in the laboratory was used to calculate the total number of fibers required in each plot. If a fiber released material at a rate of 7×10^{-5} mg/h in laboratory tests, and if materials were to be compared at an approximate rate of 5 mg/h/ha, then a total of 5,250 fibers would be needed for each test plot. The fibers would be evenly distributed among the 3 release stations in each of the 50 m^2 subplots. If another treatment were to involve this material at 10 mg/h/ha, then the number of fibers at each release station could be doubled.

Replicates and Check Plots

Disorientation in the small-plot tests is determined by a comparison of trap catch in treated and check plots. The % disorientation is calculated by the following formula:

$$\frac{\text{Catch in check plot - catch in test plot}}{\text{Catch in check plot}} \times 100$$

It is important in these tests to have the treated and check plots in areas of similar population densities. However, trap catch among areas within an orchard can be quite variable due to local "hot spots" and to "edge effects" generated by pest reservoirs located outside of the test orchard.

In an attempt to overcome the problems of a comparable check plot, we paired each treated plot with a separate check plot. Additionally, we established a protocol in which all plots were marked in advance and a pheromone trap placed in the center of each plot. After the initiation of moth flight, the traps were checked and plots were paired on the basis of similar trap catches, rather than strictly on their juxtaposition in the orchard. Treatments then were randomly distributed to one of the plots in each pair.

Studies involving several different test plots and their corresponding check plots consume a large area in the unsprayed orchards. We attempt to set up 3 replicates of all plots, but sometimes the pest population availability does not allow this. In some tests, a large number of fibers were involved to obtain high release rates. We were unable to set up more than 2 replicates, and so replicates were obtained by moving all test materials to new locations several times during the flight. Each time the new pair of treated and check plots would be established by monitoring trap catch in all plots prior to their use.

The statistical analysis of disorientation data presents a number of problems, but the paired treated and check plot arrangement is set up for using an analysis of covariance, which appears to be an excellent method for analyzing these types of data.

SMALL-PLOT TESTS WITH REDBANDED LEAFROLLER MOTHS

Three types of potential disruptants have been tested in our standard small plots with the redbanded leafroller moth (RBLR) (Fig. 3). These included (a) the pheromone component Z11-14:Ac alone and in varying mixtures with the other primary pheromone component, E11-14:Ac; (b) a parapheromone, Z11-13:Ac; (c) several compounds that decrease trap catch when added to the pheromone blend.

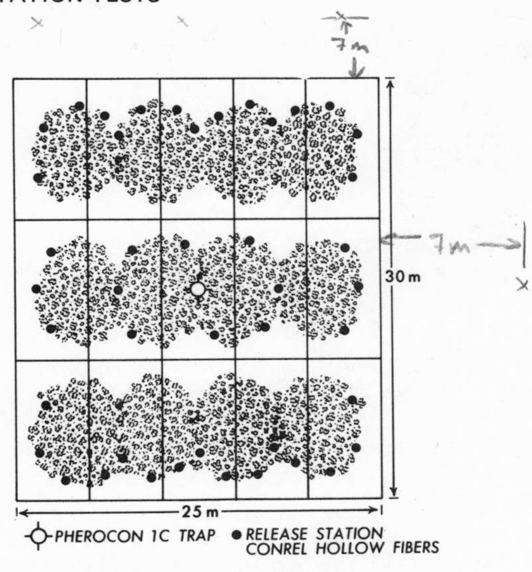

Fig. 1. A schematic disruption plot in which the fibers are distributed evenly among the release stations in the 15 subplots to obtain an estimated mg/h/ha release rate.

Fig. 2 A bulldog clip hung 2 m from the ground in an apple tree and holding the fiber strips allotted to that particular release station.

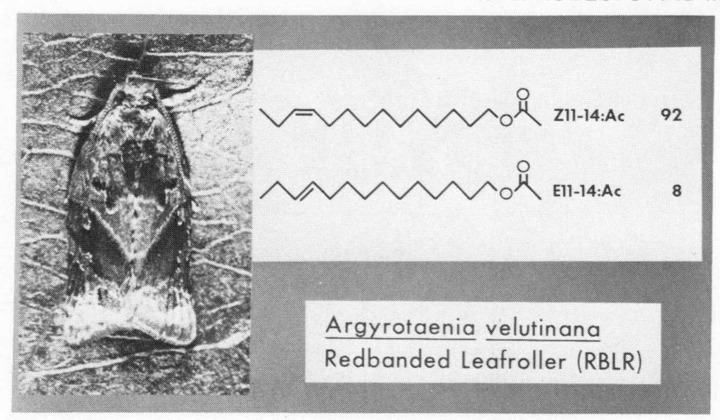

Fig. 3. Primary pheromone components identified for this species. Excess dodecenyl acetate is a secondary component.

In 1977, tests comparing Z11-14:Ac to several analogs that are effective in decreasing pheromone trap catch showed that at release rates of 5 and 10 mg/h/ha, Z11-14:Ac-treated plots exhibited 100% disorientation, whereas, the corresponding alcohol, Z11-14:OH, produced an average of 61% disorientation per plot at 10 mg/h/ha, and the corresponding formate, Z11-14:FOR, produced an average disorientation per plot of 47% at 5 mg/h/ha (Reissig et al., 1978). It could be concluded from those data that the alcohol and formate alone would not be good disruptants to use in larger test plots.

In 1978 another analog of the pheromone component was tested (Novak et al., 1979). It had been found previously that Z11-13:Ac was a parapheromone and could replace both Z11-14:Ac and dodecyl acetate in the pheromone blend used in traps (Cardé and Roelofs, 1977). In small-plot disruption tests it was found that the 13-carbon compound was as effective as the pheromone component at 3 mg/h/ha (95% and 99% disorientation, respectively). Lower release rates were not tested to determine which compound is best at the lowest rate.

In 1979 small-plot tests were conducted to determine if slight changes in the pheromone component blend (Z/E11-14:Ac in a 92:8 ratio) would give differences in disruption tests. The tests were conducted in vineyards with the same 30 x 25-m plot format as used in apple orchards. Due to the large number of treatments, however, only 2 replicates could be set up. They were re-randomized once between flights in the period from 7/9-9/27/79. This experiment was conducted in conjunction with research on the threshold hypothesis (Roelofs, 1978). The hypothesis is concerned mainly with the activation threshold at which a moth initiates a behavioral response, such as upwind flight, and the disorientation or arrestment threshold at which the moth terminates a behavior, such as upwind flight (Fig. 4). It is not known how behaviors involved in air permeation tests relate to those involved in orienting to an odor source, but there was interest in comparing results form small-plot disorientation tests to some of the proposed arrestment threshold patterns. For example, if air permeation disruption were directly related to the arrestment threshold shown in Figure 4, then the natural blend (92:8) should be disruptive at a lower release rate than pure Z11-14:Ac, or an 80:20 ratio. However, the relationship between these 2 phenomena may not be a direct one.

The design of the experiment was factorial with 4 blends of Z11-14:Ac and E11-14:Ac (100:0; 90:10; 80:20; 70:30) and 3 release rates of each blend (0.3, 1.0 and 3.0 mg/h/ha). Because these treatments are actually part of a continuum of possible blends and release rates, it is more appropriate to fit the data to a functional form than to perform a multiple comparisons analysis of variance. A regression analysis involving 2 factors (rates and blends) was used with the data transformed by $\sin^{-1}\sqrt{X}$ (angular or arcsin transformation). The results (Fig. 5) show strong linear effects for both rates (significant at $P \leq 0.002$) and blends (significant at $P \leq 0.003$). The significant linear effect for rates indicate that for each blend the disorientation increases as the release rate increases. The significant linear effect for blend indicates that for each release rate there is a change in % disorientation with a change in blend. In this test, the % disorientation increased with blends containing smaller quantities of E11-14:Ac. This suggests that of the materials tested the Z11-14:Ac alone treatment was the best disruptant. No higher order effects and no rate X blend interactions were present.

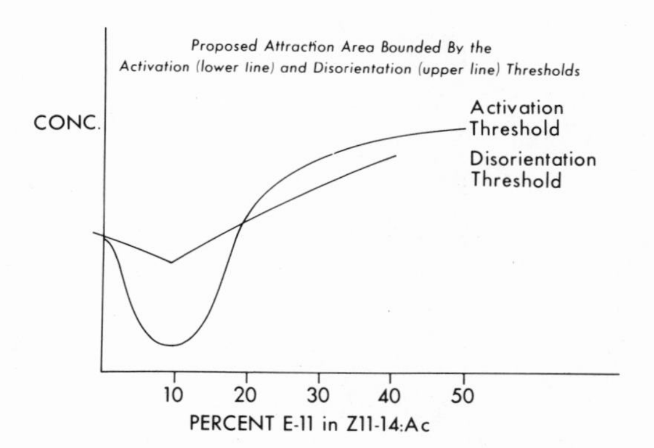

Fig. 4. A hypothetical depiction of thresholds that form the upper and lower boundaries for attraction of redbanded leafroller moths with various pheromone component ratios and release rates. (From Roelofs, 1978.)

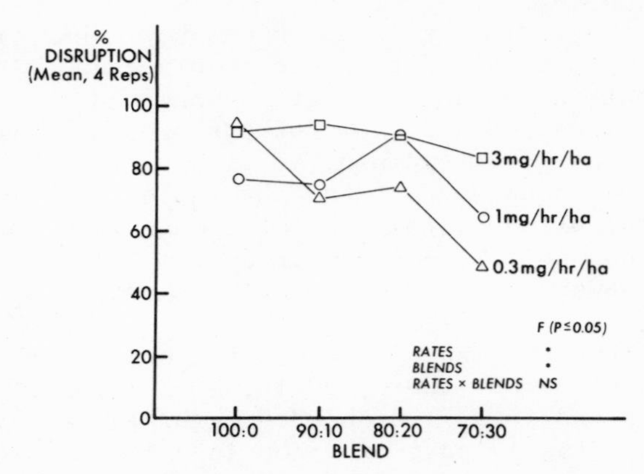

Fig. 5. Data from a small-plot disruption test with 4 blends of pheromone components at 3 release rates each. Test was conducted in vineyards in 2 flights (7/9-9/27/79). Data subjected to regression analysis (rates/blends) and transformed by $\sin^{-1}\sqrt{X}$.

The data could also be subjected to a multiple regression analysis, because there were no higher order effects indicating a non-linear relationship between release rates and blends. Using the experimental means of the 12 treatments, the equation of a plane was calculated, with the axes as blend, release rate, and % disruption. Line equations could then be calculated to predict release rates needed to achieve various levels of disruption with the different blends (Fig. 6). For example, a release rate of less than 1 mg/h/ha of Z11-14:Ac could result in 90% disorientation, whereas, it would take a release rate of at least 5 mg/h/ha of the 70:30 blend to achieve 90% disorientation.

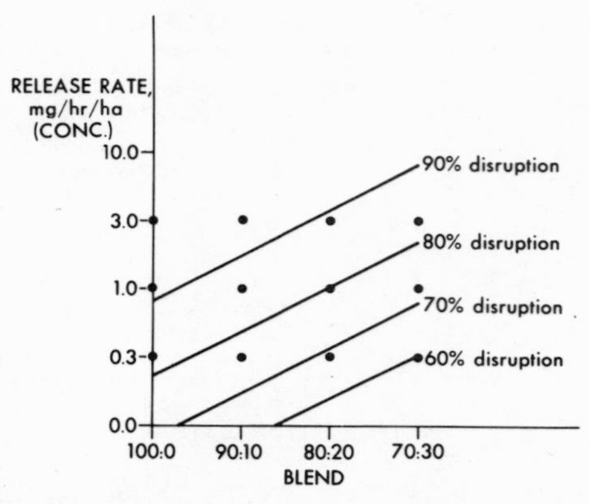

Fig. 6. Line equations predicting release rates needed to achieve various levels of disruption with different blends. (Data taken from Figure 5.)

SMALL-PLOT TESTS WITH OBLIQUEBANDED LEAFROLLER MOTHS

The obliquebanded leafroller moth (OBLR), Choristoneura rosaceana, uses a pheromone blend (Hill and Roelofs, 1979) consisting of Z/E11-14:Ac in a 95:5 ratio and ca. 5% Z11-14:OH (Fig. 7). There is not much difference in the acetate ratios used by the obliquebanded and redbanded leafrollers, and so it was surprising to find that Z11-14:Ac was not a good disruptant for obliquebanded leafroller moths in small-plot tests. At the 5 mg/h/ha level, the Z11-acetate, corresponding alcohol and corresponding formate only gave 64, 63, and 54% disorientation in 1977 (Reissig et al., 1978). In 1978, the Z11-14:Ac treatment at 5 mg/h/ha showed only 73% disorientation and the Z11-13:Ac gave an extremely poor 35% disorientation. The latter compound is not a parapheromone for obliquebanded leafroller moths as it is for redbanded leafroller moths, and so it may not have as much input to the obliquebanded leafroller moth central nervous system.

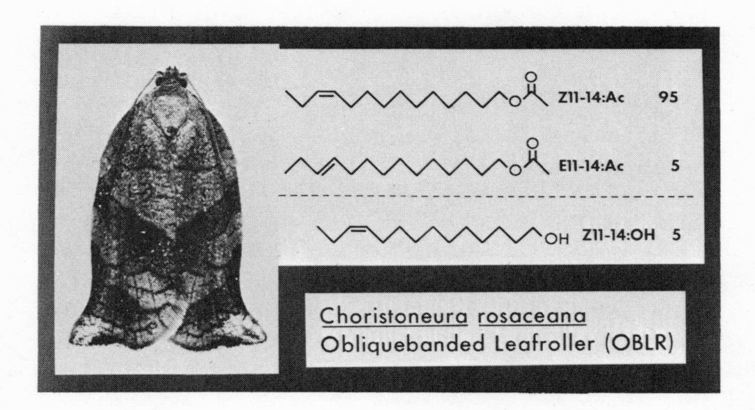

Fig. 7. An obliquebanded leafroller moth and the pheromone components identified for this species.

In another test (Table 1), it was found that neither the acetate pheromone components alone nor combinations with the corresponding alcohols provided good orientation activity with obliquebanded leafroller moths. We then tried tests with mixtures that are closer to the natural pheromone component ratios.

Table 1. Small-plot disorientation tests for oblique-banded leafroller males, 8/23-9/25/78. No significant differences among treatments. Six replicates per treatment.

Treatment 5 mg/h/ha each compound	Mean % disorientation[1]
1) Z11-14:Ac	66 b
2) E11-14:Ac	60 b
3) Z11-14:Ac + Z11-14:OH	84 a
4) E11-14:Ac + Z11-14:OH	53 b
5) E11-14:Ac + Z11-14:OH + E11-14:OH	70 ab

[1]Mean % disorientation determined from season totals of each replicate; analysis of covariance based on daily catches; data transformed by $\sqrt{x + 0.5}$; Duncan's new multiple range test; means followed by same letters do not differ at the 5% level.

Table 2. Small-plot disorientation tests for oblique-banded leafroller moths.

Compounds	% Disorientation
Test A (6/16-7/8/79)[1,2]	
Z-11-14:Ac	
5 mg/h/ha	53 b
10 "	41 b
20 "	54 ab
Z/E11-14:Ac + Z11-14:OH (90:5:5)	
5 mg/h/ha	77 a
10 "	82 a
20 "	94 a
Test B (8/16-10/8/79)[1,3]	
Blend (10 mg/h/ha total release)	
Z/E-14:Ac + Z11-14:OH	
(90:5:5)	92 a
" (1:1:1)	69 b
Z11-14:Ac + Z11-14:OH (1:1)	69 b
Z/E11-14:Ac + Z/E11-14:OH (1:1:1:1)	54 c
Z11-14:Ac + E11-14:Ac (1:1)	53 c

[1]Mean % disorientation determined from season totals of each replicate; analysis of covariance based on daily catches; data transformed by $\sqrt{x + 0.5}$; Duncan's new multiple range test; means followed by same letters within each test do not differ at the 5% level.

[2]Four replicates for each treatment; 2-4 trap observations over flight period.

[3]Three replicates for each treatment; 12 trap observations over flight period.

Natural Blend as a Disruptant

In 1979 2 sets of tests were run which compared the action of the natural pheromone blend for OBLR to Z11-14:Ac alone and to various combinations of the 3 pheromone components. Previous data from small-plot disorientation tests have been analyzed by means of an analysis of variance on the calculated % disorientation and a multiple comparisons procedure, such as Duncan's new multiple range test (Reissing et al., 1978). However, a more appropriate analysis for our small-plot design is an analysis of covariance based on a comparison of test and check plot trap catches, rather than on a calculated % disorientation. The data were transformed by $\sqrt{x + 0.5}$ and means were separated by using Waller and Duncan's Multiple Comparisons Procedure.

The first field test compared Z11-14:Ac along with the natural blend of 90:5:5 of Z11-14:Ac/E11-14:Ac/Z11-14:OH at release rates of 5, 10, 20 mg/h/ha. The results (Table 2, test A) show that the natural blend is a better disruptant than Z11-14Ac.

In the next test (Table 2, test B), the 3-component natural blend (90:5:5) was compared to the same 3 components in a 1:1:1 ratio at a release rate of 10 mg/h/ha. Included in the test were treatments consisting of a 1:1 ratio of the 2 acetates, a 1:1 ratio of the Z-acetate and Z-alcohol, and a 1:1:1 ratio of both acetates and both alcohols. Only 3 replicates could be set up, but these data show that the natural blend was the best disruptant in this small-plot test.

CONCLUSIONS

A standard format for small-plot tests has allowed us to compare various possible disruptant materials in a screening program prior to large-field projects. The small-plot tests do not test for mating disruption, but indicate chemicals, blends, and concentrations that are ineffective at the first-level of testing. Only materials giving 95% disorientation or better would be considered for further testing. The small-plot disorientation tests described in this paper involve the following general format: a) all 25 x 30-m plots to be used are marked and monitored with a pheromone trap prior to the test; b) pairs of plots are determined according to similar trap catch; c) treatments are randomly distributed to one plot of each pair; d) trap catch data from the treated plot and check plot are subjected to an analysis of covariance to determine significant differences among treatments. The standard format also can be used for testing various formulations and different spacings among the release points.

The studies conducted to date show that at this first level of testing, the redbanded and obliquebanded leafrollers react differently to the various disruptants, even though their pheromone blends are quite similar. The redbanded leafroller moth is easily disoriented by the main pheromone component, Z11-14:Ac, whereas, the obliquebanded leafroller has been disoriented in small-plot tests only by the 90:5:5 natural blend of pheromone components. Various analogs that act as trap catch decreasers have been found to be poor disruptants so far in the small-plot tests. Additional testing will be conducted in attempts to find compounds that are as good or better than the pheromone components. The small-plot tests are extremely useful for screening the new materials when insect populations and finances are limited.

ACKNOWLEDGEMENTS

We thank Dr. E. F. Taschenberg for his generous assistance in conducting small-plot tests in vineyards, and Drs. J. Barnard and L. Bjostad for conducting and discussing the statistical analyses.

REFERENCES

Cardé, R. T., and Roelofs, W. L., 1977, Attraction of redbanded leafroller moths, Argyrotaenia velutinana, to blends of (Z)- and (E)-11-tridecenyl acetates, J. Chem. Ecol., 3: 143.

Cardé, R. T., Trammel, K., and Roelofs, W. L., 1975, Disruption of sex attraction of the redbanded leafroller (Argyrotaenia velutinana) with microencapsulated pheromone components, Environ. Entomol., 4: 448.

Hill, A. S., and Roelofs, W. L., 1979, Sex pheromone of the obliquebanded leafroller, J. Chem. Ecol., 5: 1.

Novak, M. A., Reissig, W. H., and Roelofs, W. L., 1978, Orientation disruption of Argyrotaenia velutinana and Choristoneura rosaceana (Lepidoptera: Tortricidae) male moths, J. N.Y. Entomol. Soc., 4: 311.

Reissig, W. H., Novak, M., and Roelofs, W. L., 1978, Orientation disruption of Argyrotaenia velutinana and Choristoneura rosaceana male moths, Environ. Entomol., 7: 631.

Roelofs, W. L., 1976, Communication disruption of pheromone components, in "Insect Pheromones and Their Applications," Proc. Symp. Nagaoka and Tokyo, Japan, Dec. 8-11, pp. 123.

Roelofs, W. L., 1978, Threshold hypothesis for pheromone perception, J. Chem. Ecol., 4: 685.

Roelofs, W. L., 1979, "Establishing Efficacy of Sex Attractants and Disruptants for Insect Control," Entomol. Soc. Am., 97 pp.

Roelofs, W., Cardé, R., Taschenberg, E., and Weires, R., Jr., 1976, Pheromone research for control of lepidopterous species in New York, Am. Chem. Soc. Symp. Ser. 23: 75.

Taschenberg, E. F., Cardé, R. T., and Roelofs, W. L., 1974, Sex pheromone mass trapping and mating disruption for control of redbanded leafroller and grape berry moths in vineyards, Environ. Entomol., 3: 239.

Taschenberg, E. F., and Roelofs, W. L., 1976, Pheromone communication disruption of the grape berry moth with microencapsulated and hollow fiber systems, Environ. Entomol., 5: 688.

FIELD AND LABORATORY EVALUATION OF MATING DISRUPTANTS OF *HELIOTHIS ZEA* AND *SPODOPTERA FRUGIPERDA* IN FLORIDA

John R. McLaughlin

Everett R. Mitchell

John H. Cross

Insect Attractants, Behavior, and
Basic Biology Research Laboratory
AR-SEA, USDA
P. O. Box 14565
Gainesville, FL 32604 USA

INTRODUCTION

The corn earworm (cotton bollworm, tomato fruit worm), *Heliothis zea* (Boddie), is one of the most destructive pests of many valuable crops in the United States and in Central and South America. Estimated losses within the United States attributable to this pest exceed $500 million annually (unofficial estimates of USDA entomologists per authors' survey).

The fall armyworm (maize whorlworm), *Spodoptera frugiperda* (J. E. Smith), is a subtropical insect of the Western Hemisphere that lacks any diapause mechanism. It is a sporadic pest that overwinters in the United States in Florida and a few southern coastal areas. Each year it spreads to infest crops as far north as Canada and west to Montana. Corn, peanuts, sorghum, and Bermudagrass are the primary economic hosts. During the particularly severe outbreaks in 1975, 1976, and 1977, losses in the southeastern United States were estimated at from $30 million to several $100 million annually (Hunt, 1978).

These insects are the major pests of corn throughout the United States where approximately 250,000 ha of sweet corn, worth $158 million, are produced each year. Florida growers annually produce

20,000 ha of sweet corn worth $40-50 million (Koehn, 1978). The corn earworm is the primary pest in early season and both pests restrict production in mid- and late summer. As a result Florida's farmers make 14-28 applications of insecticides each season, at a cost of over $6.5 million (based on data from Brooke, 1976). The increasing costs of such control, the very real threat of the development of resistance to the materials being used, the environmental risks, and the rapidly diminishing list of insecticides that are legally available have created a need to develop alternative methods of control.

The air permeation technique for disrupting pheromonal mating communication is being explored at the Insect Attractants Laboratory as a control for these 2 pests. This program has been underway for about 7 years as 3 sub-projects--identification and selection of disruptants and dosages, investigation and selection of formulation and delivery systems, and development of pest monitoring and pest population-host relationships. We summarize here some of the research dealing with selection of disruptants and dosages, and formulation and delivery systems.

DISRUPTANT CHEMICALS

The feasibility of using the air permeation technique for mating control of the corn earworm and fall armyworm was demonstrated by Mitchell et al. (1974, 1975, 1976) who used (*Z*)-9-tetradecen-1-ol formate (Z9:14F) or (*Z*)-11-hexadecenal (Z11:16Al) against the corn earworm and (*Z*,*E*)-9,12-tetradecadien-1-ol acetate (Z9,E12:14Ac) against the fall armyworm. [Z9:14F is not a sex pheromone component but can substitute for Z11:16Al in attractant mixtures for *Heliothis virescens* (F.) (Mitchell et al., 1978); Z9,E12:14Ac is a part of the sex pheromone of several *Spodoptera* spp., but it does not attract fall armyworm males (Mitchell and Doolittle, 1976).] These materials were tested concurrently for their impact on the mating of tethered or clipped-wing virgin females in the field and on male captures in female-baited traps. The reductions in mating and in trap captures were in close agreement (Mitchell et al., 1976).

Corn Earworm

The 4 chemicals, Z11:16Al, Z9:16Al, Z7:16Al and hexadecanal (16Al) are present in a ratio of 87:3:2:8 in extracts of the ovipositors of female corn earworms and presumed to be components of the sex pheromone (Klun et al., 1979, 1980). The Z11:16Al and Z9:16Al are essential to attraction of corn earworm males; the role of the 2 other chemicals is not clear (Klun et al., 1980). Because Roelofs et al. (1974) and Sekul et al. (1975) had earlier identified only Z11:16Al from the corn earworm, the reports by Klun et al. (1979, 1980) prompted us to investigate the disruptive potential of the 3 additional aldehydes and to compare the components with Z9:14F.

Small (100 m^2 or 300 m^2) plots 45 m apart were established the summer of 1979 in a peanut field adjacent to ca. 4 ha of field corn. A cone trap (Hartstack et al., 1979) baited with a 6.45 cm^2 Hercon® bait containing the 4-component blend was placed at plant height in the center of each plot. The test materials were evaporated from 3.2-cm^2 Hercon plastic laminates attached to wooden stakes. When 2 chemicals were investigated, they were evaporated at ca. equal rates (same area of laminate), with the exception that the 4-component blend was formulated in the naturally-occurring ratio (Klun et al., 1980). A treatment consisted of 16 evaporators spaced at equal intervals throughout a plot. Two plots were established for each chemical or chemical combination using a randomized complete block design. The treatments were moved each day to the adjacent plot within a block to reduce location effects. Each test was conducted for 5 or 6 nights.

The results are summarized in Table 1. The major ovipositor component, Z11:16Al, was an effective communication disruptant when used alone. Its activity did not appear to increase with the addition of the other pheromonal components. Neither Z9:16Al, which is necessary for trap capture, or Z7:16Al was as effective a disruptant as Z11:16Al. The effect of Z9:14F was not as great as that of Z11:16Al or of the 2 pheromonal combinations used in test 3.

The capture of males in hexadecanal-permeated plots was significantly greater ($p \leq 0.01$) than in control plots (test 1). Preliminary observations from another experiment (unpublished) indicated that a trap in a hexadecanal permeated plot did not attract more males than a trap in a non-permeated area but that more males that approached the trap entered it.

We therefore concluded that the disruptant of choice for the corn earworm is Z11:16Al or a mixture of Z11:16Al-Z9:16Al. However, aldehydes are not easily utilized in large field tests. They are quite labile when exposed to the air and sunlight and we have had poor results from hollow fiber formulations of them. Hollow fibers have performed well in dispensing the formate (Mitchell, Tingle, and McLaughlin, unpublished observations).

In August 1979 a test was conducted to evaluate Hercon flake formulations of potential corn earworm disruptants dispensed by aircraft with equipment designed by Thoughts Development Corp., Sun City, AZ. The materials were applied at a rate of 12.3 g of active chemical per ha to 0.81-ha plots of commercial sweet corn arranged in a randomized complete block design with 4 blocks. Treated plots were separated by at least 137 meters. The 4 treatments were Z9:14F, Z11:16Al, a 1:1 mixture of Z11:16Al and Z9:16Al, and a 16:1 mixture of Z11:16Al and Z9:14Al (virelure). A cone trap at the center of 2 plots of each treatment was baited with 3 virgin female corn earworms and the traps in the other 2 plots were baited with synthetic corn earworm sex pheromone.

Table 1. Mean (± SE) captures of male corn earworms in traps baited with synthetic pheromone and placed in plots permeated with sex pheromone components.

Permeating chemical	Plot size (m^2)	
	100	300
	Test 1 (July 16-20)	
Control	6.4 ± 3.0	--- - ---
Z11:16Al	0.3 ± 0.1	--- - ---
Z9:16Al	3.7 ± 0.9	--- - ---
Z7:16Al	4.5 ± 4.8	--- - ---
16Al	15.7 ± 4.8	--- - ---
	Test 2 (August 1-6)	
Control	43.8 ± 15.6	--- - ---
Z11:16Al	6.6 ± 3.5	--- - ---
Z11: & Z9:16Al	9.0 ± 5.4	--- - ---
Z11: & Z7:16Al	9.6 ± 5.8	--- - ---
	Test 3	
	(July 27-31)	(August 9-13)
Control	13.3 ± 2.5	37.5 ± 8.5
Z11:16Al	1.2 ± 0.3	5.4 ± 1.8
Z11: & Z9:16Al	0.8 ± 0.3	2.7 ± 1.5
CEW-4 component	0.8 ± 0.4	3.0 ± 0.8
Z9:14F	2.1 ± 2.5	12.7 ± 3.5

The results of tests with both types of trap bait were statistically identical and are summarized in Table 2. Despite the wide separation of plots, there was a good deal of overspray into control plots, and some cross-contamination possibly occurred in adjacent field treatments. The Z9:14F was as good a communication disruptant as the *Heliothis* aldehydes when used in an aerially broadcast formulation.

Table 2. Captures of male corn earworm in pheromone traps placed in 0.81-ha plots of sweet corn treated by air with mating disruptants.

Chemical	Night after day of treatment of 1st moth capture	$\bar{X}$/trap/night (4 plots - 14 nights)
Z11:16Al & Z9:16Al	3	2.8
Z11:16Al & Z9:14Al	2	0.8
Z9:14F	14	0.4
Z11:16Al	11	1.1
Control	1	4.7

Fall Armyworm

The complete sex pheromone of the fall armyworm is not yet identified. Sekul and Sparks (1967) reported (Z)-9-tetradecen-1-ol acetate (Z9:14Ac) in extracts of female abdominal tips, but this chemical is not an effective trap bait in the field. Subsequently, Mitchell et al. (1975) and Mitchell (1975) reported that Z9:14Ac is a mating disruptant of the fall armyworm (female-baited traps). Also, Mitchell et al. (1974) determined that Z9:12Ac is a field attractant of fall armyworm males and a disruptant of mating communication (female-baited traps). Jones and Sparks (1979) reported that Z9:12Ac and Z9:14Ac in mixtures of 100:2 or 100:10 were more attractive to males than Z9:12Ac alone. Mitchell (1974) determined that Z7:12Ac and Z9,E12:14Ac each inhibit male response to Z9:12Ac and that Z9:12Ac-baited traps in Z7:12Ac-permeated plots capture more males than traps in control plots.

Thus, there appear to be 3 candidate fall armyworm disruptants, Z9:14Ac, Z9:12Ac, and Z9,E12:14Ac. However, Z9,E12:14Ac has not been readily available in quantities necessary for large-scale field tests and it is somewhat unstable in the field.

An experiment, identical in design to that of the corn earworm aerial test, but with only 2 blocks, was conducted in August 1979 to evaluate Z9:12Ac and Z9:14Ac as fall armyworm disruptants. The applications consisted of 7.0 g of active material/ha in Conrel® fibers. Disruption was measured with female-baited traps. The results are summarized in Table 3. The Z9:14Ac was effective in reducing captures of males, and the Z9:12Ac was not at all effective. Males were captured each night in the Z9:14Ac-permeated plots indicating that the application rate (or rate of evaporation from the

Table 3. Captures of male fall armyworm in female-baited traps placed in 0.81-ha plots of sweet corn treated by air with mating disruptants.

Chemical	$\bar{X}$/trap/night (2 plots - 9 nights)
Control	21.1 ± 4.3
Z9:12Ac	20.8 ± 4.6
Z9:14Ac	3.2 ± 1.4

Table 4. Rate of evaporation[a] of Z9:14F and Z9:14Ac from commercial formulations.

Age of formulation (days)	Evaporation rate (mg/h)/g of formulation	
	Z9:14F	Z9:14Ac
	Hercon (0.05 g/cm^2 of laminate)	
0.8	17.1	3.9
3.0	27.0	6.9
6.7	21.3	6.6
8.9	21.9	6.2
	Conrel (357 fibers/g)	
0.8	203.8	69.1
3.8	62.8	73.9
6.7	62.6	64.1
8.9	50.3	43.7

[a]For method see Cross et al., Insect Attractants, Behavior and Basic Biology Research Laboratory, Abstract No. 10, 176th National American Chemical Society Meeting, Division of Pesticide Chemistry, September 11, 1978.

formulation) must be increased to obtain complete communication disruption. By the 13th night of the test, the captures in the Z9:14Ac plots were equal to those in the control plots.

DOSAGE

One very important parameter of any formulation is the evaporation rate of the disruptant. Work on the disruptants of the corn earworm and fall armyworm has been done at the Attractants Laboratory by J. H. Cross (unpublished data). The approximate rates he has obtained for the disruptants that will be used in the 1980 testing are shown in Table 4. On the basis of these figures, the actual release rate for the aerially applied Z9:14F in the 1979 trials was 6-7 mg/ha/h.

FORMULATIONS AND DISPENSING SYSTEMS

A large-scale experiment in a field crop requires a broadcast formulation for the disruptant chemical. Locally, pesticides and fertilizers are applied to sweet corn by aircraft, and grower acceptance of a new pest control material would be most likely if it could be applied by the existing method. This is particularly true in sweet corn where elimination of hard pesticides is not likely given present market standards. Therefore, in 1979 we also evaluated the existing aerial application systems for mating disruptants developed by Thoughts Development Corp. for dispensing Hercon flakes and by Conrel for dispensing their hollow fibers.

A 10.9-ha field of sweet corn was divided into ca. 2 equal plots and sprayed with Z9:14F and Z9:14Ac on an as needed basis (determined by a system of female-baited, pheromone-baited, and light traps). This experiment lasted from mid-April to early June. Five Conrel treatments totaling 504 g of Z9:14F and 221 g of Z9:14Ac (4670 g and 1890 g, respectively, of fiber) were applied to one plot. Six applications of Hercon flake were attempted in the other plot. However, on one occasion the Hercon units malfunctioned, and no material was delivered, and on 2 occasions the calibration was very poor and only ca. 0.3 of the plot was treated. Also, during all Hercon applications there was severe overspray which hampered proper calibration and contaminated control areas. The attempted rate was 1800 g active flake per application.

The Conrel applications resulted in greatly reduced captures of male corn earworm and fall armyworm in female- and pheromone-baited traps. The Hercon applications resulted in erratic reductions. In-depth sampling of all stages of each pest and of damage throughout the plots revealed no major difference between the pheromone-treated area and an adjacent untreated area of the same size, corn variety, and maturation dates. However, approximately the same number of pesticide applications were made to both areas; thus, no pest control effects could be attributed to the pheromone treatment.

The 1980 field test will involve 24.3 ha of commercial sweet corn using Conrel equipment and formulations of Z9:14Ac and Z9:14F.

Pesticide applications will be controlled by the pheromone management team, not the grower.

CONCLUSION

The disruptant of choice for the corn earworm would be Z11:16Al or a mixture of Z11:16Al - Z9:16Al. However, these materials cannot be reliably dispensed for a major control study at this time. Fortunately, Z9:14F appears to be a substitute disruptant.

The chemical Z9:14Ac is the fall armyworm disruptant of choice at present. A defined pheromone for this species is needed.

We have a formulation and delivery system adequate for a major study in 1980 to assess suppression of corn earworm and fall armyworm in sweet corn. However, as stated in 1972 in an early account of a mating disruption study, "The development of a suitable delivery system for a volatile chemical which interferes with sex pheromone communication becomes a problem for the engineer once the appropriate biological parameters are known (McLaughlin et al., 1972)." In many cases today the pheromonal disruptants and required effective doses are known. The development and evaluation of mating disruption is still severely hampered by the lack of reliable delivery systems.

REFERENCES

Brooke, D. L., 1976, Costs and returns from vegetable crops in Florida, season 1974-75 with Comparisons, Univ. of Florida, Economic Information Report 49, March 1976, 27 pp.

Hartstack, A. W., Witz, J. A., and Buck, D. R., 1979, Moth traps for the tobacco budworm, *J. Econ. Entomol.*, 72:519.

Hunt, T. N., 1978, Insect detection, evaluation, and prediction committee; report of the Southeastern Branch of the Entomol. Soc. Am. Dept. Entomol., NC State Univ., Raleigh, NC 27650.

Jones, R. L., and Sparks, A. N., 1979, (*Z*)-9-Tetradecen-1-ol acetate a secondary sex pheromone of the fall armyworm, *Spodoptera frugiperda* (J. E. Smith), *J. Chem. Ecol.*, 5:721.

Klun, J. A., Plimmer, J. R., Bierl-Leonhardt, B. A., Sparks, A. N., and Chapman, O. L., 1979, Trace chemicals: The essence of sexual communication systems in *Heliothis* species, *Science*, 204: 1328.

Klun, J. A., Plimmer, J. R., Bierl-Leonhardt, B. A., Sparks, A. N., Primiani, M., Chapman, O. L., Lee, G. H., and Lepone, G., 1980, Sex pheromone chemistry of female corn earworm moth, *Heliothis zea*, *J. Chem. Ecol.*, 6:165.

Koehn, M. L., 1978, Agricultural Statistics, 1978, USDA Data Services Branch, Economics, Statistics, and Cooperatives Service, U. S. Gov't Printing Office Stock No. 001-000-03775-7, 605 pp.

McLaughlin, J. R., 1972, Sex pheromones of Lepidoptera. XXXI. Disruption of sex pheromone communication in *Pectinophora gossypiella* with hexalure, *Environ. Entomol.*, 1: 645.

Mitchell, E. R., Copeland, W. W., Sparks, A. N., and Sekul, A. A., 1974, Fall armyworm: Disruption of pheromone communication with synthetic acetates, Environ. Entomol., 3: 778.

Mitchell, E. R., 1975, Disruption of pheromonal communication among coexistent pest insects with multichemical formulations, Bioscience, 25: 493.

Mitchell, E. R., Baumhover, A. H., and Jacobson, M., 1976, Reduction of mating potential of male *Heliothis* spp. and *Spodoptera frugiperda* in field plots treated with disruptants, Environ. Entomol., 5: 484.

Mitchell, E. R., and Doolittle, R. E., 1976, Sex pheromones of *Spodoptera exigua*, *S. eridania*, and *S. frugiperda*: Bioassay for field activity, J. Econ. Entomol. 69:324.

Mitchell, E. R., Copeland, W. W., Sparks, A. N., and Sekul, A. A., 1974, Fall armyworm: Disruption of pheromone communication with synthetic acetates, Environ. Entomol., 3: 778.

Mitchell, E. R., Jacobson, M., and Baumhover, A. H., 1975, *Heliothis* spp.: Disruption of pheromonal communication with (*Z*)-9-tetradecen-1-ol formate, Environ. Entomol., 4: 577.

Mitchell, E. R., Tumlinson, J. H., and Baumhover, A. H., 1978, *Heliothis virescens*: Attraction of males to blends of (*Z*)-9-tetradecen-1-ol formate and (*Z*)-9-tetradecenal, J. Chem. Ecol., 4: 709.

Roelofs, W. L., Hill, A. S., Cardé, R. T., and Baker, T. C., 1974, Two sex pheromone components of the tobacco budworm moth, *Heliothis virescens*, Life Sciences, 14: 1555.

Sekul, A. A., and Sparks, A. N., 1967, Sex pheromone of the fall armyworm moth: Isolation, identification, and synthesis, J. Econ. Entomol., 60: 127.

Sekul, A. A., Sparks, A. N., Beroza, M., and Bierl, B. A., 1975, A natural inhibitor of the corn earworm moth sex attractant, J. Econ. Entomol., 68: 603.

EVALUATION OF MICROENCAPSULATED FORMULATIONS OF PHEROMONE COMPONENTS OF THE EGYPTIAN COTTON LEAFWORM IN CRETE

D. G. Campion

L. J. McVeigh

P. Hunter-Jones

Centre for Overseas Pest Research
College House, Wrights Lane
London, W8 5SJ England

D. R. Hall

R. Lester

B. F. Nesbitt

Tropical Products Institute
56-62 Grays Inn Road
London WC1X 8LU England

G. J. Marrs

M. R. Alder

ICI Ltd.
Plant Protection Division
Jealotts Hill, Research Station
Bracknell, Berkshire RG12 6EY England

INTRODUCTION

The female sex pheromone of the Egyptian cotton leafworm, *Spodoptera littoralis* (Boisd.), was originally identified by Nesbitt et al. (1973) using insects obtained from Cyprus. The female moth was found to produce a complex of 4 components: tetradecyl acetate (I), (*Z*)-9-tetradecenyl acetate (IIA), (*E*)-11-tetradecenyl acetate (IIB) and (*Z,E*)-9,11-tetradecadienyl acetate (III). Field experiments conducted by Campion et al. (1974, 1980b) in Cyprus and Crete showed that of the 4 components isolated, the diene (III) was a long-range attractant, while the monoene (IIA) was a powerful attractant inhibitor in that, when dispensed in traps together with (III) or virgin female moths, catches of male moths were reduced almost to the level of catches in unbaited traps.

Therefore, the possibility exists of controlling S. littoralis by means of communication disruption between male and female moths using either IIA or III so as to prevent mating and hence subsequent larval infestations. To achieve control by communication disruption, an effective pheromone concentration in the field has to be maintained over a period of several weeks. Campion et al. (1976) reported results obtained by spraying formulations of IIA microencapsulated within a polyurea or polyurea/polyamide shell. A number of formulations with different release rates were prepared by changing the permeability of the capsule. Polyurea-based microcapsules, average size 2 to 3 µm, applied at the rate of 100 g/ha^{-1} caused 97% disruption for 1 week and a level of 80% for a further 2-week period. This effect was measured by comparing catches in traps baited with synthetic attractant located in the treatment areas with catches in traps in control areas 200 m distant. On the basis of these experiments, larger-scale applications were made in a semi-isolated area of 50 ha in northwest Crete. The persistence of the formulation was measured both by the trap catches and by subsequent chemical analyses. The rate of loss of the pheromone measured chemically was faster than had been anticipated from laboratory studies of the formulation with most of the active ingredient being lost after 6 or 7 days. In spite of these limitations it appeared from extensive larval sampling that some reduction in the population had been achieved (Beevor and Campion, 1979; Campion et al., 1980a).

This paper describes further tests made both in the laboratory and in the field with the aim of developing a more persistent formulation employing either polyurea/polyamide or gelatin-based microcapsules. Evidence reported by Campion et al. (1976) from trials using polythene dispensers suggested that confusion disruption using the pheromone attractant (III) was likely to be more effective than disruption based on pheromone inhibition. The performance of microencapsulated formulations of III was, therefore, also investigated.

MATERIALS AND METHODS

Several polyurea/polyamide based microencapsulated formulations of IIA prepared by methods described previously (Campion et al., 1976) were supplied by Imperial Chemical Industries Ltd. (ICI). The essential characteristics of formulations are listed in Table 1. They were formulated in aqueous dispersions containing 4% wt/vol active ingredient. The capsule size averaged

2 to 3 μm except for formulation JP 6344 for which the average capsule size was 20 μm. Applications were made at 100 g/ha^{-1} using high volume or ultra-low volume spraying equipment. In a further series of tests a conventional UV screener was added to one of the formulations (JF 6209) prior to spraying at a concentration of 5% (of formulation).

Table 1. Disruption of pheromone communication in _S. littoralis_ following application of polyurea/polyamide microcapsules of the pheromone inhibitor IIA on lucerne.*

Code	Microcapsule characteristics	Reps	% Disruption averaged over 5-day periods		
			1-5	6-10	11-15
JF 6210	Lightly cross-linked	6	76	60	48
JF 6340	Heavily cross-linked	2	43	30	10
JF 6209	No internal solvent	3	35	3	0
JF 6341	Dibutyl phthalate as internal solvent	1	12	40	30
JF 6344	Large diameter capsules	3	67	50	15
JF 6209	No internal solvent plus UV screener added after preparation	2	51	41	5

*Application rate 100 g ha^{-1}. Standard internal solvent of the microcapsules dodecyl benzene.

Polyurea-based microcapsules were also prepared (JF 5999) containing III and applied at rates ranging from 1 to 20 g/ha^{-1} using a high volume applicator.

Gelatin-based microcapsules containing IIA were supplied by Food Industries Ltd (FIL). The formulations contained 0.4% active ingredient and were diluted with water saturated with xylene and amyl acetate. The capsules, which ranged in size from 50-100 μm, were suspended in a cellulose thickener plus spreading agent and a latex sticker. The surface of the capsules had been hardened by treatment with formaldehyde followed by resorcinol so as to provide an outer coating of "Bakelite." The 3 formulations provided differed in the degree of thickness of this outer coating. Applications were made at 100 g/ha^{-1} using a high volume applicator.

The treatment areas consisted of lucerne fields sited mainly along the north coast of Crete, which varied in size from 0.7 to 6.2 ha. For initial experiments, 1 to 4 standard water traps (WT), baited with polythene vials containing 1 mg III, as described by Campion et al. (1974), were located in each plot; the number used depended on plot size. Control plots were sited at least 500 m from the treatment areas and contained pheromone traps only. Funnel traps (FT) (McVeigh et al., 1979) were used in later experiments.

Daily records of catch were made after each spraying for periods up to 15 or 20 days. The difference in catch between control and treatment areas was taken to indicate the level and persistence of disruption achieved. The number of replicates for each formulation varied from 1 to 6 depending on availability of material.

The persistence of the disruptant deposited in the crop was also measured chemically. Filter papers (9 cm-diam. Whatman® PS Silicone treated papers) were stapled to the crop before spraying began. Two batches of 20 papers were collected immediately after spraying and similar batches were collected at daily intervals for periods of up to 6 days.

Filter papers were also exposed on boards (2 x 2 m) disposed within the lucerne subsequently sprayed with formulation (JF 6209). The papers were sampled at 12-h intervals corresponding to day and night periods to compare rate of loss of pheromone during the 2 periods.

Similar tests of persistence were also carried out in London. Filter papers sprayed with the formulations were pinned to boards sited on the laboratory roof. The papers were positioned either face upward (exposed) or face downward (shielded) and were sampled at regular intervals for periods of up to 13 days.

Filter paper samples were assayed by chloroform extraction followed by GC analysis of residual pheromone.

Tests were also carried out in Crete to determine the persistence of some of the formulations when exposed or shielded from sunlight by measuring inhibition of trap catch in a "surround" experiment. For each test 8 filter paper discs sprayed with formulation were positioned on boards in a circle 1 m from the center of a standard WT trap baited with III, with 1 disc opposite each corner and 1 opposite each face. The height of the boards and papers corresponded to the trap height of 35 cm. For each formulation tested, one series had the discs on top of the boards and exposed to sunlight, and the other series had the discs positioned underneath the boards and therefore shielded from the effects of sunlight. Daily records of moth catch were made, together with catches from control traps not less than 100 m distant. After periods of 8 to 20 days the discs were removed and subsequently analyzed by GC to determine the amount of residual disruptant.

RESULTS

The level and persistence of the disruption effect of IIA formulated in polyurea/polyamide and gelatin microcapsules. The level and persistence of disruption of polyurea/polyamide formulations of IIA, JF 6209, JF 6210, JF 6340, and JF 6344 are shown in Table 1. Relatively low levels of disruption were achieved for all formulations tested ranging from 12 to 76% for initial periods of 5 days after application. The levels of disruption of subsequent intervals of 10 and 15 days were lower. There was some indication that the addition of the UV screener had slightly improved the persistence of formulation JF 6209. A fast rate of pheromone loss with a half-life of approximately 2 days was confirmed by GC analysis of recovered filter papers.

Similar low levels of disruption were achieved using gelatin-based formulations of IIA (Table 2). GC analysis of residual IIA on recovered filter papers again indicated rapid loss of pheromone with a half-life for formulations of approximately 2 days. It was concluded that the formulations tested were, in general, less effective than those used for earlier trials by Campion et al. (1976).

The level and persistence of the disruptive effect of III formulated in polyurea-based microcapsules. The results of applying polyurea-based microcapsules containing III at doses ranging from 1 to 20 g/ha^{-1} are shown in Table 3. Relatively high rates of disruption in excess of 90% were achieved for the first 5 days after application at rates of 10 and 20 g/ha^{-1}. Amounts of active ingredient to achieve disruption are therefore considerably lower than rates necessary for applications of IIA both in the present series of experiments and those reported earlier by Campion et al. (1976).

Table 2. Disruption of pheromone communication in S. littoralis following application of gelatin-based microcapsules containing IIA on lucerne.*

Code	Characteristics	Reps	% Disruption averaged over 5-day periods		
			1-5	6-10	11-15
FIL/1	Capsules with a thick polymer outer coat	1	55	60	64
FIL/2	Capsules with a thin polymer outer coat	1	69	54	17
FIL/3	Capsules with a medium polymer outer coat	5	58	32	25

*Application rate 100 g ha^{-1}.

* * *

Table 3. Disruption of pheromone communication in S. littoralis following application of polyurea-based microcapsules (JF 5999) containing the pheromone attractant (III) on lucerne.

Rate	% Disruption averaged over 5-day intervals			
(A.I. in g/ha^{-1})	1-5	6-10	11-15	16-20
1	64	51	30	0
5	85	36	44	25
10	94	78	70	0
20	98	57	57	32

The effects of sunlight on the persistence of IIA in polyurea/polyamide and gelatin-based microcapsules assayed by GC. The persistence of IIA in formulations JF 6210 and JF 6340, when sprayed on filter papers and then maintained in shielded and exposed positions in London compared with the persistence on filter papers sprayed in a lucerne field in Crete is shown in Table 4. When shielded from the direct rays of the sun in London, formulation JF 6210 exhibited a very slow rate of release, with a half-life of approximately 12 days, whereas under exposed conditions a very fast rate of loss occurred with a half-life of approximately 1 day. This is compared with the field situation in Crete where a half-life of approximately 2 days was found. Similar results were obtained for formulation JF 6340. Under shielded conditions only 10% of the disruptant was released from the filter papers after 6 days, compared with a half-life of approximately 1 day from the filter papers in an exposed situation and 2.5 days from papers collected from a sprayed lucerne field in Crete. A similar effect was also found for gelatin-based formulation of IIA. Because of temperature similarities for exposure of filter papers above and below test boards, the results strongly suggest that sunlight is involved in causing excessive loss of disruptant either by pheromone degradation and/or rupture of the capsule walls. This effect of sunlight thus overrides differences in composition, wall thickness and the level of cross-linking in capsule polymers; differences in release rate because of such parameters are only apparent when the capsules are protected from sunlight either in shielded positions out-of-doors or in laboratory wind-tunnel experiments.

Comparison of loss of IIA from microcapsules during day and night. The losses of IIA from formulation JF 6209 under day and night conditions (12-h intervals) assayed by GC analysis of filter papers attached to boards in an exposed position sited in a lucerne field in Crete which had been sprayed at the rate of 100 g ha^{-1} are shown in Figure 1. This probably represents extreme conditions of exposure that presumably will occur on the upper leaf surfaces. It can be calculated from the losses shown that only between 17 and 31% of the sprayed formulation was released during the night. Furthermore, since up to 75% of the active chemical had been lost during the first 24 h after spraying, the rate of loss for the following 3 nights was only 45-170 mg ha^{-1} h^{-1}. Since approximately 70% of the sprayed formulation is likely to be deposited on the upper leaf surfaces, only the remaining 30% deposited on the under-surface of the leaves is likely to contribute to persistence beyond the first day.

Persistence of disruption under shielded and exposed conditions measured biologically. The results of tests showing the persistence of disruption using formulations JF 6209, JF 6210, and JF 6340 under shielded and exposed conditions, when compared with JF 5887, that was used in the large-scale trial described by Beevor and Campion

(1979), and Campion et al. (1980a) are shown in Table 5. The residual amount of disruptant remaining on the filter papers after 9 days' exposure in the field is also given.

Table 4. Persistence of IIA in microencapsulated formulations on filter papers measured by GC assay after spraying in the field in Crete and also in London where the papers were arranged in shielded and exposed conditions.

Formulation	Position of sprayed filter papers	% IIA remaining on days postspray					
		1	2	3	4	5	6
JF 6210	Shielded (London)	95	105	82	-	75	-
	Exposed (London)	46	27	7	-	4	-
	Field (Crete)	52	37	25	-	22	-
JF 6340	Shielded (London)	-	96	95	-	-	90
	Exposed (London)	-	42	35	-	-	2
	Field (Crete)	75	56	-	35	34	30

As expected, the heavily cross-linked formulation JF 6340 lost virtually no IIA when shielded and hence only low levels of disruption were achieved; 89% of JF 6209 and 78% of JF 6210 were lost in exposed conditions but only 11 and 5%, respectively, when shielded. Interpretation of trap catch reductions therefore must be considered in relation to fast early losses of IIA from exposed papers with possibly excessively slow rates of loss under shielded conditions which would be insufficient to cause appreciable disruption. JF 5887 was marginally the best formulation in this series, having the fastest release rate under shielded conditions.

These results led to the development of a technique for distinguishing between pheromone release and degradative loss. This involved incorporating a known amount of a stable saturated compound, tetradecyl acetate (I), into the formulation. Because its volatility was closely similar to IIA, comparison of the amount of IIA and of (I) remaining after different periods of exposure gave a measure of the amount of pheromone degradation. Using this technique to assess stability, a formulation JF 6869 was developed in which pheromone degradation was markedly reduced by incorporation of various stabilizing agents. The results of analyses of filter papers sprayed with this formulation and exposed above and below boards are given in Table 6. Degradation of IIA is ca. 10% after nearly 2 weeks' exposure.

Table 5. Persistence of microencapsulated formulations of IIA when sprayed on filter papers and used in "surround" experiments in exposed and shielded positions.

	% Disruption in days*				% IIA remaining
Formulation	1-2	3-4	5-6	7-8	at 9 days
			Exposed		
JF 5887	89	34	63	13	33
JF 6209	86	58	44	10	11
JF 6210	86	32	35	45	22
JF 6340	73	22	22	22	56
			Shielded		
JF 5887	86	76	88	69	87
JF 6209	82	50	48	24	89
JF 6210	100	63	15	10	95
JF 6340	62	18	0	0	100

*As measured by catches of S. littoralis males in pheromone-baited traps within the "surround" area when compared with catches in similarly baited traps in the same field.

Persistence of biological activity of the formulation was also established by a "surround" experiment, the results of which are shown in Table 7. The degree of disruption produced and its persistence was similar for exposed and shielded formulation. However, the percentage of disruption was relatively low, and this attributed to the actual release of pheromone from the formulation being too slow.

A formulation of similar composition having faster release characteristics by virtue of having thinner capsule walls (JF 6956) was therefore prepared and tested in the field. The results showed 100% disruption up to day 8 and 86% for days 9-12 when the formulation was sprayed at the rate of 100 g/ha-1. Subsequent small scale trials in Crete and Egypt have confirmed the effectiveness and persistence of similar formulations incorporating either IIA or III.

Table 6. The persistence of IIA and I in formulation JF 6869 measured by GC analysis, expressed as % remaining after spraying on filter papers and displaying in exposed or shielded positions.

Pheromone component	Position of filter papers	Days exposure of filter paper postspray 0	1	2	6	13
IIA	Exposed	100	76	74	54	36
	Shielded	100	82	85	81	81
I	Exposed	100	83	85	64	45
	Shielded	100	86	88	81	81
IIA (% degradation)		0	8	11	10	9

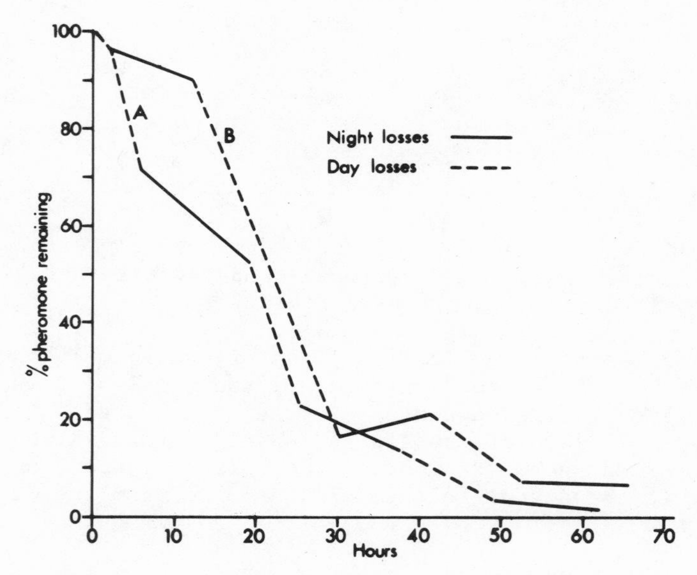

Fig. 1. Losses of IIA from formulation JF 6209 under day and night conditions assayed by GC analysis of filter papers attached to boards in an exposed position sited in a lucerne field in Crete which has been sprayed at the rate of $100g/ha^{-1}$

Table 7. Persistence of microencapsulated formulation JF 6869 when sprayed on filter papers and used in a "surround" experiment in exposed and shielded conditions.

Position of papers	% Disruption averaged over 4-day periods*				
	1-4	5-8	9-12	13-16	17-20
Exposed	62	43	62	69	86
Shielded	56	52	73	67	77

*Measured by catches of S. littoralis males in pheromone traps within the "surround" area when compared with pheromone-baited traps in the same field. The results are measured for 2 replicates.

DISCUSSION

Campion et al. (1976, 1978) had shown in the laboratory that differences in the release rate of pheromone from polyurea and polyurea/polyamide-based microcapsules could be achieved by varying wall thickness and the level of polymer cross-linking of the capsules as well as by changes in temperature and wind-speed. However, in the field, at least in the case of S. littoralis pheromones, the overruling effect is now shown to be the degradation of the pheromone within the capsules when directly exposed to sunlight. Thus, the pheromone from heavily cross-linked capsules (JF 6340) was lost as quickly as from lightly cross-linked polymer (JF 6210) or indeed as from gelatin (FIL/3). Changes in the internal solvent within the capsule (JF 6209, JF 6341) or increased capsule size (JF 6344) similarly made no essential difference. Other release characteristics could only materially influence release from that proportion of the sprayed capsules landing in a shielded position that might occur on the undersurfaces of the leaves.

The persistence of the biological effect of the formulations in the field as measured by trap catches had in earlier experiments been greater than had been found from GC analysis of filter papers collected from the sprayed fields. This difference presumably reflected the release of pheromone from shielded lower-leaf surfaces when opposed to losses measured chemically from filter papers generally secured at the top of the plants. Before the importance of UV degradation had been appreciated, efforts to slow down apparently fast rates of release by thickening the capsule walls or increased polymer cross linkage had merely served to reduce almost completely the actual release of pheromone from the sprayed capsules landing in shielded positions. This probably explains, at least in part, the relatively poor performance of formulations

JF 6209 to JF 6344 when compared with some of those reported earlier by Campion et al. (1976).

The results also suggest that the actual amounts of pheromone inducing disruption are very much lower than actually applied in g/ha. Furthermore, assuming these problems have been overcome and that larval population reductions were achieved earlier in Crete as the result of spraying a relatively inefficient formulation of IIA (Beevor and Campion, 1979; Campion et al., 1980a), the future possibilities for control by this method seem good. The pheromone attractant (III) has also been shown to possess much greater disruptant activity than IIA; further large-scale tests with III would therefore seem to be desirable.

ACKNOWLEDGMENTS

We thank Mr. W. H. Nightingale and Mr. D. W. Hartley of Food Industries, Ltd. for the gelatin-based microencapsulated formulations. The field work described here formed part of a collaborative project between the Centre for Overseas Pest Research and the Greek Ministry of Agriculture. Therefore, we thank Mr. N. Psylakiis, Director of the Institute of Subtropical and Olive Trees, Chania, Crete for all his help and for providing us with work facilities.

REFERENCES

Beevor, P. S., and Campion, D. G., 1979, The field use of 'inhibitory' components of lepidopterous sex pheromones and pheromone mimics, in "Chemical Ecology: Odour Communication in Animals," F. J. Ritter, ed., Elsevier/North-Holland Biomedical Press, Amsterdam.

Campion, D. G., Bettany, B. W., Nesbitt, B. F., Beevor, P. S., Lester, R., and Poppi, R. G., 1974, Field studies of the female sex pheromone of the cotton leafworm *Spodoptera littoralis* (Boisd.) in Cyprus, *Bull. Entomol. Res.*, 64: 89.

Campion, D. G., McVeigh, L. J., Murlis, J., Hall, D. R., Lester, R., Nesbitt, B. F., and Marrs, G. J., 1976, Communication disruption of adult Egyptian cotton leafworm *Spodoptera littoralis* (Boisd.) (Lepidoptera: Noctuidae) in Crete using synthetic pheromones applied by microencapsulation and dispenser techniques, *Bull. Entomol. Res.*, 66: 335.

Campion, D. G., Lester, R., and Nesbitt, B. F., 1978, Controlled release of pheromones, *Pestic. Sci.*, 9: 434.

Campion, D. G., McVeigh, L. J., Hunter-Jones, P., Hall, D. R., Lester, R., Nesbitt, B. F., and Marrs, G. J., 1980a, Reductions in larval populations of Egyptian cotton leafworm by spraying a microencapsulated formulation of a pheromone component (Z)-9-tetradecenyl acetate in a semi-isolated area in Crete, *Bull. Entomol. Res.*, (In press).

Campion, D. G., Hunter-Jones, P., McVeigh, L. J., Hall, D. R., Lester, R., and Nesbitt, B. F., 1980b, Modification of the attractiveness of the primary pheromone component of the Egyptian cotton leafworm *Spodoptera littoralis* (Boisd.) (Lepidoptera: Noctuidae) by secondary pheromone components and related chemicals, *Bull. Entomol. Res.*, (In press).

McVeigh, L. J., Paton, E. M., and Hall, D. R., 1979, Factors affecting the performance of pheromone traps for male *Spodoptera littoralis*, *in* "Proceedings 1979 British Crop Protection Conference, Pests and Diseases," Brighton, 409-419.

Nesbitt, B. F., Beevor, P. S., Cole, R. A., Lester, R., and Poppi, R. G., 1973, Sex pheromones of two noctuid moths, *Nature New Biol.*, 244: 208.

PINK BOLLWORM AND TOBACCO BUDWORM MATING DISRUPTION STUDIES ON COTTON

Thomas J. Henneberry

Louis A. Bariola

Hollis M. Flint

Pete D. Lingren

Western Cotton Research Laboratory
AR-SEA, USDA, Phoenix, AZ 85040 USA

Janice M. Gillespie

Herculite Products, Inc.
2321 North 27th Street
Phoenix, AZ 85006 USA

Agis F. Kydonieus

Herculite Products, Inc.
1107 Broadway
New York, NY 10010 USA

INTRODUCTION

The role of insect pheromones in mediating behavior of insects and the possible use of these chemical substances for control have recently attracted considerable attention. One such method of control, mating disruption by permeation of the atmosphere of crop environments with sex pheromones so as to prevent reproduction and thereby reduce infestations of economic insect pests, was first suggested by Beroza in 1960. The fact is that in most cultivated crop systems there is more than one pest species. Thus, as the technology of mating disruption has advanced, the possibility of using sex pheromones to disrupt the communication of more than one species within a same crop system has received greater attention (Mitchell, 1975; Tumlinson et al., 1976). One such complex of pest insects attacks cotton in Arizona and southern California. The most damaging species of this complex are the pink bollworm, *Pectinophora gossypiella* (Saunders); the bollworm, *Heliothis zea* (Boddie); and the tobacco budworm, *H. virescens* (F.).

Hummel et al. (1973) and later Bierl et al. (1974) identified the pink bollworm sex pheromone as a mixture of *Z*,*Z*- and *Z*,*E*-isomers of 7,11-hexadecadienyl acetate (gossyplure). Subsequently, promising results were reported when gossyplure was used to produce mating disruption for control of the pink bollworm (Boness, 1975; Shorey, 1975; Shorey et al., 1976; Gaston et al., 1977; Boness et al., 1977; Brooks and Kitterman, 1978; and Brooks et al., 1979).

Roelofs et al. (1974) and Tumlinson et al. (1975) identified (*Z*)-11-hexadecenal (Z-11-HDAL) and (*Z*)-9-tetradecenal (Z-9-TDAL) as sex pheromones (called virelure) of the tobacco budworm. However, a 16:1 mixture of the 2 chemicals as bait in traps often did not catch as many male tobacco budworms as virgin females (Tumlinson et al., 1975), which indicated that other chemical components of the pheromone existed. Klun et al. (1979) subsequently identified 5 additional chemicals and found that the 7-component mixture equaled or exceeded the attractiveness of virgin females. Sparks et al. (1979) corroborated these results. Field evaluations of virelure or of the 7-component pheromone mixture as mating disruptants for tobacco budworm control on cotton have not been conducted but Hendricks (1976) demonstrated that catches of male tobacco budworms were significantly reduced when 0.643 μl of the one component of virelure, Z-9-TDAL, was dispensed near virgin females. Thus, the compound might have potential for mating disruption.

Also, Mitchell et al. (1975) reported that when (*Z*)-9-tetradecen-1-ol formate (Z-9-TDF), a "mimic" of one of the tobacco budworm pheromone components, was dispensed from polyethylene caps at the rate of 300 ng/min in the field, catches of male tobacco budworm and bollworm moths in traps baited with virgin females were reduced. Also, combinations of Z-9-TDF and Z-9-TDAL used as bait in traps (16:1 or 32:1 ratios) caught as many tobacco budworm males as virelure or as 3 virgin females (Mitchell et al., 1978). Further, when Z-9-TDF was evaporated from 1.25 ml polyethylene vials arranged in a checkerboard square with dispensers 3 m apart in cotton or corn plots, they found reduced catches of male moths and reduced mating of bollworms and tobacco budworms. Likewise, Z-11-HDAL, evaporated from plastic laminates, reduced mating of bollworms, and Z-9-TDAL, evaporated from polyethylene vials, reduced mating of tobacco budworms (Mitchell et al., 1976).

The availability of gossyplure, virelure, 1-tetradecenal formate (TF), and Z-9-TDF in polymeric membrane formulations

(Hercon®) and the need for efficient, effective control methods for the pink bollworm and tobacco budworm in cotton production systems in the Southwest prompted us to evaluate the effect of combinations of these materials on pink bollworm and tobacco budworm populations in 1979 in Rainbow Valley, AZ. The present paper is a report of this work.

METHODS AND MATERIALS

The studies were conducted in 17 cotton fields, either adjacent to or within 1/4 km of one another. Disruption treatments were (1) gossyplure plus TF, (2) gossyplure plus virelure, (3) gossyplure plus Z-9-TDF, and (4) the control, fields treated with commercial applications of insecticide as recommended by grower-employed field scouts. Each treatment was replicated 4 times (fields ca. 16 ha each). In one additional 4-ha field (untreated check), no insecticides or pheromones were applied. Polymeric plastic laminated flakes containing pheromone and pheromone mimics were produced by the Hercon Group, Herculite Products, Inc., New York, NY. The gossyplure, plus TF, virelure, or Z-9-TDF, was suspended in an acrylic adhesive carrier (Phero-Tac 5®) and applied every 13-19 days from May 16 until September 27 using the aerial flake dispensing unit manufactured for the Herculite Products, Inc. and described by Thoughts Development, Inc., Sun City, AZ. Beginning August 23, additional applications of TF, virelure, and Z-9-TDF were made weekly until October 4, except that Z-9-TDF was applied for the last time on September 20.

Results were monitored with pheromone traps, by picking cotton bolls, by operating flight traps, by egg and larval counts on terminals, and by nocturnal collections.

Four Delta-type traps baited with a rubber septa containing 1 mg of a 1:1 ratio of *Z*,*Z*- to *Z*,*E*-isomers of gossyplure were placed at the tops of cotton plants in each quadrant of the 17 experimental fields on April 4. On May 15, 2 of these traps, those in diagonal quadrants, were replaced with Lingren live traps (Lingren et al., 1980) containing the same bait formulation. Traps and all baits were replaced every 2-4 weeks or as needed throughout the season. Also, 2 cone traps (Hartstack et al., 1979), baited with Hercon-laminated virelure baits, were placed one each in a diagonal quadrant in each field on April 18. However, on June 16, the virelure bait in one of the cone traps in each field was replaced with the Hercon 7-component bait (Klun et al., 1979), and a 3rd cone trap, baited with 5 live virgin female moths, was placed in another quadrant of each field on May 10.

The live female moths used as bait were produced at the USDA Western Cotton Research Laboratory on wheat germ diet, sexed as pupae and then held at 24±1°C for emergence. When they were ca. 2-5 days old, they were placed in 5 x 20 cm screen cylinders into which a 5 cc glass vial containing 10% sugar water was inserted. Initially (May-June), the cylinders were placed in the cone traps in the morning, but during July-September, they were placed in the traps in late afternoon to prevent death from high midday temperatures. Females were placed centrally about 5 cm below the skirt of the cone trap, and the trap was placed 0.5-1 m above the cotton plants early in the season and at the tops of the plants during mid- and late season. All male tobacco budworm moths captured in the Delta-type traps, Lingren live traps, and cone traps were counted twice each week, and the live females were replaced during these observations. The laminated baits were replaced every 2-4 weeks or as needed throughout the season. All traps were removed from the fields on October 15.

Samples of cotton bolls (25) were picked twice weekly from July 9 to October 9 from each of 4 locations in each field to determine the effect of the treatments on pink bollworm infestations. Fifty of the 100 were cut open and examined for pink bollworm larvae; the other 50 were incubated for 14 days at 26°C, after which all emerging pink bollworms were counted.

One flight trap (Butler, 1966) was operated from June 14 to October 1, 1979 in each test field to obtain information on the mating of female pink bollworms in pheromone-treated and control fields. Pink bollworm moths were collected from the traps 1-2 times a week, the numbers were recorded, and the females were dissected to determine the presence of spermatophores.

Twice a week beginning June 7, 1979 each of 25 cotton terminals at 4 locations in each field was examined for eggs and larvae of *Heliothis* spp. All eggs and larvae found were returned to the laboratory and reared to adults to identify the species of *Heliothis* present.

Plots were searched with the aid of 6-V headlamps for 8 nights from June 19 through September 12 to collect individual moths and mating pairs of tobacco budworms and bollworms. The procedure was as follows: From 6 to 10 observers searched from 3 to 5 fields for 3 to 10 h/night. Searches were concentrated in control fields and in fields treated with gossyplure plus virelure, but 2 of the fields receiving gossyplure plus Z-9-TDF and gossyplure plus TF were searched for 2 to 8 nights. All searches were made 1 day before treatment and the night of treatment. Species were identified, numbers of single moths and mating moth pairs were recorded, and females were dissected and examined for the presence of spermatophores.

RESULTS

Pheromone Applications

The total amounts of gossyplure, TF, virelure, and Z-9-TDF applied per ha for the season were 46.2, 269.6, 158.7, and 142.7 g, respectively. Amounts applied on a given date ranged from 3.5 to 4.9, 12.4 to 30.1, 7.4 to 14.8, and 8.7 to 19.8 g/ha for gossyplure, TF, virelure, and Z-9-TDF. Gas chromatographic analysis of gossyplure, TF, and virelure residues in flakes under similar circumstances indicated that ca. 29, 49, 69, and 78% of the gossyplure was evaporated in 7, 14, 21, and 28 days, respectively. Similarly, 22-84% of the TF and 66-92% of the virelure evaporated from plastic laminate flakes during the same period (Henneberry et al., 1980).

Effect on Catches of Pink Bollworm Males

The effects of gossyplure plus TF, virelure, or Z-9-TDF on catches of male moths in gossyplure-baited traps were similar so they were combined in Figure 1. From the time of the first application on May 16 and continuing to mid-July, catches in all fields were low. Then when populations began to increase from July 14 to September 11, captures in the treated fields were significantly reduced an average of 93% (range 78-99).

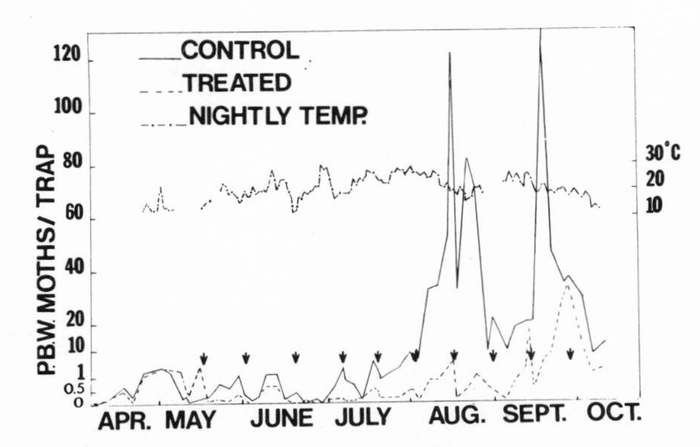

Fig. 1. Mean number of male pink bollworm moths/trap in gossyplure-treated and control cotton fields (means of 12 replications for gossyplure-treated and 4 replications for control, 4 traps/replication), Rainbow Valley, AZ, 1979. (Arrows indicate pheromone treatment dates.)

Apparently, catches of male moths were also affected in the control fields; this was a relatively small experimental area. For example, the average catches of male moths per trap/night in the treated fields for 2-3 days before and for the day after treatments were 2.9 and 0.3, respectively. In the control fields, the average catches of male moths per trap/night on the same occasions were 115.1 and 70.4, respectively. Catches of male moths in the additional untreated check field increased throughout the season, and a peak catch of 260 male moths/trap was recorded in late September (Fig. 2).

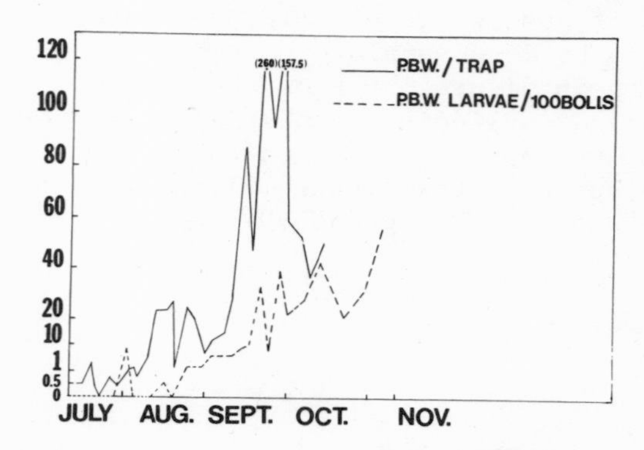

Fig. 2. Mean number of male pink bollworm moths/trap and numbers of pink bollworm larvae/100 bolls in samples from the untreated check cotton field, Rainbow Valley, AZ, 1979.

<u>Effect on Boll Infestations by Pink Bollworms</u>

Infestations in bolls developed slowly in all fields (Table 1) but began to increase in late August. More larvae were found in bolls from the control fields than from the treated fields, and the differences were statistically significant on six of the sampling dates. Over the season, the total numbers of larvae found in bolls were 68% lower in the treated fields than in the control fields.

In the single untreated check field, the first infested bolls were found July 30 (8 larvae/100 bolls, Fig. 2). Thereafter, until late August, infestations were variable. Beginning in late August, populations increased steadily to the end of the season and reached a high of 58 larvae/100 bolls with 38% of the bolls infested.

Table 1. Mean[a] number of pink bollworm larvae/100 cotton bolls from pheromone-treated and control fields, Rainbow Valley, AZ, 1979.

Sampling date	Gossyplure plus			Control
	TF	Virelure	Z-9-TDF	
July 9-Aug. 16[b]	< .1	.2	< .1	.7
Aug. 21	1.0 b	1.8 b	2.8 b	6.0 a
23	2.3	1.0	2.0	4.0
27	10.8	1.8	3.3	5.8
30	1.5 b	4.3 b	4.0 b	15.8 a
Sept. 6[c]	9.3 b	3.0 b	7.8 b	20.8 a
10	3.5	3.0	4.0	5.3
13	1.3	.5	3.3	6.8
17	1.5 b	.8 b	1.3 b	14.8 a
21	1.0	.3	3.8	3.8
25	1.8 b	2.5 b	2.0 b	10.5 a
27	.8	.8	2.0	3.3
Oct. 3	5.8	3.8	1.8	3.0
9	2.5	5.0	.3	1.5
18	4.0	4.3	1.5	2.8
25	4.3	3.5	.8	1.3
30	5.8	2.3	3.3	1.0

[a]Means of 4 replications. Means in the same row not followed by the same letter are significantly different by Duncan's multiple range test ($P = 0.05$).

[b]Means of 12 sampling dates during indicated period.

[c]First insecticide applications Aug. 31 to 2, 1, 1, and 3 fields of gossyplure-TF, gossyplure-virelure, gossyplure-Z-9-TDF and control, respectively. Average number of treatments for the season were .8, 1.0, 1.8, and 2.8, respectively.

Pink Bollworm Adults in Flight Traps

The catch of pink bollworm adults in flight traps showed more pink bollworm moths, more spermatophores/female, and higher percentages of mated females in control fields than fields receiving 3 treatments (Table 2). Differences were not always statistically significant for these individual treatments, but the comparison between treated fields vs. control fields showed that significantly more pink bollworm males were caught in the control fields (149 vs. 62), that percentages of mated females were higher (64 vs. 35%), and that the numbers of spermatophores/female (1.1 vs. 0.5) were reduced 45% in treated fields.

Table 2. Number of pink bollworm moths caught in flight traps in pheromone-treated and control fields, percentages of females mated, and spermatophores/female, Rainbow Valley, AZ, 1979.

Treatment	No. moths caught/ treatment[a,b]		Females	
	Males	Females	% Mated	Spermato- phores/♀
Gossyplure plus				
TF	63 a	32 a	33 ab	0.5 a
Virelure	38 a	19 a	18 b	0.3 a
Z-9-TDF	84 a	51 a	55 a	0.8 a
Control	149 a	69 a	64 a	1.1 a

[a]Total moths caught/4 fields/treatment for the season.

[b]Means/4 replications. Means within a column not followed by the same letter are significantly different. Duncan's multiple range test (*P* = 0.05).

Effect on Catches of Tobacco Budworm Males

Prior to pheromone applications, catches of male tobacco budworm moths ranged from 0.3 to 9/trap and averaged 2.2 moths/trap in the 17 cotton fields before the disruptant treatments began (Fig. 3). (The first applications were made May 16, but there were no significant differences in catches between treated and control fields through July 19.)

Beginning on July 20, after the fifth application, catches of male moths in fields treated with gossyplure plus virelure were lower on all but one sampling date than in control fields, and the difference was significant on 21 of the 30 sampling dates. The impact of gossyplure plus TF and gossyplure plus Z-9-TDF on male catches in pheromone-baited traps was more variable, but catches were reduced in treated fields on 10 of 30 sampling dates. On most dates, the effects of the 3 treatments were not significantly different. The dramatic reductions in catches after each treatment were of short duration; catches increased within 2-3 days.

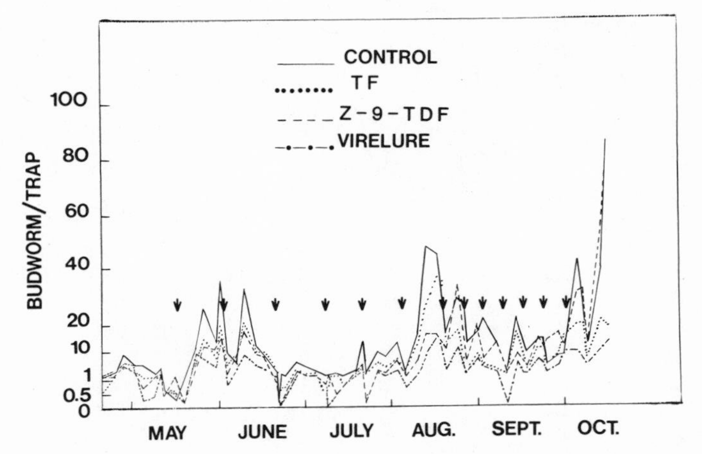

Fig. 3. Mean number of male tobacco budworm moths/trap in gossyplure plus TF, gossyplure plus virelure, gossyplure plus Z-9-TDF, and control cotton fields (means of 4 replications, 4 traps/replication), Rainbow Valley, AZ, 1979. (Arrows indicate pheromone treatment dates.)

Heliothis Eggs and Larvae

Few *Heliothis* spp. eggs and larvae were found on cotton terminals throughout the season and almost none in any fields through August 14 (Table 3). In August, fewer *Heliothis* eggs were found in terminals in fields treated with gossyplure plus virelure on some sampling dates, but differences were not significant; and during September, egg counts were similar in all fields. On October 4 and 8, significantly higher numbers of eggs were found in control fields than in treated fields.

Table 3. Mean[a] number of *Heliothis* spp. eggs and larvae in pheromone-treated, control and untreated check cotton plots, Rainbow Valley, AZ, 1979.

Sampling date	Gossyplure plus TF		Gossyplure plus Virelure		Gossyplure plus Z-9-TDF		Control		Untreated check	
	E[b]	L[c]	E	L	E	L	E	L	E	L
June 7–Aug. 14[d]	<1	0	<1	0	<1	0	1	0	<1	0
17	1	0	7	0	10	0	12	0	1	0
20	8	0	14	0	24	<1	20	0	12	0
23	6	<1b	12	<1b	18	1b	15	4a	11	0
27	1	1	0	1	6	8	5	7	4	1
30	3	6	1	3	3	1	2	5	4	5
Sept. 5	1	1	1	1	1	<1	<1	1	0	2
10	0	<1	1	0	1	1	1	0	2	0
17	6	1	12	2	5	2	7	2	4	1
20	5	3a	11	1ab	5	3a	4	1b	5	2
24	6	1	6	3	13	2	5	2	1	5
26	5	3	3	2	2	0	3	0	4	6
Oct. 1	13	4	6	2	6	4	18	2	2	0
4	9b	1	12b	1	31a	2	34a	2	4	2
8	7c	6	7c	2	17b	1	28a	4	2	2
11	14	4	15	2	15	1	26	4	-	-

[a]Means of 4 reps, 100 terminals/rep. Means for Aug. 23, Sept. 20, Oct. 4 and 8 within the same row and data comparison not followed by the same letter are significantly different.

[b]E = eggs. [c]L = larvae. [d]Means of 4 reps on 11 sampling dates.

There were no significant differences in populations of larvae in treated and control fields except on August 23 when more larvae were found in the control fields than in any treated fields. *Heliothis* populations in the untreated check fields developed slowly and never reached damaging numbers (Fig. 4).

Egg hatch from field-collected bolls ranged from 13 to 100% for all fields on all sampling dates and was not affected by treatments. On all sampling dates, except August 14, 75-100% of the eggs collected and reared to adults were tobacco budworms.

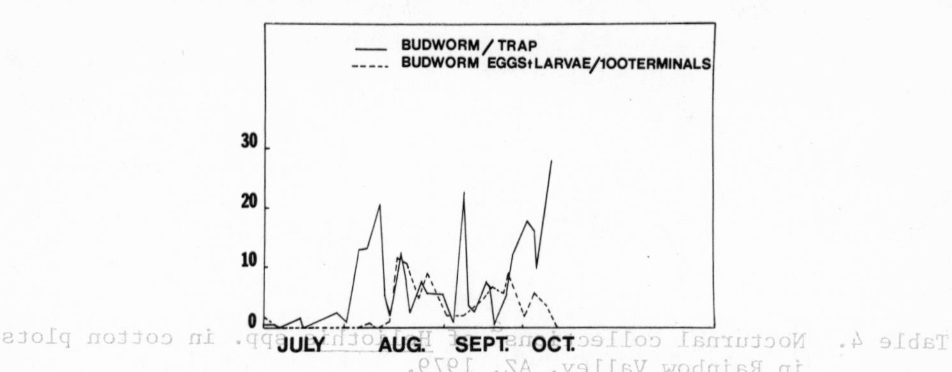

Fig. 4. Mean number of male tobacco budworm moths/trap and Heliothis spp. eggs and larvae/100 terminals in the untreated check cotton field, Rainbow Valley, AZ, 1979.

Nocturnal Collections of Heliothis Moths

The results of the nocturnal collections of Heliothis spp. moths are summarized in Table 4. Peak collections were made August 23 when 4.9 moths/man-hour were captured. Overall, about 2X more tobacco budworms were captured than bollworms. The high female-to-male sex ratio of the tobacco budworm population may be related to the removal of tobacco budworm males from what was a low density population by the 3 pheromone traps present in each field throughout the season.

The fewest number of unmated females and the largest number of spermatophores/female for both tobacco budworms and bollworms were recorded at the time of the peak population. In fact, the high percentages of mating pairs and the increase in number of spermatophores/female between August 15-22 suggests moth emergence during that time. The results are similar to those of Lingren et al. (1979) and Raulston et al. (1979). Both groups found that emergence and mating of tobacco budworm occurs within a few days and that aging of the population could be indexed by a preponderance of multiple matings.

Mating of both tobacco budworms and bollworms was reduced in the fields treated with virelure during the first night after application (Table 5), but the percentages of females mating on the night before application and the night after (Table 5), and the number of spermatophores/female in treated and control fields (Table 6) were not affected. Possibly rapid dissipation of the pheromone or adaptation of the males is responsible, but rapid dissipation is suggested since a dramatic drop in male moth catches occurred after each treatment (Fig. 3).

Table 4. Nocturnal collections[a] of Heliothis spp. in cotton plots in Rainbow Valley, AZ, 1979.

Sample date	Man-hr[b]	No. of Heliothis captured BDW[c] ♀	♂	BLW[d] ♀	♂	% Unmated BDW	BLW	Spermatophores/♀ BDW	BLW	% As mating pairs BDW	BLW
June 19	<1	0	1	0	1	-	-	-	-	-	-
Aug. 15	1	13	3	4	7	0	25	2	1	13	18
16	1	4	3	4	2	25	25	3	1	43	33
22	4	26	16	9	10	4	0	3	2	30	16
23	5	59	14	20	13	2	5	3	2	5	15
Sept. 5	1	1	2	0	0	100	-	0	-	0	0
6	1	8	2	2	0	50	50	1	1	0	0
12	<1	3	0	3	0	33	33	1	1	33	0
Total		114	41	42	33	-	-	-	-	-	-

[a]Total numbers collected in treated and untreated plots.

[b]From 6-10 workers searching and capturing all Heliothis spp. at night in 1-3 reps of 4 treatments before and after applications of pheromone.

[c]BDW = budworm. [d]BLW = bollworm.

Table 5. Mean number of *Heliothis* spp. females collected[a] and percent mating the night before and night of virelure treatment, Rainbow Valley, AZ.

	Control		Virelure-treated[b]	
	No.	% Mating	No.	% Mating
	Tobacco budworm			
Night--				
before application	13	15	43	26
of application	18	22	58	5
Total and average over all nights	31	19	101	15
	Bollworm			
Night--				
before application	4	100	14	21
of application	7	29	19	16
Total and average over all nights	11	36	33	18

[a]One to 3 ha cotton fields searched in each treatment by 1-5 people for 3-9 h/night over 6 nights from Aug. 15-Sept. 6.

[b]In combination with gossyplure.

Table 6. Average numbers of spermatophores/female tobacco budworm captured in control fields and in virelure-treated fields in Rainbow Valley, AZ.

Sampling date	Control No. spermatophores/♀	Virelure No. spermatophores/♀
June 19	-	-
Aug. 15	1.9	2.0
16	3.7	0
22	-	3.3
23	2.4	3.5
Sept. 5	-	0
6	1.3	0.3
12	-	3.0

Pesticide Applications

The applications of permethrin plus chlordimeform on August 31, 1979 to 2, 1, and 3 fields of gossyplure-TF, gossyplure-virelure, gossyplure-Z-9-TDF, and the control treatments, respectively, reduced the average infestation of pink bollworms in the control fields in mid-September (Table 2) and prevented the populations from reaching high numbers. For example, an average of 14 larvae/100 bolls was found on September 6 in insecticide-treated control fields. The single untreated field had 33 larvae/100 bolls on September 17, 2 days before it was treated with insecticide. The insecticides also suppressed development of *Heliothis* populations. Thus, significant differences in oviposition between the control fields and the treated fields did not occur until the 1st week of October, as previously noted.

DISCUSSION

The pioneering research of McLaughlin et al. (1972), Shorey et al. (1974), Shorey et al. (1976), and Gaston et al. (1977) demonstrated that the sex pheromone mating disruption method was a feasible and potential method for suppressing pink bollworm populations. In fact, as the data obtained in the present studies show, catches of male pink bollworms were reduced in gossyplure-baited traps as a result of gossyplure disruptant treatment. Although some moths continued to be caught, catches generally did not exceed 1 moth/trap each night. These results are similar to those of Shorey et al. (1974), Boness (1977), and Brooks et al. (1979). Possibly, mating interactions between male and female moths are density dependent and may not depend totally on long-range pheromone communication between sexes (Smith et al., 1978). Also, male pink bollworm moths, like the gypsy moth, *Lymantria dispar* (L.), (Cameron, 1973), may be stimulated to intensify searching activities in the presence of the sex pheromone or, like the European corn borer, *Ostrinia nubilalis* (Hübner), (Klun and Robinson, 1971), inhibition of flight orientation occurs but not inhibition of precopulatory behavior.

Certainly boll infestations and captures of mated females were reduced by the treatments. Also, captures of male tobacco budworms in pheromone and live-female traps were reduced on several sampling dates after the fields were treated. However, the results were more variable than they were for the pink bollworm and there were no reductions in egg hatch or egg and larval populations. However, mating of tobacco budworms and bollworms was reduced the night of the treatments so the pheromone apparently dissipated rapidly. Thus, mating disruption may have potential for control of *Heliothis* spp. if a formulation can be perfected that will lengthen its effect.

The polymeric membrane flake formulation of gossyplure proved to be an effective vehicle for dissemination of the pheromone in the field. Further, application of the gossyplure flakes in combination with TF, virelure, or Z-9-TDF formulated as flakes did not seem to reduce or increase the effectiveness of gossyplure. Still further studies will be needed to clarify this point since interspecific disruption of insect sex pheromone communication has been shown for closely related pheromones, pheromone mimics, and closely related chemicals (Kaae et al., 1972; 1974; McLaughlin et al., 1972; Mitchell et al., 1975; 1976). However, Coudriet and Henneberry (1976) did not detect any effect of virelure on male pink bollworm moth catches in gossyplure-baited traps.

REFERENCES

Beroza, M., 1960, Insect attractants are taking hold, Agric. Chem., 15: 37.

Bierl, B. A., Beroza, M., Staten, R. T., Sonnet, P. E., and Adler, V. E., 1974, The pink bollworm sex attractant, J. Econ. Entomol., 67: 211.

Boness, M., 1975, Field trials with the synthetic sex pheromone of Pectinophora gossypiella, Pflanzenschutz-Nachrichten Bayer, 28: 156.

Boness, M., Eiter, K., and Disseln-Kotter, H., 1977, Studies on sex attractants of Lepidoptera and their use in crop protection, Pflanzenschutz-Nachrichten Bayer, 30: 213.

Brooks, T. W., and Kitterman, R. L., 1978, Gossyplure H. F. - pink bollworm population suppression with male sex attractant pheromone released from hollow fibers, 1976 experiments, Proc. Beltwide Prod. Mech. Conf. Jan. 10-13, 1977, Atlanta, GA, pp. 79-82.

Brooks, T. W., Doane, C. C., and Staten, R. T., 1979, Experience with the first commercial pheromone communication disruptive for suppression of an agricultural insect pest, in: "Odour Communication in Animals," F. J. Ritter, ed., Elsevier/North Holland Biomedical Press, Amsterdam, pp. 375-388.

Butler, G. D., Jr., 1966, An insect flight trap for dry areas, J. Econ. Entomol., 59: 1030.

Cameron, E. A., 1973, Disparlure: A potential tool for gypsy moth population manipulation, Bull. Entomol. Soc. Am., 19: 15.

Coudriet, D. L., and Henneberry, T. J., 1976, Captures of male cabbage loopers and pink bollworms: Effect of trap design and pheromone, J. Econ. Entomol., 69: 603.

Gaston, L. K., Kaae, R. S., Shorey, H. H., and Sellers, D., 1977, Controlling the pink bollworm by disrupting sex pheromone communication between adult moths, Science, 196: 904.

Hartstack, A. W., Witz, J. A., and Buck, D. R., 1979, Moth traps for the tobacco budworm, J. Econ. Entomol., 72: 519.

Hendricks, D. E., 1976, Tobacco budworm: Disruption of courtship behavior with a component of the synthetic sex pheromone, Environ. Entomol., 5: 978.
Henneberry, T. J., Gillespie, J. M., Bariola, L. A., Flint. H. M., Butler, G. D., Jr., Lingren, P. D., and Kydonieus, A. F., (1980), Mating disruption as a method of suppressing pink bollworm and tobacco budworm populations in cotton, in: "Insect Suppression Controlled Release Pheromone Systems," A. F. Kydonieus and M. Beroza (eds.), (In press).
Hummel, H. E., Gaston, L. K., Shorey, H. H., Kaae, R. S., Byrne, K. J., and Silverstein, R. M., 1973, Clarification of the chemical status of the pink bollworm sex pheromone, Science, 181: 873.
Kaae, R. S., McLaughlin, J. R., Shorey, H. H., and Gaston, L. K., 1972, Sex pheromones of Lepidoptera XXXII, Disruption of intraspecific pheromone communication in various species of Lepidoptera by permeation of the air with looplure or hexalure, Environ. Entomol., 1: 651.
Kaae, R. S., Shorey, H. H., Gaston, L. K., and Hummel, H. E., 1974, Sex pheromones of Lepidoptera: Disruption of pheromone communication in Trichoplusia ni and Pectinophora gossypiella by permeation of the air with non-pheromone chemicals, Environ. Entomol., 3: 87.
Klun, J. A., and Robinson, J. F., 1971, European corn borer moth: Sex attractant and sex attraction inhibitors, Ann. Entomol. Soc. Am., 64: 1083.
Klun, J. A., Plimmer, J. R., Bierl-Leonhardt, B. A., Sparks, A. N., and Chapman, O. L., 1979, Trace chemicals: The essence of sexual communication systems in Heliothis species, Science, 204: 1328.
Lingren, P. D., Raulston, J. R., Sparks, A. N., and Proshold, F. I., 1979, Tobacco budworm: Nocturnal behavior of laboratory-reared irradiated and native adults in the field, USDA Tech. Bull. ARR-W-5, 17 pp.
Lingren, P. D., Burton, J., Shelton, W., and Raulston, J. R., 1980, Design, evaluation, and comparative efficiency of a pheromone trap for capturing live male pink bollworms, J. Econ. Entomol., (In press).
McLaughlin, J. R., Gaston, L. K., Shorey, H. H., Hummel, H. E., and Stewart, F. D., 1972, Sex pheromones of Lepidoptera, XXXIII. Evaluation of the disruptive effects of tetradecyl acetate on sex pheromone communication in Pectinophora gossypiella, J. Econ. Entomol., 65: 1592.
McLaughlin, J. R., Shorey, H. H., Gaston, L. K., Kaae, R. S., and Stewart, F. D., 1972, Sex pheromones of Lepidoptera, XXXII. Disruption of sex pheromone communication in Pectinophora gossypiella with hexalure, Environ. Entomol., 1: 645.
Mitchell, E. R., 1975, Disruption of pheromonal communication among coexisting pest insects with multichemical formulations, BioScience, 25: 493.

Mitchell, E. R., Jacobson, M., and Baumhover, A. H., 1975, Heliothis spp.: Disruption of pheromonal communication with Z-9-tetradecen-1-ol formate, Environ. Entomol., 4: 577.

Mitchell, E. R., Baumhover, A. H., and Jacobson, M., 1976, Reduction of mating potential of male Heliothis spp. and Spodoptera frugiperda in field plots treated with disruptants, Environ. Entomol., 5: 484.

Mitchell, E. R., Tumlinson, J. H., and Baumhover, A. H., 1978, Heliothis virescens: Attraction of male to blends of Z-9-tetradecen-1-ol formate and Z-9-tetradecenal, J. Chem. Ecol., 4: 709.

Raulston, J. R., Lingren, P. D., Sparks, A. N., and Martin, D. F., 1979, Mating interaction between native tobacco budworms and released backcross adults, Environ. Entomol., 8: 349.

Roelofs, W. L., Hall, A. S., Cardé, R. T., and Baker, T. C., 1974, Two sex pheromone components of the tobacco budworm moth, Heliothis virescens, Life Sciences, 14: 1555.

Shorey, H. H., 1975, New advances in pink bollworm control, Proc. Beltwide Cotton Prod.-Mech. Conf., New Orleans, LA, pp. 24-25.

Shorey, H. H., 1976, Application of pheromones for manipulating insect pests of agricultural crops, in "Insect Pheromones and Their Applications," Proc. Symp., Tokyo, Japan, 97-108.

Shorey, H. H., Kaae, R. S., and Gaston, L. K., 1974, Sex pheromones of Lepidoptera: Development of a method for pheromonal control of Pectinophora gossypiella in cotton, J. Econ. Entomol., 67: 347.

Shorey, H. H., Gaston, L. K., and Kaae, R. S., 1976, Air permeation with gossyplure for control of the pink bollworm, in "Pest Management with Insect Attractants," M. Beroza, ed., Am. Chem. Soc. Symp. Series, 23: 67.

Smith, R. L., Flint, H. M., and Forey, D. E., 1978, Air permeation with gossyplure: Feasibility studies on chemical confusion for control of the pink bollworm, J. Econ. Entomol. 71: 257.

Sparks, A. N., Raulston, J. R., Lingren, P. D., Carpenter, J. E., Klun, J. A., and Mullinix, B. G., 1979, Field response of male Heliothis virescens (F.) to pheromonal stimuli and traps, Bull. Entomol. Soc. Am., 25: 268.

Tumlinson, J. H., Hendricks, D. E., Mitchell, E. R., Doolittle, R. E., and Brennan, M. M., 1975, Isolation, identification, and synthesis of the sex pheromone of the tobacco budworm, J. Chem. Ecol., 1: 203.

Tumlinson, J. H., Mitchell, E. R., and Chambers, D. L., 1976, Manipulating complexes of insect pests with various combinations of behavior-modifying chemicals, in: "Pest Management with Insect Sex Attractants and Behavior Controlling Chemicals," M. Beroza, ed., Am. Chem. Soc. Symp. Series, 23: 53.

RESEARCH AND DEVELOPMENT OF PHEROMONES FOR INSECT CONTROL WITH EMPHASIS ON THE PINK BOLLWORM

Charles C. Doane

Thomas W. Brooks

Albany International
Controlled Release Division
Post Office Box 537
Buckeye, AZ 85326 USA

INTRODUCTION

Commercial development of pheromones for mating disruption of insects has proceeded at a brisk pace in recent years, largely because a number of unique problems encountered in research and development have been solved. In 1978 the U. S. Environmental Protection Agency granted the first commercial registration for gossyplure H. F. (now Nomate PBW®) for suppression of pink bollworm, [Pectinophora gossypiella (Saunders)], infestations in growing cotton. This product is a hollow fiber formulation containing gossyplure, the male sex attractant, and is designed to disrupt mating communication by the air permeation technique. This was the first commercial use of a pheromone for direct control of an important agricultural pest. Several seasons of experience with the Nomate PBW system, both in the southwestern desert areas of the United States and in a number of cotton growing countries around the world, have demonstrated the practical value of Nomate PBW in protecting cotton from attack by the pink bollworm.

There is a great need for a noninsecticidal, alternative method of control for this devastating cosmopolitan pest. The eggs and larvae are generally well protected by the cotton fruiting bodies and are difficult to reach with conventional insecticides. Control is accomplished through repeated applications of insecticides against the adult moths. Besides the high direct costs of the insecticides, there are other indirect penalties. Early application of insecticide removes nontarget beneficial insects that hold such important pests as *Heliothis virescens* (F.) under control. The consequences are made more serious because of the marked levels of resistance to insecticides in this species. The impact of heavy insecticide applications on the ecosystem is also a serious problem, particularly in countries of the world where residual insecticides are still in use.

It is increasingly evident that pheromone systems, when joined with appropriate biological and cultural methods, can form the basis of successful integrated pest management programs. This paper briefly surveys the development of the hollow fiber control system and discusses some of the features unique to pheromone mating disruption of pink bollworm.

EARLY HISTORY OF AIR PERMEATION TECHNIQUES

Mating communication between the sexes of pink bollworm is mediated by gossyplure, a 1:1 mixture of (*Z*,*Z*)- and (*Z*,*E*)-7,11-hexadecadienyl acetate (Hummel et al., 1973; Bierl et al., 1974). Flint et al. (1979) have shown that in most countries around the world most pink bollworm males were caught in traps baited with the 1:1 mixture, although baits with 55 or 60% of the *Z*,*Z*-isomer caught equal or, in some cases, greater numbers of moths.

Shorey and his colleagues first demonstrated that air permeation with pheromones could be effective in reducing mating frequency and thus in controlling the pink bollworm in cotton fields. McLaughlin et al. (1972) found that hexalure, *cis*-7-hexadecenyl acetate, a compound that is attractive to pink bollworm males, could disrupt mating communication. Males were unable to locate females releasing pheromone from traps set in arrays. Treatment of small fields with hexalure did not appear to reduce mating in wild females. However, further tests with hexalure with continuous release into the air throughout the cotton growing season did demonstrate clearly that mating was disrupted; this resulted in a reduction in larval boll infestations comparable to that obtained by commerical insecticide applications (Shorey et al., 1974). Interestingly, they also measured a 90% reduction in resident male and female moths in the hexalure-treated fields. This supports the contention that large numbers of moths are not attracted to treated fields from surrounding areas.

Further tests with the true pheromone, gossyplure, in hollow fibers established the concept of air permeation for control of the pink bollworm (Shorey et al., 1976; Gaston et al., 1977). These same tests also demonstrated that with use of pheromones for control of the pink bollworm, greater reductions in seasonal use of insecticides are attained.

COMMERCIAL DEVELOPMENT OF HOLLOW FIBERS

The hollow fiber system for controlled release of pheromones has been under investigation for commercial development since 1974 and release characteristics have been discussed in detail (Ashare et al., 1976; Brooks et al., 1977). While some air permeation tests have been conducted with long fibers in loops (Gaston et al., 1977), most tests have employed single chopped fibers. Baits for monitoring in traps consisted of parallel fibers fastened to a sticky tape backing.

The basic strategy for control is to apply the Nomate PBW formulation about 2 weeks before the first susceptible pin squares form on cotton. This presupposes that mated females may be present at the time of treatment but that few will survive until susceptible pin squares are available for deposition of their eggs. Those females not mated at the time of first application will not be able to attract males for mating. At this time of season the population density of moths surviving in planted cotton fields is low, and fields with high densities of moths should not be selected for a pheromone program. Repeated applications of Nomate PBW over the growing season maintain air permeation at an effective concentration and suppress the natural rate of increase of the moth, thus maintaining the infestation at levels below the economic threshold. Classic suppression patterns of pink bollworm in cotton may be seen in Shorey et al. (1974).

The first commercial application of the mating disruption system was made in 1976 in Arizona and California using the chopped fiber formulation. These were Celcon® hollow fibers measuring 1.75 cm in length with a 200 micron I.D. capillary. The fibers were sealed at random distance from the open fiber ends. The pheromone diluted with hexane was loaded into the fibers to produce a formulation of 6.8% active pheromone. This was designed to produce a field life of approximately 14 days at 37.7° C. Applications were broadcast both with ground equipment and air equipment at the rate of 2.4 to 7.4/g/ha. Applications were repeated at intervals through the growing season. Roughly 1,214 ha in 4 widely-separated locations were treated with the pheromone, and these were compared with 350 ha of fields treated with conventional insecticides. The fibers were applied with a polybutene sticker to fasten them to the foliage. All fields were monitored for disruption of mating by daily readings of monitor traps and twice-weekly samples of bolls. Good results were obtained in terms of reduction of moths caught in traps and in boll infestation compared with conventional insecticide-treated fields. Most of the pheromone-treated fields were maintained at less than 1% pink bollworm infestation through the season. Conventional practice fields required earlier insecticide treatment by at least one complete generation. Reduction in insecticide applications in pheromone fields ranged from 33% to a high of 68% compared to conventional fields.

In the 1977 growing season, 9 test areas comprising 9311 ha in Arizona and California were treated with hollow fibers (Brooks et al., 1979). The same basic system was used although application equipment was altered and improved. Treatment programs that year were under supervision of agricultural chemical fieldmen and consulting entomologists who were being introduced to pheromone technology for the first time.

At the start of the season, plans were made to carefully monitor the effectiveness of the program by using monitor traps baited with hollow fiber lures, by monitoring the infestation in the bolls and by checking for the presence of other cotton pests that might become injurious. Results again were very satisfactory although some cases of poor control were experienced. Problems with control were usually traced to one or more factors such as poor isolation in areas with high pink bollworm pressure; poor timing of the 1st treatment of Nomate PBW or improper spacing of application.

The results again clearly demonstrated that satisfactory control of pink bollworm can be accomplished with pheromones and, at the same time, the amount of hard chemical applications can be greatly reduced. The value of delaying the 1st applications of insecticides also was demonstrated. In both the Palo Verde and Imperial Valleys of California heavy outbreaks of <u>H. virescens</u> occurred following heavy rains, and pressure was aggravated by heavy insecticide use during early season. The fact that many pheromone-treated fields yielded more cotton than conventional practice fields was partly due to better natural suppression of <u>H. virescens</u> by predators.

In 1978 approximately 20,000 ha were placed under commercial treatment in the desert southwestern U.S.A. Commercial applications had been under experimental label granted by the United States Environmental Protection Agency. In March of that year a full registration was granted by the same agency. The pink bollworm was extremely abundant, particularly in Arizona, where the winter was mild and stub or ratoon cotton was cultivated after years of regulated early-plowdown to destroy ratoon cotton. The pheromone system performed well over all of the areas but because of the extremely high populations of pink bollworm, it often became necessary to terminate pheromone programs in either July or August and begin insecticide treatments. This most often occurred when dispersal of moths began after heavy populations built up in nearby fields (Brooks et al., 1979). Careful monitoring of traps and boll infestation made it possible to accurately predict the time when the change was necessary, and it allowed proper timing to avoid economic loss. Once populations of the pink bollworm reach high densities, the pheromone system will no longer prevent mating, presumably because males simply make too many chance encounters with females as they search the tops of cotton plants where females are usually located.

By contrast, the 1979 populations of pink bollworm in Arizona and California were very low due to cold, wet weather, and commercial applications performed well through the entire season. The exponential growth patterns in untreated cotton did not develop until the end of the season when applications of pheromones had ceased.

Experimental Studies in 1978-1979

Replicated field plots were established to examine the direct effect of the pheromone system on mating (Brooks et al., 1979). Previously, there had been no direct proof that the hollow fiber system actually disrupted mating in the pink bollworm. Many of the benefits of pheromone programs had been attributed to the indirect benefits of pheromone application, particularly to beneficial effects on parasites and predators that control insecticide-resistant pests such as *H. virescens*. Mating table studies were conducted by placing 25 virgin pink bollworm females on observation tables during the night. Several bouquets of cotton terminals placed in water on the tables and partial clipping of the wings helped keep the females on the tables. In early morning the females were collected and later examined for the presence of spermatophores. From 13 nights of replicated observations there was a 97% reduction in mating frequency compared with that in the control (Brooks et al., 1979). The small amount of mating that did take place occurred during peak emergence periods of the adults as measured by trap catches in untreated fields.

Monitoring Traps and Pheromone Concentrations

Pink bollworm males appear to continue searching behavior in fields permeated with normal background concentrations of pheromone capable of producing effective mating disruption. In a commercial field treated on June 13 with Nomate PBW at the rate of 4.6 g of actual gossyplure/ha, a trapping experiment was initiated on June 18 comparing catch of males in delta traps baited with various numbers of fibers (Table 1). The standard lure for small delta traps used in monitoring treated fields is 10 fibers, so 1/10, 1/2, 1 lure, 2, 4, and 8 lures were used. The test was replicated 5 times and readings were made over a period of 12 days. The test was set up in the pheromone-treated field and in a nearby untreated field. The order of the traps within replicates was changed at each reading.

There was a significant increase in catch between 1 lure and 2 lures and a clear indication, that within certain ranks, as pheromone concentration increases in traps, the chances for capture of males flying in gossyplure-treated fields also increases. In the untreated control field only one fiber caught effectively and there was little difference in catch between traps baited with 5 to 80 fibers. The interpretation of these data is that male moths are continuously searching in fields treated at normal commercial levels of gossyplure, which range from 2.8 to 5.6 grams of actual/ha. If a trap is releasing sufficient pheromone to create a concentration gradient

Table 1. Mean Catch of Pink Bollworm Males in Traps Baited with Fibers and Placed in Gossyplure-Treated and Untreated Fields.

No. of Fibers	Nomate PBW	Untreated
1	3.8 bc[a]	63 b
5	13.4 b	102 a
10	12.0 b	116 a
20	32.2 a	114 a
40	30.8 a	130 a
80	47.8 a	124 a
0	0.4 c	3.6 c

[a]Means followed by the same letter in column are not significant at 0.05% level. Student-Newman Keuls multiple range test.

with the background concentration in the field, then a searching male will be able to follow this and locate the trap. In the gossyplure-permeated field, traps baited with 2-8 lures created a gradient with the background concentration, and thereby caught more male moths. Conversely, in untreated cotton fields, even low concentrations produced effective downwind gradients. It should also be noted that the higher concentrations tested in the untreated pheromone field did not produce a depression in male catch. Males approaching the traps were apparently not deterred by the higher concentrations.

It is important to balance the lure concentrations in delta traps with the background level of gossyplure when monitoring a treatment for its effectiveness in mating disruption. Considerable observation of large-scale applications show that when traps baited with 1 lure of 10 fibers do not catch, then mating does not occur, or at least, occurs at very low levels as implied by boll infestation.

The above experiment was run following several instances in

which growers reported that, after regular application of Nomate PBW at 37 g/ha, the monitor traps in treated fields were still catching several moths/trap/night. They assumed that the pheromone system was not producing disruption of mating. After careful sampling showed that the boll infestations were at, or near zero, we found that the monitor traps that caught heavily were being baited with cotton wicks treated with pure gossyplure. Nearby delta traps baited with one lure of 10 fibers did not catch. Thus, it is clear that a proper balance must be maintained between lure concentration of traps and the background concentration of gossyplure in the field in order to monitor efficacy.

Shorey et al. (1976) discussed the process of communication disruption and suggested that the major factor in disruption was the inability of the male to detect an odor trail of the female. Their suggestion that males, continuously exposed to the odor, may become habituated and unable to search is more questionable for pink bollworm. The data here suggest that a large number of male moths continue to search in the gossyplure-treated fields but are unable to detect the weaker pheromone trails of the females. Cardé (1979) has recently discussed literature pertaining to the relationship between pheromone dispersal and trap catch. Response to concentration tends to vary between species of insects, and it is obvious that generalizations among species must be avoided. Detailed studies should be made with each insect species. The ideal concentration of lure and the correct trap design should be tested under natural conditions. Then the behavior of the species should be compared under atmospheric permeation with an effective dose of the pheromone. As suggested here, the behavior of the male in the 2 settings will probably be different.

DEMONSTRATION STUDIES IN FOREIGN COUNTRIES

Bolivia

Until late 1978 the Nomate PBW system had been tested only in the arid, hot, cotton-growing areas of the southwestern U.S.A. A Bolivian test farm 45 kilometers east of Santa Cruz was selected to test Nomate PBW under humid, tropical conditions (Brooks et al., 1980). The farm, comprising 8 cotton fields totalling 320 ha, was completely isolated from other cotton-growing areas by dense jungle. However, pink bollworm pressure was present since the farm had been heavily infested during the previous 4 years. Mating suppression of the pink bollworm was excellent (Brooks et al., 1980). For most of the season the boll infestation in the treated field remained at zero and rose only briefly to 7% at the end of the season after picking had started. By comparison, the infestation in the conventional practice fields began the somewhat typical rapid increase about midseason and it eventually rose to 100% infestation with an average of

3 larvae/boll. Rain was very abundant and caused the bottom crop of bolls to rot. This made the middle and top crop more important for yield, but it was badly damaged in the conventional practice field by the high infestation of pink bollworm. The conventional practice grower made repeated applications of insecticide for control of _H. virescens_ and _Alabama argillacea_ (Hübner), but did not apply insecticide specifically for control of the pink bollworm. The Nomate PBW-treated fields also suffered from attacks by _H. virescens_ and _A. argillacea_. Spot treatments of insecticide and 1 treatment with _Bacillus thuringiensis_ were applied early in the season to avoid decimation of the parasites and predators.

Although the total applications, including pheromone, were about the same in the 2 comparison farms, the reduction in insecticide use on the Nomate PBW fields amounted to 67%. Costs on the Nomate PBW fields were higher but the yield and fiber quality were far superior in the pheromone demonstration where there was a yield advantage of $390/ha. In addition, the grower sold all of his cottonseed for seeding purposes since it was among the cleanest in the country. In spite of heavy rains and very humid conditions the performance of Nomate PBW was outstanding.

There was a heavy dispersal phase in the spring before and at planting of the test farm. During that season we assumed that this was local dispersal but recent observations in the current season show that early dispersal occurred in newly broken land in the jungle many miles from the nearest planted cotton. This land has never had cotton planted on it so it is likely that there are alternate hosts of the pink bollworm in the jungle.

The progress of the demonstration was monitored by scientists appointed by the Government. Following approval of the final report on the project, full registration of Nomate PBW for control of the pink bollworm was granted by the Bolivian authorities. The 1979-80 cotton-growing season is now in progress and approximately 20% of the total cotton land is being treated with the hollow fibers.

Colombia

Colombian agriculturists have made great progress toward reducing their dependence on hard chemicals in cotton production and have advanced considerably in their use of biological and alternative methods of control. For example, there are some 40 _Trichogramma_ producers in the country. There are strict regulatory and cultural controls involved with suppression of the pink bollworm. These are basically directed toward reducing numbers of the insect and eliminating the need for early application of insecticides. Colombians have experienced heavy losses in cotton from insecticide-resistant _Heliothis_ in the recent past and have dramatically reduced the number of insecticide applications/season so as to avoid destruction of parasites and predators that effectively hold *H. virescens* in check.

With these conditions prevailing, the value of a control system based on mating disruption by broadcast of pheromone was readily accepted, and demonstration plots were established in 2 localities in the Cauca Valley during the 1979 cotton-growing season. The demonstration was established in cooperation with Colombian authorities and was monitored by scientists from Instituto Colombiano Agropecuario (ICA).

Two experimental application plots were established in the Buga region of the Cauca Valley and matched against 2 sets on other farms that were handled in the conventional manner. In the Roldanillo region, 1 demonstration field was established and matched against 2 conventional practice fields. There were 197 ha of cotton treated with Nomate PBW compared to about 167 ha of cotton treated with conventional pest control. The temperatures throughout the Cauca Valley are cool with average daily temperatures of about 24°C during the growing season. Planting dates for the localities, which are carefully regulated, were mostly from January 30 to February 25. Applications of pheromone began at 40 days after planting. A total of 7 applications were made by air at 21-day-intervals at the rate of 53 g/ha. Delta traps were established in all fields at the rate of 1 trap/8 ha and were used to monitor population levels and to assess the efficacy of the pheromone treatments. Boll samples were collected twice a week at the rate of 25 bolls for each 8 ha of cotton and examined for early instar larvae. Infestations of other pests such as *H. virescens* and *A. argillacea* as well as numbers of predators were also monitored.

The trapping data show that population densities in early-to mid-season were low in both test areas (Figs. 1 & 2). These low trap catches of >0.5 moths/trap/day were reflected in low boll infestations until mid- or late-season. Boll infestations at Buga in the conventional practice fields remained low until late June and then rose in the typical rate of increase until the end of the season (Fig. 3). Boll infestations at Roldanillo began to increase toward the end of May in the conventional practice fields and in both areas, differences between the pheromone fields and conventional practice fields became apparent (Fig. 4). Although the population pressure was low at the start of the season, the increase in density later was typical of that observed in many other field demonstrations in other countries. For example, they are very much like those shown in Shorey et al. (1974).

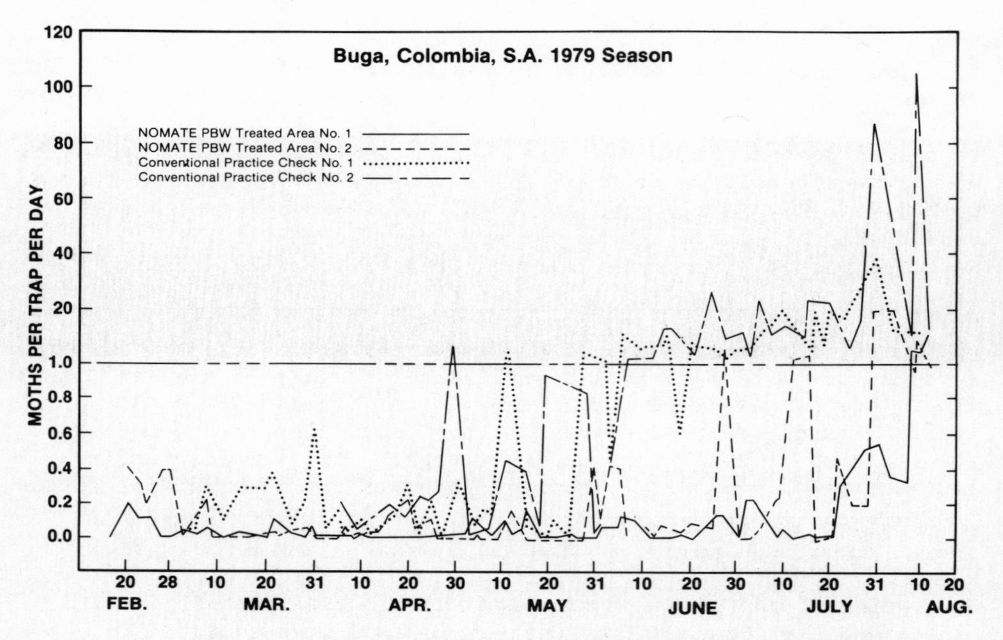

Fig. 1 Mean number of pink bollworm males caught in delta traps in 2 gossyplure-treated farms and in 2 conventional-practice farms near Buga, Colombia.

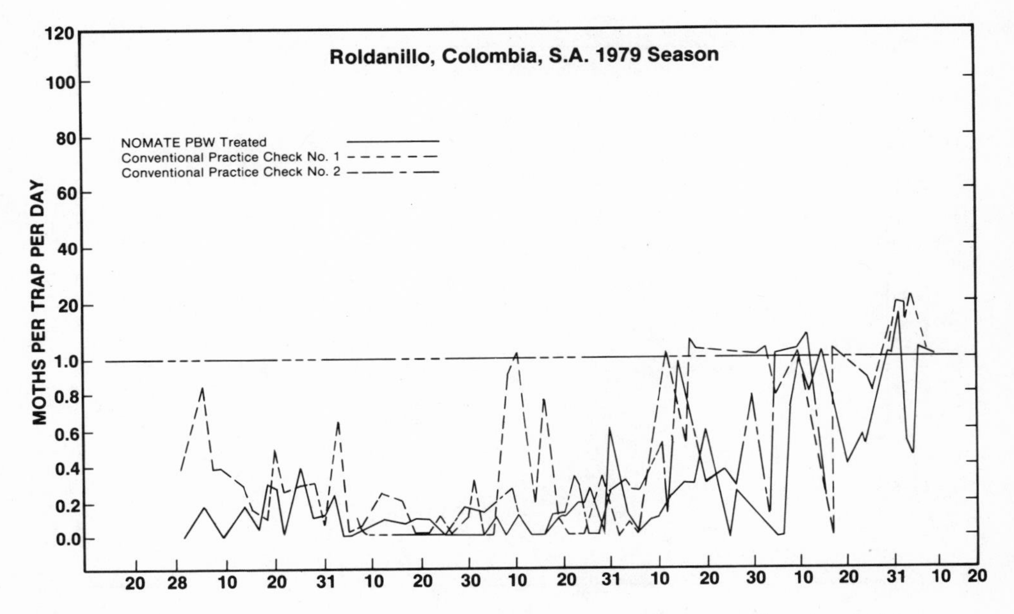

Fig. 2 Mean number of pink bollworm males caught in delta traps in a gossyplure-treated farm and in 2 conventional-practice farms near Roldanillo, Colombia.

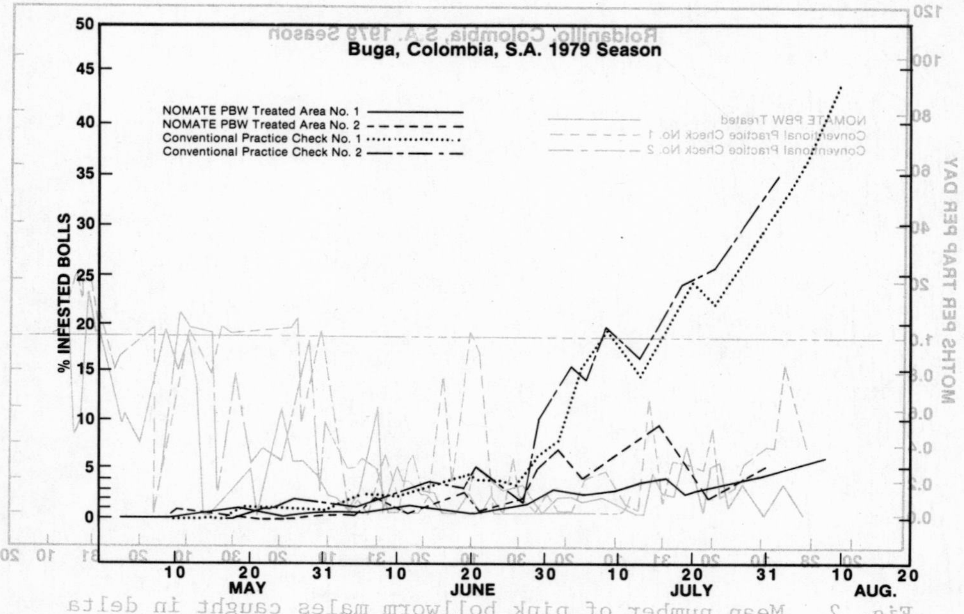

Fig. 3 Percent boll infestation in cotton on 2 farms treated with gossyplure and 2 conventional practice farms near Buga, Colombia.

At the close of the season, a month and a half after the last Nomate PBW application, the conventional practice fields in the Buga area reached about 40% infested bolls as compared to about 5% in the 2 pheromone fields. The infestation rate at Roldanillo was lower than that in Buga but the rates of increase after June 30 were remarkably similar.

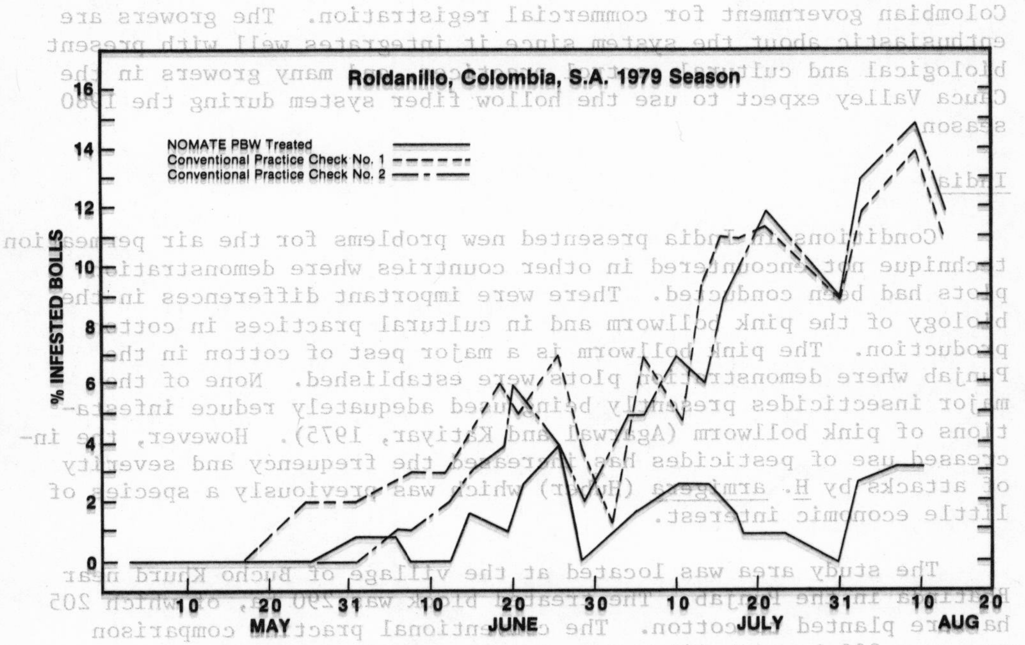

Fig. 4. Percent boll infestation in cotton on a farm treated with gossyplure and on 2 conventional-practice farms near Roldanillo, Colombia.

In general, the usual pests did not present a severe problem, mainly due to good cultural practices, use of natural beneficial insects, and restricted but timely use of insecticides. Early releases of Trichogramma were made in both pheromone and conventional practice fields. Control of A. argillacea was accomplished through application of arsenicals and Bacillus thuringiensis. H. virescens became a problem mainly in April and May and was controlled through the use of synthetic pyrethroids. Trichogramma was reintroduced when necessary. The growers and their advisors throughout the valley were aware of the need to apply as little insecticide as late as possible so that the predators and parasites would be of greatest value in controlling secondary pests. There was little difference in insecticide use between the pheromone and the conventional practice fields. Yield of seed cotton was about the same although final grades have not been received.

The results of these trials were considered highly successful in terms of suppression of the pink bollworm and data along with other results from one other region have been submitted to the Colombian government for commercial registration. The growers are enthusiastic about the system since it integrates well with present biological and cultural control practices, and many growers in the Cauca Valley expect to use the hollow fiber system during the 1980 season.

India

Conditions in India presented new problems for the air permeation technique not encountered in other countries where demonstration plots had been conducted. There were important differences in the biology of the pink bollworm and in cultural practices in cotton production. The pink bollworm is a major pest of cotton in the Punjab where demonstration plots were established. None of the major insecticides presently being used adequately reduce infestations of pink bollworm (Agarwal and Katiyar, 1975). However, the increased use of pesticides has increased the frequency and severity of attacks by *H. armigera* (Hübер) which was previously a species of little economic interest.

The study area was located at the village of Bucho Khurd near Bhatinda in the Punjab. The treated block was 290 ha, of which 205 ha were planted to cotton. The conventional practice comparison area was 223 ha with 141 ha planted in cotton, and this area was situated about 0.5 kilometers from the treated area.

Farms are small with an average holding of 4 ha. Cotton was planted from April 21 to June 10 with most of the planting taking place in May. May and June are the hottest months with high temperatures averaging between 40-45°C. The monsoon begins the first of July bringing cooler, more humid weather.

While several factors, such as planting of infested cottonseed contribute to high populations of pink bollworm, there is one major practice that makes control most difficult. In the fall, after cotton harvest, the stalks are cut and removed from the fields prior to wheat planting. The stalks are then piled in the villages and used for fuel. When such piles were examined in the village of Bucho Khurd, the seeds were heavily infested with diapausing pink bollworm larvae. The pattern of emergence of the pink bollworm in India much different than that in North or South America. In the Americas the major emergence and dispersal phase occur before or soon after cotton planting time. In India the major emergence is delayed until the monsoon rains break diapause in early July (Fig. 5). By July many of the heavily infested piles of cotton stalks remain unused and although it is known that these initiate the pink bollworm infestation each year, little appears to have been done to correct the problem (Sukhija and Sidhu, 1975).

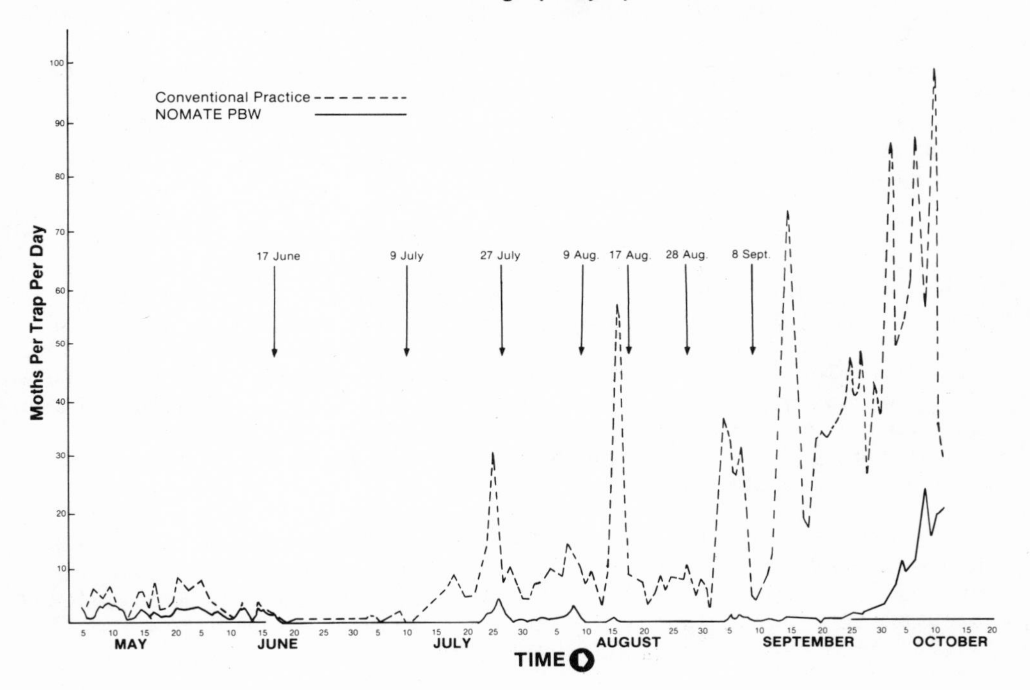

Fig. 5 Mean numbers of pink bollworm males caught in delta traps in gossyplure-treated fields and in conventional-practice fields near Bucho Khurd in India. Arrows indicate treatment dates with Nomate PBW.

Monitoring began at the start of the planting season using pheromone-baited delta traps at densities of one trap/6-8 ha. Traps were examined 6 times a week throughout the season. Further monitoring included twice-weekly sampling of flower blooms and bolls for infestation. Pheromone applications were applied from the air at rates of 50 to 75 grams of Nomate PBW/ha. Seven applications were made beginning June 18 and ending September 8. Most insecticide applications in the pheromone fields were organophosphate systemics to control cotton jassids. Other organophosphate sprays were directed toward pests other than pink bollworm. The conventional practice fields were also treated for jassid control and later with organophosphates and chlorinated hydrocarbons for control of the pink bollworm.

The trapping results show that the major emergence did not occur until early July, a few days after the first rain (Fig. 5). Although the first pheromone applications were applied in June, it is now doubtful whether an application was necessary until the first of July. First, few of the moths emerge early. Second, under prevailing conditions of inadequate irrigation water and exceedingly hot weather during May and June the cotton plants shed almost all the fruiting bodies. In fact, at this time of year, the buds are termed "pseudo-squares" since almost all of them drop to the ground. The main crop begins to develop in July when the weather cools and precipitation increases.

Traps placed in the village of Bucho Khurd and out in the fields near the village measured heavy dispersal, especially toward the pheromone-treated fields. As a result, some of the border plots in the pheromone treatment became heavily infested. Since the traps indicated effective disruption and because of the pattern of dispersal in the fields, it appears that the pheromone system was effective and that the infestation occurred largely from mated moths dispersing into the plots. Typical peaks representing adult emergence and increasing population densities were observed in traps in the conventional practice fields.

The boll infestations in the pheromone-treated fields remained approximately half that in the conventional fields throughout the season (Fig. 6). The differences were not as great as expected and this is attributed to the dispersal of so many adult moths into the fields from the piles of cotton stalks.

The seed cotton yield in the pheromone test fields was significantly greater than that from plots in the conventional practice fields. The average number of cotton plants/ha was estimated by sampling and was about identical in both areas. Using randomly selected subsampling plots, an average of 15.8 kg/10m^2 were taken from the pheromone-treated fields compared with 11.9 kg/10m^2 from the grower practice fields. This represented a 33.7% increase in yield over the grower practice fields.

In spite of the adverse conditions encountered, the Nomate PBW system performed well. On the basis of farming practices and a different life cycle of the pink bollworm, it should be possible to delay first treatment of cotton in the Punjab until just before the first rains in July. Emergence during the extremely hot, dry months of May and June is limited and most fruiting bodies drop from the plants so that the moth could not effectively reproduce. Further observations of the applications also indicate that at least one other late application of pheromone could also be eliminated. Sanitation, leading to destruction of piles of cotton stalks before monsoon rains begin, will be necessary for economic application of pheromone.

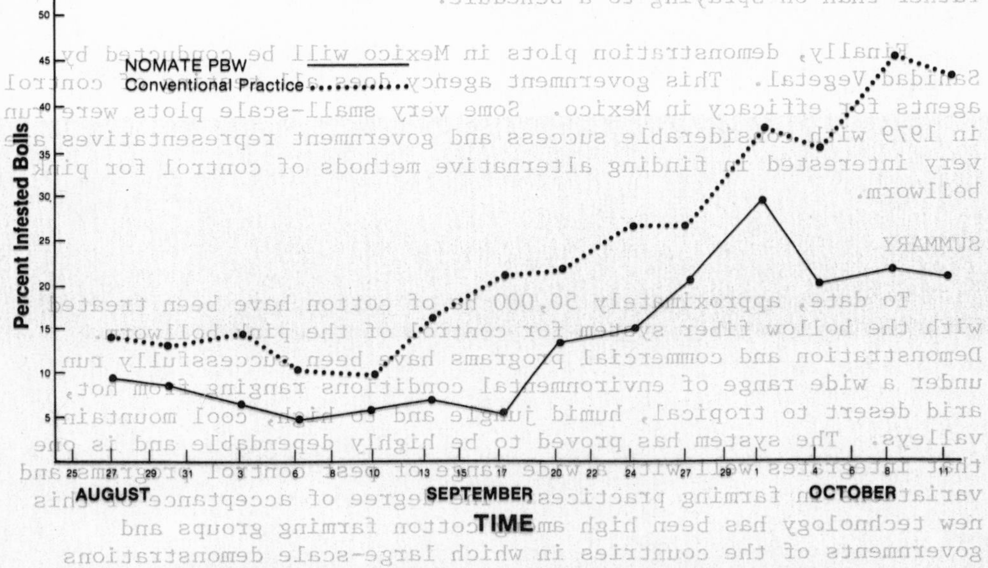

Fig. 6 Boll infestation from pink bollworm in gossyplure-treated fields and in the conventional-practice fields near Bucho Khurd, India.

Other Countries

Large scale demonstrations of efficacy of the hollow fiber system against pink bollworm are in progress or completed in several countries other than those discussed above. During the 1978-79 season a demonstration was run in Chaco province in Argentina with rather poor results. This was largely due to poor isolation of the test fields coupled with a severe drought that occurred about 2 months after planting and lasted for at least 6 weeks. The plants stopped growth and shed most of the fruit with some cotton actually dying from lack of water. The cotton was badly damaged and yields were very low. It was uneconomical to treat the cotton during the dry period but applications had to be made since the system demands that constant air permeation be maintained.

A large demonstration plot is just nearing completion in Brazil and although final results are not in, they are very promising. Control of the pink bollworm was very effective although there was only a light population in the area until the latter part of the growing season. A major factor in the demonstration was the striking reduction in the use of pesticides in the pheromone-treated fields. Careful monitoring of the fields allowed for the reduction in hard chemicals, and applications during the season were based on need rather than on spraying to a schedule.

Finally, demonstration plots in Mexico will be conducted by Sanidad Vegetal. This government agency does all testing of control agents for efficacy in Mexico. Some very small-scale plots were run in 1979 with considerable success and government representatives are very interested in finding alternative methods of control for pink bollworm.

SUMMARY

To date, approximately 50,000 ha of cotton have been treated with the hollow fiber system for control of the pink bollworm. Demonstration and commercial programs have been successfully run under a wide range of environmental conditions ranging from hot, arid desert to tropical, humid jungle and to high, cool mountain valleys. The system has proved to be highly dependable and is one that integrates well with a wide range of pest control programs and variations in farming practices. The degree of acceptance of this new technology has been high among cotton farming groups and governments of the countries in which large-scale demonstrations have been conducted.

REFERENCES

Agarwal, K. A., and Katiyar, K. N., 1975, Effects of insecticide against jassid and pink bollworm in cotton, Pesticides, 9: 30.

Ashare, E., Brooks, T. W., and Swenson, D. W., 1976, Controlled release from hollow fibers, in: "Controlled Release Polymeric Formulations," Paul, D. R. and Harris, F. W. eds., ACS Symp. Series 33, American Chemical Society, Washington, DC.

Bierl, B. A., Beroza, M., Staten, R. T., Sonnet, P. E., and Adler, V. E., 1974, The pink bollworm sex attractant, J. Econ. Entomol., 67: 211.

Brooks, T. W., Ashare, E., and Swenson, D. W., 1977, Hollow fibers as controlled release devices, in: "Textile and Paper Chemistry and Technology," J. C. Arthur, ed., ACS Symp. Series 49, American Chemical Society, Washington, DC.

Brooks, T. W., Doane, C. C., and Staten, R. T., 1979, Experience with the first commercial pheromone communication disruptive for suppression of an agricultural pest, *in*: "Chemical Ecology: Odour Communication in Animals," F. J. Ritter, ed., Elsevier/ North-Holland Biomedical Press, Amsterdam.

Brooks, T. W., Doane, C. C., and Haworth, J. K., 1980, Suppression of *Pectinophora gossypiella* with sex pheromone, Proceedings, 1979 British Crop Protection Conference - Pests and Diseases, The Boots Co. Ltd., Nottingham.

Cardé, R. T., 1980, Behavioral responses of moths to female-produced pheromones and the utilization of attractant-baited traps for population monitoring, *in*: "Highly Mobile Insects: Concepts and Methodology in Research," R. L. Rabb and G. G. Kennedy, eds., N. C. State Univ. Press, Raleigh.

Flint, H. M., Balasubramanian, M., Campero, J., Strickland, G. R., Ahmad, Z., Barral, J., Barbosa, S., and Khalil, A. F., 1979, Pink bollworm responses of native males to ratios of *Z*,*Z* and *Z*,*E*-isomers of gossyplure in several cotton growing areas of the world, *J. Econ. Entomol.*, 72: 758.

Gaston, L. K., Kaae, R. S., Shorey, H. H., and Sellers, D., 1977, Controlling the pink bollworm by disrupting sex pheromone communication between adult moths, *Science*, 196: 904.

Hummel, H. E., Gaston, L. K., Shorey, H. H., Kaae, R. S., Byrne, K. J., and Silverstein, R. M., 1973, Clarification of the chemical structure of the pink bollworm sex pheromone, *Science*, 181: 873.

McLaughlin, J. R., Shorey, H. H., Gaston, L. K., Kaae, R. S., and Stewart, F. D., 1972, Sex pheromones of Lepidoptera. XXXI. Disruption of sex pheromone communication in *Pectinophora gossypiella* with hexalure, *Environ. Entomol.*, 1: 645.

Shorey, H. H., Gaston, L. K., and Kaae, R. S., 1976, Air-permeation with gossyplure for control of the pink bollworm, *in*: "Pest Management with Insect Sex Attractants," M. Beroza, ed., ACS Symposium Series 23, American Chemical Society, Washington, DC.

Shorey, H. H., Kaae, R. S., and Gaston, L. K., 1974, Sex pheromones of Lepidoptera. Development of a method for pheromonal control of *Pectinophora gossypiella* in cotton, *J. Econ. Entomol.*, 67: 347.

Sukhija, H. S., and Sidhu, A. S., 1975, Studies on the carryover of pink bollworm *Pectinophora gossypiella* in the Punjab, *Indian J. Ecol.*, 2: 51.

FORMATE PHEROMONE MIMICS AS MATING DISRUPTANTS OF THE STRIPED RICE BORER MOTH, CHILO SUPPRESSALIS (WALKER)

P. S. Beevor

Tropical Products Institute
56/62 Gray's Inn Road
London, WCIX 8LU England

V. A. Dyck and G. S. Arida

International Rice Research Institute
Post Office Box 933
Manila, Philippines

INTRODUCTION

The striped rice borer moth, Chilo suppressalis (Walker), is a severe pest of rice throughout the Far East and Asia. Extensive damage is caused by the larva boring into the stem of the plant where it feeds, thus preventing the panicles from filling to form rice grains. Without close inspection, damage only becomes apparent when unfilled panicles show up as white heads near harvest. Resistant varieties of rice afford only partial protection against this pest, and the use of pesticides can lead to insect resistance and environmental problems. Therefore, identification and synthesis of the sex pheromone of this species was undertaken in the hope that an alternative means of control by mating disruption could be found.

The female sex pheromone of C. suppressalis was identified as a mixture of (Z)-11-hexadecenal (HDA) and (Z)-13-octadecenal (ODA) (Nesbitt et al., 1975). Field trials in the Philippines, Korea, and Iran demonstrated the attractiveness of synthetic HDA and ODA to male moths (Beevor et al., 1977). Further trials in the Philippines showed both the pheromone components and their structurally-related but chemically more stable analogues, (Z)-9-tetradecenyl formate (TDF) and (Z)-11-hexadecenyl formate (HDF), to be efficient communication disruptants (Beevor et al., 1977).

In 1978, mating disruption experiments in $100m^2$ field cages demonstrated that TDF and HDF pheromone mimics can disrupt mating in this species by 100% at rates of loss in excess of 2 g/ha/day (Beevor and Campion, 1979). This paper describes further field cage experiments to determine the effects of lower concentrations of these compounds on mating disruption of *C. suppressalis* in 16 m^2 field cages.

MATERIALS AND METHODS

Field Cages

A rice variety highly susceptible to stem borer attack, IR 127-80-1, was transplanted on the experimental farm at the International Rice Research Institute. Thirty days after transplant and before natural infestation by *C. suppressalis*, cages (4 m x 4 m x 2 m high) consisting of a wooden frame covered by nylon mesh were constructed within the rice paddy. In this way, a nearly normal crop was grown in isolation from immigrant moths.

TDF and HDF Dispensers

TDF and HDF were dispensed from polyethylene vials in a ratio of 4.5:1 at total loadings of 0.1, 0.5, and 1.0 mg/vial and containing equal amounts of 2, 6-di-*tert*-butyl-p-cresol as antioxidant (Beevor et al., 1977).

As communication disruption studies (unpublished data) indicated that the disruptive source height was important for efficient disruption, vials were placed at 2 levels in the cages, 10 cm and 30 cm below the top of the flag leaf. They were arranged in an 0.5 m^2 lattice, and one additional row of vials at the 2 heights was placed encircling the cages to permeate air entering the cage. The vials were placed in position 40 days after transplant. The cages containing 0.1 and 0.5 mg vials and the untreated control cages were replicated once. The cage containing 1.0 mg vials was not replicated since previous studies (Beevor and Campion, 1979) had shown this treatment causes 100% mating disruption up to 18 days.

To determine the rate of loss for TDF and HDF from the vials, additional vials were placed in the open field at a short distance from the cages and duplicate samples were taken at intervals during the experiment. These were analyzed for residual formate by extraction with chloroform followed by gas chromatographic analysis of the extract.

Insect Material

Pupae obtained by dissection of field-collected, infested rice stems were sexed and reared separately in the laboratory. Between

1 and 4 pairs of male and female C. suppressalis moths thus obtained were introduced into the cage on alternate days. Each cage received 40 pairs during the 30-day experiment. The first introduction was one day after placement of the TDF and HDF vials; this delay ensured that the vials had reached their full emission level prior to introduction of the moths.

Throughout the experiment each cage was searched for egg masses on alternate days. The eggs were collected and incubated at 28°C and 80% RH in the laboratory; the number of emerging larvae provided a measure for the degree of mating in each of the treatments.

RESULTS AND CONCLUSIONS

Rates of Loss for TDF and HDF

Gas chromatographic analysis of TDF and HDF remaining in 1 mg vials after field exposure showed a rapid loss of material between Days 0 and 4, with an average rate of loss of 8.04 g/ha/day decreasing to 2.16 g/ha/day between Days 4 and 8. Between Days 8 and 32 the rate slowed markedly to 0.94 g/ha/day. The relative rate of loss was similar with 0.1 and 0.5 mg vials. These data are presented in Figure 1.

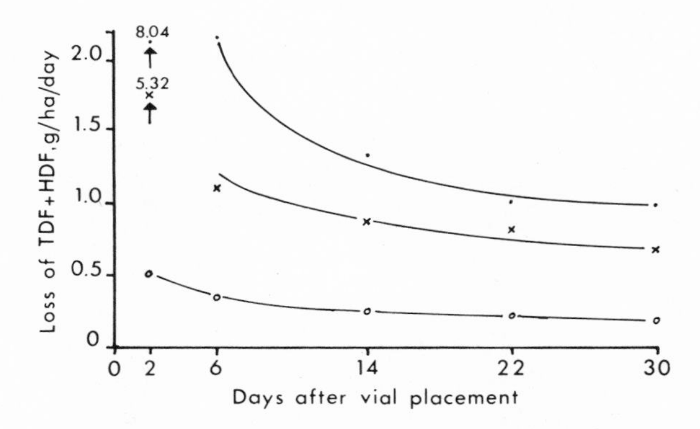

Fig. 1 Rate of loss of TDF and HDF from vials at three loadings after field exposure.

Rates of loss were calculated from the mean of each replicated sample and the average loss of TDF and HDF over 2 days before and 2 days after the sample day. These data agree with rate of loss studies carried out in a laboratory wind-tunnel (Beevor et al., 1977) and indicates that loss of TDF and HDF is due to release into the atmosphere and not to degradation on exposure to tropical field conditions.

Mating Disruption

The number of larvae emerging from eggs collected from each of the cages are presented in Figure 2. The average numbers of larvae emerging from each treatment were 384.5 (0.1 mg vials), 135.5 (0.5 mg vials), 60 (1 mg vials, no replication) and 1601.5 (untreated controls). On this basis the levels of mating disruption achieved throughout the experiments were 76, 92 and 96%, respectively.

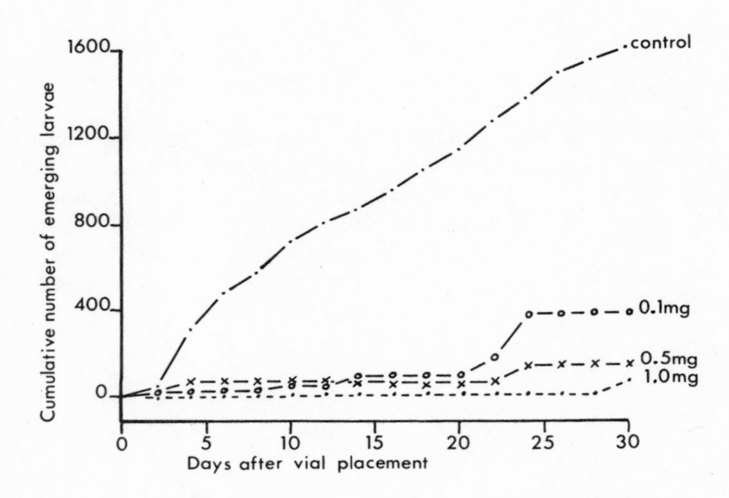

Fig. 2 Effect of vial loading of TDF and HDF on oviposition by C. suppressalis in 16 m^2 field cages.

A comparison of the rate of loss for TDF and HDF and viable egg laying by C. suppressalis in the field cages with the 3 different treatment levels is presented in Table 1. Increasing rate of loss for TDF and HDF from the 0.1, 0.5 and 1.0 mg vials resulted in increasing levels of mating disruption. Decreasing loss of TDF and HDF from Days 14 to 30, when less than 1.1 g/ha/day, resulted in decreasing levels of disruption. These data are presented in Figure 3.

Table 1 Effect of TDF and HDF rate of loss on mating by *C. suppressalis* in 16 m^2 field cages.

Initial vial loading		0.1mg	0.5mg	1.0mg
Day 14	rate of loss[+]	0.24	0.87	1.34
	% disruption[‡]	91	93	100
Day 22	rate of loss[+]	0.22	0.81	1.08
	% disruption[‡]	86	95	100
Day 30	rate of loss[+]	0.18	0.67	0.94
	% disruption[‡]	76	92	96

[+] g/ha/day

[‡] based on emerging larvae

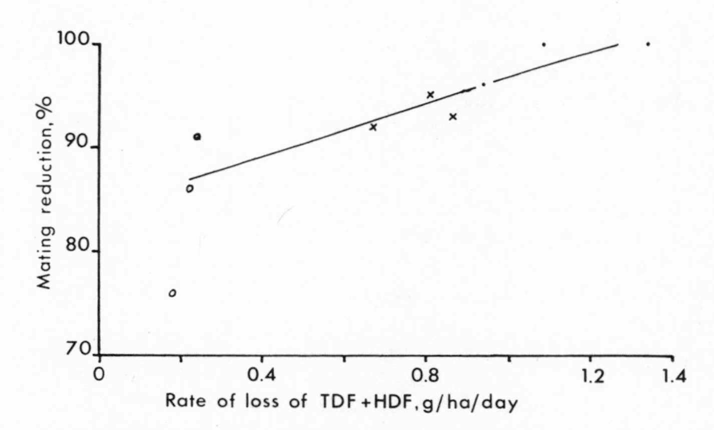

Fig. 3 Effect of TDF and HDF rate of loss on mating by *C. suppressalis* in 16 m^2 field cages.

This shows a marked correlation of TDF and HDF rates of loss between 0.25 and 1.10 g/ha/day and mating disruption between 90 and 100%. Thus, an estimate can be made of the amount of TDF and HDF needed to reduce the number of viable eggs and subsequent larval populations to levels which are not of economic significance.

DISCUSSION

These experiments have demonstrated the feasibility of disrupting mating in C. suppressalis using the formates TDF and HDF, compounds which are structurally-related to the natural pheromone components but which have greater stability under field conditions. The experiments were carried out in cages set in growing rice to avoid immigration of mated female moths into the test areas, and the conducting of open field trials is a necessary next step. However, the results from the control cages show that C. suppressalis mates readily in the cage environment when inhibitory chemicals are absent. This encourages us to believe that the results obtained in the cages will be reproducible in the open field, provided a sufficiently large area is treated to overcome the problem of gravid female immigration.

The experiments described have defined the amount of the 2 formates, which need to be released daily into the atmosphere to effect a high percentage of mating disruption. Interestingly, this figure of approximately 1 g/ha/day for the C. suppressalis pheromone mimics is of the same order as the amount 1.6-0.4 g/ha/day of inhibitory pheromone needed to disrupt mating of Diparopsis castanea (Hmps.) in a large field cage (Marks, 1976). It is also comparable with that (0.4-0.8 g/ha/day) of synthetic attractant pheromone reported as necessary to markedly reduce Pectinophora gossypiella (Saunders) mating in the open field (Brooks et al., 1979).

The C. suppressalis populations in our experimental cages were higher than normal field infestation in this part of the Philippines. It is not known if the mating disruption achieved is population density dependent or whether percentage of mating disruption would fall at still higher moth populations. Nor is it known whether this effect (if it occurred) could be countered by higher atmospheric concentration of the formates. However, if these compounds are to be used to control C. suppressalis in the field, application starting early in the season when populations are lowest would almost certainly provide the best chance of success.

REFERENCES

Beevor, P. S., and Campion, D. G., 1979, The field use of 'inhibitory' components of lepidopterous sex pheromones and pheromone mimics, in: "Chemical Ecology: Odour Communication in Animals", F. J. Ritter, ed., Elsevier/North Holland Biomedical Press, Amsterdam.

Beevor, P. S., Hall, D. R., Nesbitt, Brenda, F., Dyck, V. A., Arida, G. S., Lippold, P. C., and Oloumi-Sadeghi, H., 1977, Field trials of the synthetic sex pheromones of the striped rice borer, Chilo suppressalis (Walker) (Lepidoptera: Pyralidae), and related compounds, Bull. Entomol. Res., 67:439.

Brooks, T. W., Doane, C. C., and Staten, R. T., 1979, Experience with the first commercial pheromone communication disruptive for suppression of an agricultural insect pest, in: "Chemical Ecology: Odour Communication in Animals", F. J. Ritter, ed., Elsevier/North Holland Biomedical Press, Amsterdam.

Marks, R. J., 1976, The influence of behaviour modifying chemicals on mating success of the red bollworm, *Diparopsis castanea* Hmps., (Lepidoptera, Noctuidae) in Malawi, *Bull. Entomol. Res.* 66: 279.

Nesbitt, Brenda F., Beevor, P. S., Hall, D. R., Lester, R., and Dyck, V. A., 1975, Identification of the female sex pheromones of the moth *Chilo suppressalis*, *J. Insect Physiol.* 21: 1883.

DISRUPTION OF SEX PHEROMONE COMMUNICATION IN *CHILO SUPPRESSALIS* WITH PHEROMONE AND ANALOGS

Sadahiro Tatsuki

Laboratory of Insect Physiology & Toxicology
The Institute of Physical & Chemical Research
Wako, Saitama 351, Japan

Hiroo Kanno

Laboratory of Applied Entomology
Hokuriku National Agricultural Experiment Station
Joetsu, Niigata 943-01, Japan

INTRODUCTION

Chilo suppressalis (Walker) is the most important species of several rice borers in the genus *Chilo* (Pyralidae) in Asia. The sex pheromone of this species was previously identified from females as a mixture of 2 homologous aldehydes, (*Z*)-11-hexadecenal (Z-11-HDAL), and (*Z*)-13-octadecenal (Z-13-ODAL) (Nesbitt et al., 1975; Ohta et al., 1976). We then initiated research on the application of the sex pheromone for control of this pest. Factors affecting the attractiveness of the synthetic pheromone have been examined in the field (Tatsuki et al., 1979), and pheromone release rate was shown to be a critical factor for attraction of males. Namely, rubber septa loaded with 1 or 2 mg of the pheromone (5:1 mixture of Z-11-HDAL and Z-13-ODAL) were attractive, while 0.1 or 10 mg dispensers showed almost no activity. These results indicated that the release rate of pheromone for male attraction was narrow, and suggested that mating disruption may be a profitable approach for direct control of *C. suppressalis*. Because aldehydes are very unstable in the field, we attempted to identify disruptants with more stability than the pheromone components from structurally related compounds. We synthesized the sex pheromone and 12 analogs including compounds having the same alkyl moieties as the pheromone components and other functional end groups; i.e., alcohol, acetate, and methyl groups.

From the results of preliminary field trials and electroantennogram (EAG) tests, we screened out (Z)-5-hexadecene (Z-5-HD) as a possible substitute for the pheromone component Z-11-HDAL (Kanno et al., 1978). This paper deals with the disruptive effect of Z-5-HD and other pheromone-related compounds on sex pheromone communication of C. suppressalis and the possible mechanism of this disruption.

DISRUPTION OF SEX PHEROMONE COMMUNICATION

Inhibitory activities of the sex pheromone components and 12 structurally-related compounds (Table 1) were tested in paddyfields in Niigata Prefecture, Japan in June 1977 against orientation of C. suppressalis males to female-baited traps. Each compound was tested individually on a cotton wick placed in a trap with 2 virgin females. The effect of each compound was shown by comparing the total number of males caught in treated traps with those captured in control traps baited with females only. As shown in Table 1, all of the compounds tested had considerable inhibitory effect on male orientation to virgin females. The sex pheromone components, as well as their E-isomers and alcohol, acetate and methyl substitutes were very effective. (Z)-5-Hexadecene and (Z)-5-octadecene (Z-5-OD), although of extremely less polarity than the pheromone components and the other derivative groups, were also potent inhibitors of male attraction.

In some Lepidoptera, non-pheromone chemicals that inhibit male orientation to pheromone-baited traps have little or no inhibitory effect when dispensed into the atmosphere surrounding the traps (Mitchell et al., 1974; Rothschild, 1974). Therefore, we conducted another field test in which the compounds were evaporated over rather wide areas around pheromone traps. The pheromone components and 8 pheromone-related compounds that showed relatively potent inhibition in the previous test were used. Twelve polyethylene vials, each containing 100 mg of the test chemical, were evenly spaced in 2 concentric circles of 1 and 2 m radius (Test A), or one circle of 5 m radius (Test B) with a virgin female-baited trap at the center. The results are shown in Table 2. Relatively good inhibition of male attraction was achieved with most of the compounds used in Test A. However, even in Test B, in which the dispensers were more sparsely placed around the traps, considerable inhibitory effect was obtained with most of the compounds. These results suggested that sex pheromone communication of C. suppressalis could be disrupted by permeating the atmosphere with the pheromone components or related chemicals.

With the exception of the aldehydes, the inhibitory effects observed with each of the C-18 compounds were less than that of the corresponding C-16 compounds used in Test A. This might have been because of the difference in release rate between the C-16 and the

Table 1. Inhibitory Effect of Sex Pheromone Components and 12 Analogs Placed in Female-Baited Traps on Orientation of C. suppressalis males to traps.[a]

Compound	Males caught[b]	% Reduction[c]
(Z)-11-Hexadecenal	0	100 c
(Z)-13-Octadecenal	15	89 c
(E)-11-Hexadecenal	7	95 c
(E)-13-Octadecenal	7	95 c
Hexadecanal	37	73 b
Octadecanal	26	81 bc
(Z)-5-Hexadecene	13	90 c
(Z)-5-Octadecene	1	99 c
(E)-5-Hexadecene	32	77 b
(E)-5-Octadecene	35	74 b
(Z)-11-Hexadecen-1-ol	3	98 c
(Z)-13-Octadecen-1-ol	2	99 c
(Z)-11-Hexadecenyl acetate	1	99 c
(Z)-13-Octadecenyl acetate	7	95 c
Control	136	0 a

[a]Data from Kanno et al. (1978).

[b]Two replicates for 14 successive nights. Treatments were rotated each day.

[c]Percentages followed by the same letter are not significantly different (P = 0.05, Duncan's multiple range test).

C-18 compounds since release rates of the former were undoubtedly much higher than those of the latter.

The inhibitory activities of Z-5-HD and Z-11-HDAL on attraction of C. suppressalis males were compared in a field test using different quantities of each compound. Six rubber septa of each dosage (treatment) were spaced evenly in a circle (1 m radius) with a virgin female-baited trap at the center. Plots were separated by a minimum of 150 m. The results of this test are shown in Table 3. The disruptive effect of Z-11-HDAL was greater than Z-5-HD. Nevertheless, Z-5-HD is potentially useful as a mating disruptant for C. suppressalis because of its greater stability and ease of synthesis.

Table 2. Inhibitory Effect of Sex Pheromone Components and 8 Analogs Surrounding Female-Baited Traps on Orientation of C. suppressalis Males to Traps.

Compound	Test A		Test B	
	Males caught[a]	Percent reduction[b]	Males caught[a]	Percent reduction[b]
(Z)-11-Hexadecenal	1	99 a	12	91 a
(Z)-13-Octadecenal	3	98 a	35	73 a
(E)-11-Hexadecenal	17	89 ab	45	66 a
(E)-13-Octadecenal	15	90 ab	15	89 a
(Z)-5-Hexadecene	1	99 a	8	94 a
(Z)-5-Octadecene	50	67 c	39	70 a
(Z)-11-Hexadecen-1-ol	1	99 a	28	79 a
(Z)-13-Octadecen-1-ol	42	73 bc	37	72 a
(Z)-11-Hexadecenyl actate	71	54 c	44	66 a
(Z)-13-Octadecenyl acetate	148	3 d	91	31 b
Control	153	0 d	131	0 c

[a]Total of 2 replicates for 5 nights.

[b]Percentages followed by the same letter within a column are not significantly different (P = 0.05, Duncan's multiple range test).

Table 3. Comparison of Inhibitory Activities Between (Z)-11-hexadecenal and (Z)-5-hexadecene against Orientation of Male C. suppressalis to Female-Baited Traps.

Compound	Mg/ septum	Males caught[a]	% Inhibition[b]
(Z)-11-Hexadecenal	0.1	92	57 cd
	0.3	43	80 abc
	1	12	94 ab
	3	3	99 a
	10	0	100 a
(Z)-5-Hexadecene	0.1	176	17 ef
	0.3	87	59 bcd
	1	125	41 de
	3	80	62 bcd
	10	30	86 abc
Control	-	212	0 f

[a]Total of 2 replicates for 8 nights. Treatments were re-randomized every other day.

[b]Percentages followed by the same letter are not significantly different (P = 0.05, Duncan's multiple range test).

SUPPRESSION OF MATING WITH PHEROMONE

Components and Hydrocarbons

The sex pheromone components and the hydrocarbons, Z-5-HD and Z-5-OD, were field tested in June 1979 in Niigata to see if mating suppression of C. suppressalis could be achieved simultaneously with inhibition of male attraction under conditions similar to those used in the previous tests. Eight polyethylene vials (10 mg compound/vial) were spaced in a circle of 3 m radius in which 15 virgin females (evenly spaced) were tethered on a cotton thread attached to a plastic rod. The moths were replaced daily during the test period; females recovered were dissected to determine mating status. The plots were separated by a minimum of 150 m. Release rates of the compounds from the dispensers were determined by direct weighing in the laboratory.

The effect of each compound on mating is shown in Table 4. All compounds tested significantly suppressed mating. Each C-16 compound was more effective in suppressing mating than the corresponding C-18 compound. This difference in effectiveness might reflect differences in release rate. The release rate of each hydrocarbon was several times greater than that of the corresponding aldehyde.

Table 4. Suppressive Effect of Sex Pheromone Components and Hydrocarbons on Mating by *C. suppressalis*.

Compound	Females collected[a]	% Mating suppression	Release rate (μg/vial/day)[c]
(*Z*)-11-Hexadecenal	212 (8 a)	86	91
(*Z*)-13-Octadecenal	209 (26 bc)	54	40
(*Z*)-5-Hexadecene	223 (18 b)	69	536
(*Z*)-5-Octadecene	221 (32 c)	45	144
Control	218 (57 d)	0	-

[a]Total of 2 replicates for 8 nights. Treatments were re-randomized every other day, figures in ( ) = % females mated.

[b]Percentages followed by the same letter are not significantly different (*P* = 0.05, Duncan's multiple range test).

[c]Means for 8 days in the laboratory.

Large Scale Communication Disruption Tests with Z-5-HD

The potential usefulness of Z-5-HD as a mating disruptant for *C. suppressalis* was demonstrated in several tests previously described. Therefore, we attempted a large-scale disruption experiment to see whether crop damage caused by *C. suppressalis* larvae could be decreased with Z-5-HD. The test was carried out in a paddyfield at Niigata in the summer of 1979. A total of 6 experimental blocks were used, 2 of which were treated with the Z-5-HD and the other 4 were untreated controls. Each block was ca. 2000 m^2 and separated by 200 to 300 m. The 2 treated areas were ca. 600 m apart. The compound was impregnated in synthetic rubber film (40% Z-5-HD in styrene-isoprene copolymer, 60 μ thick), a side of which was covered with polypropyrene film (25 μ) to control the release rate. The other side was covered with polyester film (25 μ) as a barrier. The laminated film was cut into 1 x 4 cm strip dispensers. Release rate was estimated at approximately 600 μg/dispenser/day (range 300-1000 μg) for 24 days by direct weighing in the laboratory (at ca. 30°C).

The dispensers were attached to fishing lines at 1 m intervals. The lines were strung 20 to 30 cm above the plants throughout the test field at intervals of 1.5 m. Total number of dispensers used in each treated area was 1221 for block I (2002 m^2) and 1120 for block II (1999 m^2), respectively. The total quantity of Z-5-HD evaporated during the test period was ca. 35 g (87.5 g/ha). Treatments were in place from July 31 to August 25. This period coincided with the 2nd moth flight which usually occurs from late July to late August with the peak period around August 10. We intended to monitor the disruptive effect of Z-5-HD with female-baited traps during the test period, but we terminated the test halfway through the period because of extremely low male catches, even in the control traps. No insecticide was sprayed on the field during or after the experiment.

The effect of Z-5-HD was evaluated early in September from the degree of the larval injury in rice stubble checked immediately after harvest. The data showed significantly less injury to the plants in the Z-5-HD treated plots than in the control areas (Table 5). This suggests that mating by C. suppressalis was decreased by the treatment, despite the relatively small areas used.

Table 5. Effect of (Z)-5-Hexadecene on Injury in Rice Plants by Larvae of C. suppressalis.

		Degree of injury[a]	
Treatment	Block	% Stubble injured[b]	% Stem injured[b]
(Z)-5-Hexadecene	I	14.2 a	1.01 a
	II	12.2 a	1.04 a
Untreated	III	32.4 c	2.60 d
	IV	25.8 b	2.17 c
	V	25.2 b	1.54 b
	VI	31.6 bc	2.77 d

[a]For evaluation, 500 stubbles were randomly sampled from each block.

[b]Percentages in each column followed by the same letter are not significantly different ($P = 0.05$, χ^2 test).

LABORATORY BIOASSAY OF DISRUPTANT CHEMICALS

A wind tunnel (45 x 45 x 240 cm, air flow rate 0.3-0.4 m/sec) was used in the laboratory to observe the sexual behavior of male C. suppressalis in the presence of a pheromone-emitting female. Male moths released in the tunnel downwind of a calling female displayed typical pheromone response behavior such as upwind orientation of flight including casting and hovering, landing on the cage near the female, and wing fanning while walking around the cage (mating dance). The effects of Z-5-HD and Z-11-HDAL on the response of males to pheromone was tested by placing filter papers treated with 500 μg of chemical in the wind tunnel ca. 10 cm downwind of the female-baited cage, so that the pheromone plume would be permeated with the compound. Ten to 15 male moths were introduced into the wind tunnel from an inlet hole near the downwind end. The behavior of the males

Table 6. Effect of (Z)-11-Hexadecenal and (Z)-5-Hexadecene on Pheromone Behavior of *C. suppressalis* Males in Laboratory Wind Tunnel.

Male response[a]	No. of males responding[b]							
	Control I[c]		Control II[d]		Z-5-HD		Z-11-HDAL	
++	0[e]	0	6	6	0	3	0	0
+	0	0	0	0	0	2	1	0
-	6	6	0	0	6	1	5	6

[a]++ = Two or more males simultaneously performed "mating dance," + = individual male(s) performed "mating dance," - = no pheromone behavior was observed.

[b]Number of tests in which each response grade was observed, 6 replicates.

[c]Solvent and blank cage.

[d]Solvent and virgin female-confined cage.

[e]Left column in each treatment or control group represents 1st 5 min observation and the right column the 2nd 5 min observation (see text).

was observed and recorded for the 1st 5 min in the presence of the treated filter paper, and for an additional 5 min after the removal of the filter paper. Table 6 shows that Z-5-HD and Z-11-HDAL completely blocked the response of the males to pheromone. The difference in response to Z-5-HD and Z-11-HDAL occurred after the removal of the treated filter paper. When Z-5-HD was removed, the males began to respond to the pheromone within a few minutes, while the effect of the Z-11-HDAL remained throughout the observation period. These data suggest that the mechanism involved in the behavioral inhibition by Z-5-HD is similar to that caused by Z-11-HDAL.

ELECTROANTENNOGRAM (EAG) EXPERIMENTS

The EAG technique (Ando et al., 1980) was used to evaluate the response of C. suppressalis males to the sex pheromone components, Z-11-HDAL and Z-13-ODAL, and the hydrocarbons, Z-5-HD and Z-5-OD. The 2 pheromone components gave higher EAG responses than either of

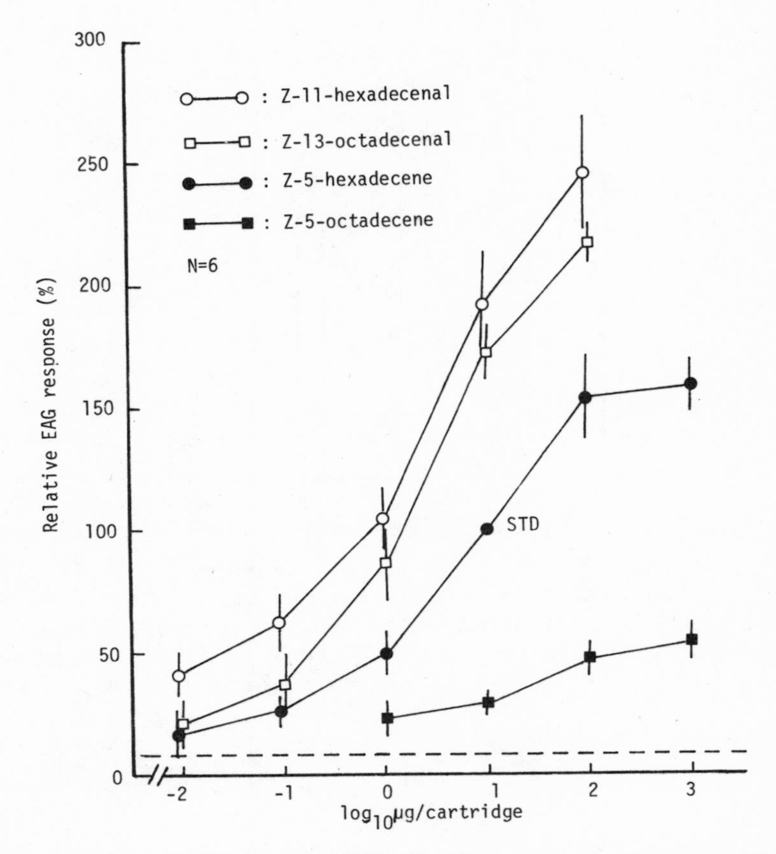

Fig. 1. EAG responses from male antennae of C. suppressalis to various concentrations of sex pheromone components and hydrocarbons. Ten μg of (Z)-5-hexadecene was used as the standard.

the analogs (Fig. 1); (Z)-5-hexadecene, which was much more active than Z-5-OD, was much less active than Z-11-HDAL. It is notable that the difference observed in EAG activity between Z-5-HD and Z-11-HDAL corresponded well with the observed inhibitory effects on male attraction in the field.

EAG responses of C. suppressalis females were also recorded. The results in Figure 2 suggest that Z-5-HD and probably Z-5-OD as well do not act on the general odor receptors, but rather on receptors on male antennae specifically adapted for pheromone perception.

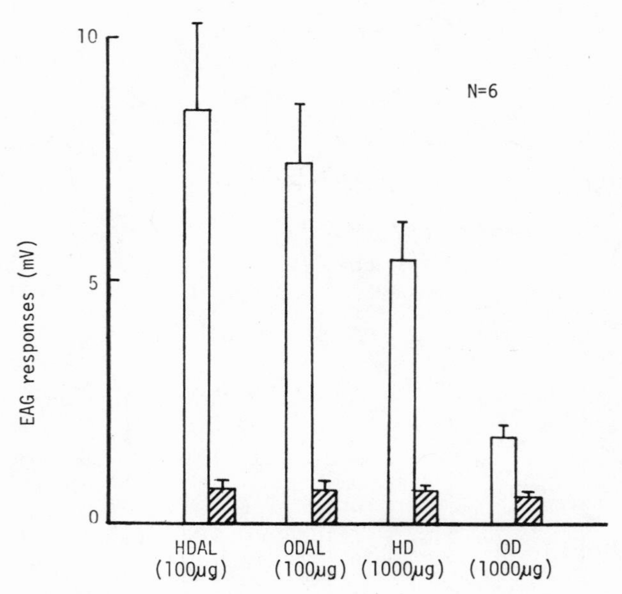

Fig. 2. Difference in EAG responses between male and female C. suppressalis to sex pheromone components and hydrocarbons. Open columns represent male responses and striped column female ones. HDAL = (Z)-11-hexadecenal, ODAL = (Z)-13-octadecenal, HD = (Z)-5-hexadecene, OD = (Z)-5-octadecene.

Receptor adaptation experiments were conducted using a modified version of the method described by Baker and Roelofs (1976). All combinations of Z-11-HDAL, Z-13-ODAL, and Z-5-HD were tested. The male antennae were exposed to a constant flow of air (3 liter/min) from a glass tube. A test compound was puffed over the antennae using a glass syringe, and the EAG response (RI) was recorded. The 2nd compound was then blown over the antennae with the constant air flow. After adjustment of the base line deflection caused by this compound, it was puffed a 2nd time and the EAG response (R2) was recorded. This response was then compared to that recorded for the 1st chemical (R1) to get a measure of the relative interaction of the compounds at the receptor level (R2 x 100/R1).

Results in Figure 3 indicate considerable interaction among these compounds at the receptor level. Pre-exposure of the antennae with either Z-11-HDAL or Z-5-HD resulted in a greatly reduced response to subsequent exposures of the same or other compounds tested. However, pre-exposure with Z-13-ODAL had no significant effect on EAG responses when followed by puffs of air containing Z-11-HDAL. These results suggest the presence of at least 2 receptor types on the male's antennae, and that sensory adaptation may be involved in the mechanism of mating disruption with Z-5-HD.

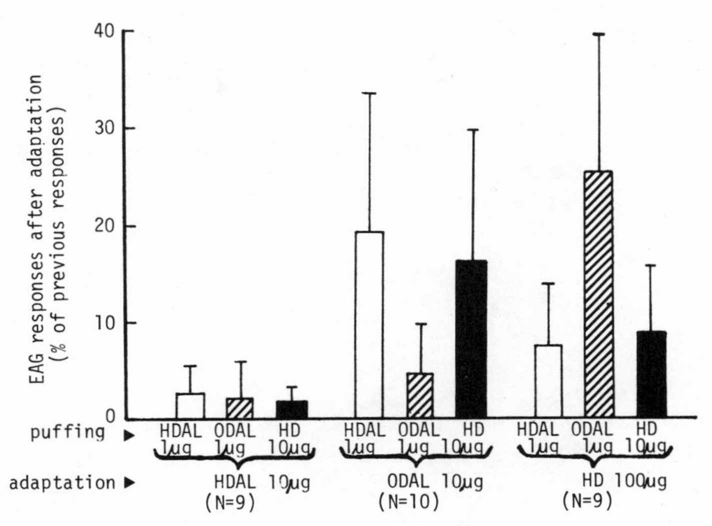

Fig. 3. Relative interaction of sex pheromone components (Z)-11-hexadecenal and (Z)-13-octadecenal, and hydrocarbon (Z)-5-hexadecene at receptor level. HDAL = (Z)-11-hexadecenal, ODAL = (Z)-13-octadecenal, HD = (Z)-5-hexadecene.

CONCLUDING REMARKS

In this paper, we emphasized the potential of Z-5-HD for control of C. suppressalis via mating disruption. The disruptive activity was demonstrated by several behavioral experiments conducted both in the field and in the laboratory. EAG experiments supported the field data with quantitative aspects of activity. Furthermore, the laboratory and EAG experiments showed that sensory adaptation may be the mechanism of disruption. However, many problems still remain before Z-5-HD can be used for control of this pest. For example, evaluation of the effectiveness of Z-5-HD under high populations of C. suppressalis is most important. Comparative studies on the effectiveness of Z-5-HD alone and in mixtures with Z-5-OD, or of the hydrocarbons and the formates (Beevor et al., 1977; Beevor and Campion,

1979) are also needed. Moreover, it would be interesting to know what effects, if any, that Z-5-HD and other hydrocarbons have on sex pheromone communication of other species that utilize aldehydes as pheromone.

ACKNOWLEDGMENT

This investigation was in collaboration with the following co-workers: K. Uchiumi, M. Kurihara, J. Fukami, Y. Fujimoto and T. Tatsuno of the Institute of Physical and Chemical Research (IPCR), M. Hattori and A. Sato of Hokuriku National Agricultural Experiment Station (HNAES). We wish to thank Nissan Chemical Co., Ltd., and Y. Hariya (IPCR) for their help in the synthesis of the compounds, Nitto Denko Co., Ltd. for formulation of the laminated dispensers, Takeda Chemical Co., Ltd. for supplying the traps, K. Tokuda, Y. Ohguchi (IPCR) and T. Hattori (HNAES) for their assistance in insect rearing. The research was partly supported by the Science and Technology Agency of Japan.

REFERENCES

Ando, T., Kishino, K., Tatsuki, S., and Takahashi, N., 1980, Sex pheromone of the rice green caterpillar: Chemical identification of three components and field tests of synthetic pheromones, Agric. Biol. Chem., 44: 765.

Baker, T. C., and Roelofs, W. L., 1976, Electroantennogram responses of the male moth, Argyrotaenia velutinana to mixtures of sex pheromone components of the female, J. Insect Physiol., 22: 1357.

Beevor, P. S., Hall, D. R., Nesbitt, B. F., Dyck, V. A., Arida, G., Lippold, P. C., Oloumi-Sadeghi, H., 1977, Field trials of the synthetic sex pheromones of the striped rice borer, Chilo suppressalis (Walker) (Lepidoptera: Pyralidae), and of related compounds, Bull. Entomol. Res., 67: 439.

Beevor, P. S., and Campion, D. G., 1979, The field use of 'inhibitory' components of lepidopterous sex pheromones and pheromone mimics, in: "Chemical Ecology: Odour Communication in Animals," F. J. Ritter, ed., Elsevier/North-Holland Biomedical Press, Amsterdam.

Kanno, H., Tatsuki, S., Uchiumi, K., Kurihara, M., Fukami, J., Fujimoto, Y., and Tatsuno, T., 1978, Disruption of sex attraction in the rice stem borer moth, Chilo suppressalis (Walker), with components of the sex pheromone and related chemicals, Appl. Entomol. Zool., 13: 321.

Mitchell, E. R., Copeland, W. W., Sparks, A. N., and Sekul, A. A., 1974, Fall armyworm: Disruption of pheromone communication with synthetic acetates, Environ. Entomol., 3: 778.

Nesbitt, B. F., Beevor, P. S., Hall, D. R., Lester, R., and Dyck, V. A., 1975, Identification of the female sex pheromones of the moth, Chilo suppressalis, J. Insect Physiol., 21: 1883.

Ohta, K., Tatsuki, S., Uchiumi, K., Kurihara, M., and Fukami, J., 1976, Structures of sex pheromones of rice stem borer, Agric. Biol. Chem. 40: 1897.

Rothschild, G. H. L., 1974, Problems in defining synergists and inhibitors of the Oriental fruit moth pheromone by field experimentation, Entomol. Exp. & Appl., 17: 294.

Tatsuki, S., Kurihara, M., Uchiumi, K., Fukami, J., Fujimoto, Y., Tatsuno, T., and Kishino, K., 1979, Factors improving field trapping of male rice stem borer moth, Chilo suppressalis (Walker) (Lepidoptera: Pyralidae), by using synthetic attractant, Appl. Entomol. Zool., 14: 95.

FIELD EXPERIMENTS TO DEVELOP CONTROL OF THE GRAPE MOTH, *EUPOECILIA AMBIGUELLA*, BY COMMUNICATION DISRUPTION

Heinrich Arn

Stefan Rauscher

Swiss Federal Research Station
for Horticulture and Viticulture
CH-8820 Wädenswil, Switzerland

Augustin Schmid

Claude Jaccard

Station Fédérale de Recherches
Agronomiques de Changins
CH-1260 Nyon, Switzerland

Barbara A. Bierl-Leonhardt

Agricultural Environmental
Quality Institute
AR-SEA, USDA
Beltsville, MD 20705 USA

INTRODUCTION

The 2 tortricoid moths attacking grapes in Europe, *Eupoecilia* (*Clysia*) *ambiguella* (Hb.) and *Lobesia* (*Polychrosis*) *botrana* Schiff., were among the first agricultural pests studied in sex pheromone research. Investigations by Götz (1941a,b; for a complete bibliography see Jacobson, 1972) already concerned laboratory bioassay of the pheromone, use of clock traps to determine diurnal rhythms, and mass-trapping for control. The work was done explicitly with today's objective: to do away with toxic chemicals. It is interesting to speculate whether the 30-year interruption in this line of work had more to do with wartime or simply with the advent of more effective insecticides.

Research toward identification of the sex pheromones in both species was initiated by Roelofs. (E)-7,(Z)-9-Dodecadienyl acetate was found in *L. botrana* (Roelofs et al., 1973; Buser et al., 1974; Lalanne-Cassou, 1977) and (Z)-9-dodecenyl acetate (Z9-12Ac) in *E. ambiguella* (Arn et al., 1976). Some confusion exists in the literature (Jacobson, 1972) between *E. ambiguella*, which belongs to the *Cochylidae* family, and *Paralobesia viteana*, the grape berry moth of North America, a tortricid. The 2 species happen to use the same main pheromone component (Roelofs et al., 1972).

Preliminary experiments in the laboratory and field indicated that a number of chemicals related to the *E. ambiguella* pheromone could be used to disrupt male orientation, but the most effective one appeared to be Z9-12Ac (Arn, 1979). In a small-scale field test using tethered females and released males from a laboratory culture, evaporation of ca. 700 mg of chemical/ha/day resulted in ca. 98% reduction of mating (Rauscher and Arn, 1979). These results called for further tests under field conditions.

THE PERROY EXPERIMENT

In 1978, a field experiment was carried out in a vineyard near Lake Geneva (Switzerland) where a sizable population of *E. ambiguella* and virtually no *L. botrana* were known to occur. Some of the data have already been reported (Arn et al., 1978; Jaccard, unpublished data).

Materials and Methods

The test site was located at the edge of several square km of vineyards, thus permitting valid check areas on 3 sides. Eighty-six trap stations were placed in a 25 x 25-m grid inside and outside the 1.5 ha plot to be treated (Fig. 1). Each trap station consisted of: 1) A pheromone trap; a Tetra trap with flaps containing a polyethylene mini-cap which was impregnated with 1 mg of a 1:5 mixture of Z9-12Ac (0.1% E) and dodecyl acetate (12Ac) (Arn et al., 1979) which was replaced every 2 weeks; and 2) A bait trap containing a mixture made from 75 liter wine, 15 liter vinegar, 10 kg sucrose and 12 liter apple juice.

Catches from bait traps were recorded weekly, those of pheromone traps, twice a week. Insects caught in bait traps were kept in Oudeman's solution (87 parts 70% ethanol, 5 parts glycerol, 8 parts glacial acetic acid) until the fall when they were sexed and females were inspected for presence of spermatophores.

On May 6, 4000 pheromone dispensers were deployed, equally spaced (ca. 2-m intervals) over the test area of 150 x 100 m. The dispensers were made by Hercon® and consisted of a 3-layered laminate made with 3 ml (0.08 mm) white acrylic polymer. Each dispenser measured ca. 2 x 5 cm and contained ca. 13.2 mg Z9-12Ac (1.5% E). The strips were attached with staples around the fence wires, alternately at heights of 1 and 1.5 m.

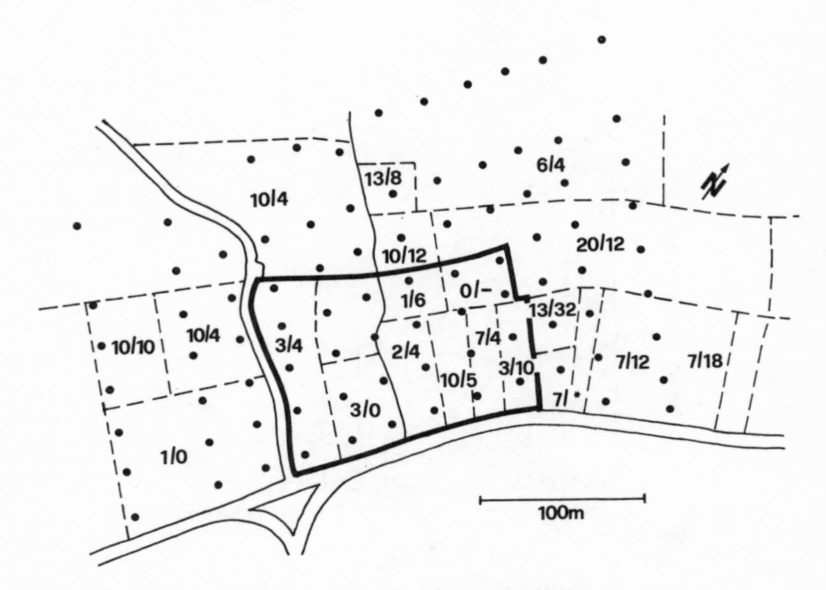

Fig. 1. Map of Perroy experiment. Pheromone-treated area is marked with heavy margin. Dots represent trap stations (pheromone + bait trap). Figures indicate % of grapes attacked by larvae of 1st/2nd generation.

The release rate of unused dispensers measured under laboratory conditions (Bierl-Leonhardt et al., 1979) was 5.6 μg/h/dispenser at 32°C and 100 ml air/min. Dispenser samples were placed in the field near the test site, retrieved at regular intervals and refrigerated until analysis. From the residual lure content (Fig. 2) the release rate of Z9-12Ac in the test plot was an average of 500 mg/ha/day in the 1st flight and 50 mg/ha/day in the 2nd.

Grapes (30 to 90/plot) were sampled for moth attack throughout the area. The 1st ovipositions were observed on June 14 and assessment of 1st generation attack was made by numbers of larval webs observed by July 3. Attack of the 2nd generation was assessed by the number of penetrations observed by August 6.

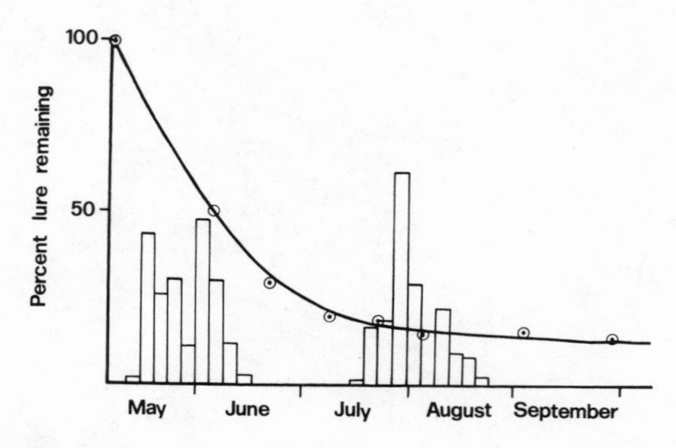

Fig. 2. Amount of (Z)-9-dodecenyl acetate remaining in dispensers during the season. Bars mark the flight curve of E. ambiguella obtained with pheromone traps, in untreated plots.

Results

Captures of males using pheromone traps are given in Table 1. Disruption was nearly total during both flights, indicating that even the low release rate during the 2nd flight was sufficient to effect male disorientation.

Table 1. *E. ambiguella* catches with pheromone traps, Perroy, 1978.

Flight	No. insects trapped in: Treated area (25 traps)	Untreated area (61 traps)	Disruption (%)
1st	2	604	99.2
2nd	4	515	98.2

Captures with bait traps are given in Table 2. Numbers of males and females caught, mated and unmated, were of the same order of magnitude. From the number of males caught in the check area with the 2 types of traps, bait trap efficiency was 6% of that of the pheromone traps in the 1st flight and 19% in the 2nd.

Table 2. *E. ambiguella* catches with bait traps, Perroy, 1978.

Date		Untreated (U) vs. treated (T) areas[1]			
		No. ♂	No. ♀	Percent ♀ mated	Number ♂/trap
5-27	(U)	27	2	50	0.44
	(T)	5	2	0	0.2
6-3	(U)	5	20	45	0.08
	(T)	5	5	20	0.2
6-10	(U)	3	1	100	0.05
	(T)	0	4	50	0
7-22	(U)	5	4	50	0.08
	(T)	1	1	0	0.04
7-29	(U)	27	42	45	0.44
	(T)	5	21	33	0.2
8-5	(U)	55	65	51	0.9
	(T)	15	39	33	0.6
8-13	(U)	13	14	71	0.21
	(T)	2	4	50	0.08

[1]Untreated area with 61 traps. Treated area with 25 traps.

It is interesting that the male/female ratio of the bait trap catches were always higher in the untreated than in the treated area. This is also true for the absolute number of males caught per trap. With the competition from the pheromone traps one would expect the opposite effect. It is tempting to speculate that the pheromone treatment drove the males out of the center area. Charmillot (1978) made observations in support of this theory when determining the trapping range of codling moth males in a disruption experiment. On the other hand, the effect might be explained by the proximity of the 2 types of traps at each trap station; in the untreated area the pheromone trap brought the males near the bait trap, while this attraction was not effective in the pheromone-treated area.

The spermatophore counts showed that matings in the treated area were reduced with respect to the check. In the treated area, 33% of the females trapped were mated; in the untreated area, 51%. The reduction was consistent throughout the trapping periods. Mating reduction could also be observed from multiple matings. Mated females in the treated area contained a maximum of 2 spermatophores, 1.4 on the average, while those in the check area contained up to 5 with an average of 1.7.

Mapping of grape moth attack shows the great variability between plots owned by different growers and maintained in different fashions (Fig. 1). On the average, 4.2% of the grapes in the treated area were attacked by the 1st generation and 4.4 by the 2nd, while in the untreated area the figures were 9.0 and 9.8%. The damage figure of the 1st generation in the treated area is low when compared to those obtained at the same place with no treatment in previous years (22, 21 and 16%, respectively, in 1975 to 1977). In other locations, the 1978 damage was the same or higher than in previous years. This indicates that the pheromone treatment did exert a suppressive effect on the grape moth.

Discussion

The results of this field test gave the first indication that matings in a field population of grape moths can be reduced with the disruption technique. Also, a moderate reduction of damage was achieved in this test which indicates that the method could be of use where populations are not too high. This is an encouragement. On the other hand, mating suppression was not as effective as might have been expected from the previous experiments. This could have been because of the following reasons:

1) Matings took place in the treated area because of the following: (a) Insufficient quantity of disruptant; our original aim was to measure mating inhibition right after dispenser placement,

but for unknown reasons the bait traps were not sufficiently effective during the spring flight. At the time of the 2nd flight, pheromone release had dropped to 1/10 of the initial rate or 14X less than in the tethered female experiment. Nevertheless, it is interesting that captures with pheromone traps were still fully suppressed at this low rate. It is conceivable that disruption of mating could have been improved by re-treating for the 2nd flight; on the other hand, the results of the 1st flight do not appear to be better than those of the 2nd. (b) Inadequate spatial distribution of disruptant; as long as no information of the places of female calling activity is available, it would seem advisable to use a formulation providing maximum homogeneity, both vertical and horizontal. (c) Deficiencies in the composition of the disruptant chemical; although Z9-12Ac gave satisfactory results with tethered females, secondary components (either those produced by the female or impurities present in the synthetic product) may be more important when examining the field strain of the insect or when both sexes are allowed to move freely. Analyses of female secretion are under way.

2) Gravid females migrated into the treated area. We have no information concerning flight distances of grape moth females. During the course of this experiment, we released ca. 4000 females in the center plot on 3 occasions but never made any recaptures, probably because of unfavorable weather conditions. Bait trap catches of wild females did not reveal a gradient of mated vs. unmated extending into the treated area. Grape moth males can be recaptured in significant numbers at distances of 150 m or greater from the release point (E. Boller, personal communication). If females fly similar distances, our test plot indeed could have been too small to obtain the desired effect.

It is interesting in this context that most of the bait-trap catches were made during the same week when fruit penetration reached a maximum. This indicated that many of the females caught had already oviposited and that dispersal may have taken place after oviposition. This might explain why the disruption effect achieved appears to be less when looking at the proportion of mated females caught as compared to the actual control figures.

FIRST FIELD TESTS WITH A ROPE FORMULATION

One of our objectives was to find a formulation applicable over a large area, giving a fairly homogeneous disruptant distribution. Aerial application did not seem suitable in grapes as a large amount of material would fall on the ground. Since grapes are often grown in lines, the principle of evaporator fences put forward by Campion et al. (1976) seemed to suit our purpose.

For our tests, Hercon manufactured 6 mm x 25 m laminated plastic strips made of 3 mil (0.08 mm) acrylic film outer layers. These "ropes" contained a 1:5 mixture of Z9-12Ac and 12 Ac. The saturated compound is a synergist both for trapping (Arn et al., 1979) and for disruption of the grape moth (Arn, 1979). Initial lure content was 120 mg of blend per meter of rope.

From the residual lure content, the rope released an average of 590 mg of Z9-12Ac/day and kilometer during the 1st 7 days of outdoor aging and 64 mg/day and kilometer after 3 weeks of aging. To obtain a release similar to that of the initial tethering experiment, the ropes, therefore, would have to be placed at distances of ca. 8 m. We expected that greater distances would be permitted because of simultaneous evaporation of the synergist.

In one experiment, an area of 12 x 12 m in the Wädenswil vineyard was fenced with Hercon rope, and 13 mating stations were placed inside the square at about equal distances from one another and from the fence. The rope was strung along the grape lines in zig-zag fashion (Fig. 3) to make the vertical distribution as even as possible. Across the lines the ropes were strung at a height of 2 m to clear the walkway but were run down each fence post to 1 m above ground. On each of 9 nights in July 1979, we released 200 males and placed fresh females in the mating stations. Pheromone traps were placed on either side of the square at 20 m distance. The test was conducted in the same place and under similar conditions to those already reported (Rauscher and Arn, 1979).

Microscopic examination showed that none of the 108 females recovered the next morning from the mating stations had mated. The center female, which was the farthest away from the pheromone sources, had passed this test during 4 nights with heavy flight (10 or more males caught in the outside trap) and 3 nights with light flight. Previous experience showed that under conditions favorable for male flight, ca. 50% of exposed females mate during the 1st night.

A similar test was done at Charrat during the flight of the field population. Hercon ropes were placed at distances of 7, 10 and 20 m (Fig. 4). Mating stations with laboratory-reared females were maintained in areas of highest and lowest dispenser density and in the check area. The figures show that of a total of 73 females exposed in the treated area, only one was mated; the latter was situated toward the end of the plot with a 20 m rope distance. It was near this female that we also made the only male catches obtained in the treated area. At the upwind end outside the treated area, 24% of the females were mated when compared with only 7% at the downwind end. This indicates that the effect of the treatment extends some distance outward of the pheromone fence. Catches with pheromone traps reflect the same phenomenon.

Fig. 3 Rope dispenser used along grape lines.

These results indicate that the ropes could provide a suitable formulation for larger-scale tests. Evidently, more extensive testing would be required to accurately determine the distances at which disruption breaks down, but it appears that satisfactory results may be obtained with spacings of the order of 10 m. Similar results were obtained with the gypsy moth where a 5-10 m spacing of dispensers gave the greatest degree of trap reduction.

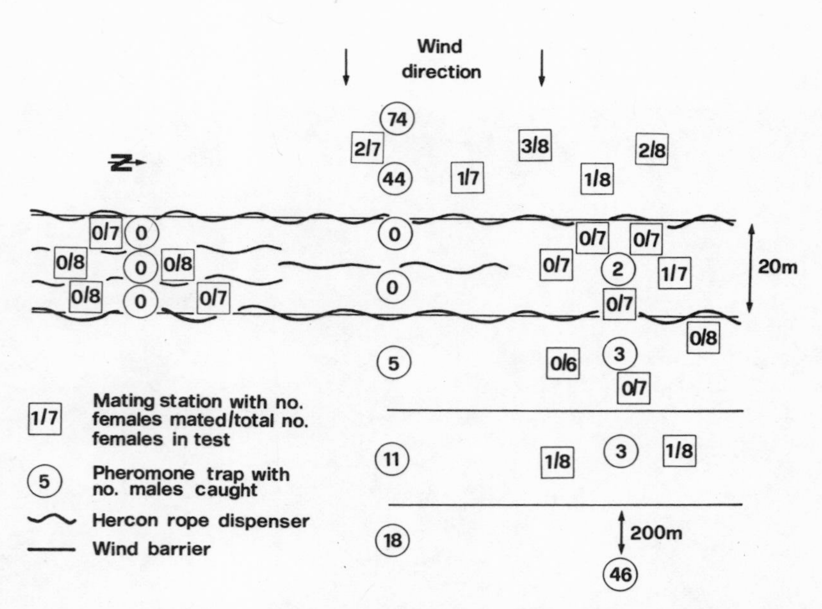

Fig. 4 Map of Charrat experiment showing disruption of *E. ambiguella* measured by trap catches and mating stations.

OUTLOOK

The results obtained in these experiments encourage further field tests to control the grape moth with communication disruption and call for more studies in pheromone chemistry and behavior. A technique for controlling *E. ambiguella* could find an immediate application in parts of Germany and Switzerland where no other insect pest occurs on grapes. To be of widespread use, the treatment would also have to be effective against *L. botrana*. First field tests in disruption of this species have given positive results (Roehrich et al., 1977), and the 2 pheromones can be used together for disrupting the 2 insects without interference (Roehrich et al., 1979). Thus, it would seem possible to incorporate them into the same formulation. Cost of the dienic compound appears to be the major holdup in large-scale experiments on *L. botrana*.

E. ambiguella and L. botrana are the only insect pests in vineyards in large parts of Europe. Control of the 2 insects with pheromones could mean a widespread elimination of insecticides in grapes. The major benefit of this measure, apart from a reduction of residues, will lie in limiting secondary ecological effects such as elimination of predators, especially Typhlodromus spp., and promotion of mites such as Panonychus ulmi and Tetranychus urticae.

The rope dispenser used in the 1979 tests was mainly conceived as a developmental formulation which could be applied by hand and provide some degree of homogeneity. Some of its properties, like tear resistance, may not have been optimal. But we feel that the principle could be of use, even in a practical situation, wherever labor cost is not limiting. This applies to grapes and other row cultures like fruit trees. The advantages of the rope over broadcast formulations are obvious: Its presence on the crop can be easily verified, it can be retrieved for performance tests and disposal, and it should leave no significant residues.

We thank Mario Menti for help in the field experiments.

REFERENCES

Arn, H., 1979, Developing insect control by disruption of sex pheromone communication: Conclusions from programs on lepidopterous pests in Switzerland, in: "Chemical Ecology: Odour Communication in Animals," F. J. Ritter, ed., Elsevier, Amsterdam, p. 365.

Arn, H., Rauscher, S., Buser, H. R., and Roelofs, W. L., 1976, Sex pheromone of Eupoecilia ambiguella: cis-9-dodecenyl acetate as a major component. Z. Naturforsch., 31c: 499.

Arn, H., Rauscher, S., Schmid, A., 1979, Sex attractant formulations and traps for the grape moth Eupoecilia ambiguella Hb., Mitt. Schweiz. Entomol. Ges. 52: 49.

Arn, H., Schmid, A., Jaccard, C., Bierl-Leonhardt, B., Rauscher, S., 1978, Mating reduction in free-living grape moths (Eupoecilia ambiguella Hb.) by disruption of pheromone communication with (Z)-9-dodecenyl acetate. Compte-rendu de la réunion sur les phéromones sexuelles des insectes et médiateurs chimiques, INRA, Antibes, p. 67.

Buser, H. R., Rauscher, S., Arn, H., 1974, Sex pheromone of Lobesia botrana: (E,Z)-7,9-dodecadienyl acetate in the female grape vine moth. Z. Naturforsch., 29c: 781.

Campion, D. G., McVeigh, L. J., Murlis, J., Hall, D. R., Lester, R., Nesbitt, B. F., Marrs, G. J., 1976, Communication disruption of adult Egyptian cotton leafworm Spodoptera littoralis (Boisd.) (Lepidoptera: Noctuidae) in Crete using synthetic pheromones applied by microencapsulation and dispenser techniques, Bull. Entomol. Res. 66: 335.

Charmillot, P. J., 1978, Influence de la diffusion de codlémone sur les déplacements du carpocapse (Laspeyresia pomonella) L.). Compte-rendu sur les phéromones sexuelles des insectes et médiateurs chimiques, INRA, Antibes, p. 75.

Götz, B., 1941a, Der Sexualduftstoff als Bekämpfungsmittel gegen die Traubenwickler im Freiland, Wein Rebe, 23: 75.

Götz, B., 1941b, Beiträge zur Analyse des Mottenfluges bei den Traubenwicklern Clysia ambiguella und Polychrosis botrana, Wein Rebe, 23: 207.

Jacobson, M., 1972, "Insect Sex Pheromones," Academic Press, New York.

Lalanne-Cassou, B., 1977, Contribution a l'étude de la communication sexuelle par phéromone chez l'Eudémis de la Vigne Lobesia botrana Schiff. Thèse Univ. Pierre et Marie Curie, Paris.

Rauscher, S., Arn, H., 1979, Mating suppression in tethered females of Eupoecilia ambiguella by evaporation of (Z)-9-dodecenyl acetate in the field, Entomol. Exp. Appl., 25: 16.

Roehrich, R., Carles, J. P., de Vathaire, M. A., 1977, La confusion sexuelle l'eudémis de la vigne au moven de phéromones de synthése: essaréalisés en 1977, Compte-rendu de la réunion sur les phéromones sexuelles des insectes, INRA, Montfavet, p. 128.

Roehrich, R., Carles, J. P., Trésor, C., de Vathaire, M. A., 1979, Essais de "confusion sexuelle" contre les Tordeuses de la grappe l'Eudémis Lobesia botrana Den. et Schiff. et la Cochylis Eupoecilia ambiguella Tr. Ann. Zool. Ecol. anim. 11: 659.

Roelofs, W., Kochansky, J., Cardé, R., Arn, H., Rauscher, S., 1973, Sex attractant of the grape vine moth, Lobesia botrana, Mitt. Schweiz. Entomol. Ges., 46: 71.

Roelofs, W. L., Tette, J. P., Taschenberg, E. F., Comeau, A., 1971, Sex pheromone of the grape berry moth: Identification by classical and electroantennogram methods, and field tests. J. Insect Physiol., 17: 2235.

DISRUPTION OF SPRUCE BUDWORM MATING--STATE OF THE ART

C. J. Sanders

Canadian Forestry Service
Great Lakes Forest Research Centre
Post Office Box 490
Sault Ste. Marie, Ontario

INTRODUCTION

As has been shown for most other Lepidoptera where this technique has been tried, the mating behavior of the spruce budworm [*Choristoneura fumiferana* (Clemens)] is profoundly disrupted in an atmosphere permeated by the synthetic attractant.

Reductions in the number of matings have been shown both in the laboratory and field (Sanders, 1974; Schmidt and Seabrook, 1979). Field application of NCR microcapsules on a moderately dense population gave better than 95% reduction in the numbers of males captured in traps baited with virgin females following the application of the synthetic attractant (Sanders, 1976). In 1977, synthetic attractant in Conrel® hollow microfibers caused a 99% reduction in the numbers of males captured in baited traps in a 250-ha stand of moderate density spruce budworm populations and also in 2 10-ha plots in low density populations--one of which had a release rate of only 3 mg/ha/h or lower.

But such results, while encouraging, are not sufficient to warrant operational use of the attractant for regulating spruce budworm outbreaks. The crux of the problem is, of course, to demonstrate that application of the attractant is followed by a reduction in population density. The 1977 trial with Conrel fibers was aimed at demonstrating this, and while there was a reduction in subsequent egg populations in the 250 ha of moderate infestation, no effect was found in the two 10-ha plots in the low density population. The reasons for the failure are not known. Application was later than originally intended and it is possible that some eggs were laid prior to the treatment; but it is unlikely. Possibly, the attractant was not as effective as the reduction in trap catches indicated and males were occasionally getting through--perhaps enough to fertilize the females--although with a 95% reduction in trap catches it seems again unlikely. The traps were placed at a height of 2 m but most spruce budworm are active in the canopy (at a height of 10-15 m in this case). Possibly, therefore, we were measuring activity in the wrong place. Possibly, too, the plots were too small; a 10-ha plot is only a little over 300 m on each side. Monitoring was carried out in the center of each of these plots in an area measuring 100 m on each side. Thus, a fertile female from outside the plot had to fly only 100 m to enter the monitoring zone, and we do not know how far ovipositing females fly.

In 1978, a more extensive trial was carried out in the Maritime Provinces. Four different application rates were applied by aircraft to 100-ha plots, each replicated twice. Again, Conrel hollow fibers were used, but unfortunately in this trial virtually all the attractant was dissipated from the fibers before the adult budworm started to emerge (Wiesner et al., in press) and again no effects on the following generations could be detected (Miller, personal communication).

In western North America, Hercon® flakes impregnated with synthetic attractant were used against the western spruce budworm, _C. occidentalis_ Freeman, in 1979. Again, considerable reduction in trap catches were recorded, but no reduction was found in egg populations (Daterman, personal communication). Since the plots in this trial were again only 10 ha, it was impossible to determine whether the failure was because of an ineffective treatment or invasion by egg-carrying females from outside the plots. Where then does this leave us? Clearly, we still have to demonstrate that the synthetic attractant can reduce population density. If it can, we then need to know how much of the attractant is needed to cause a given amount of population reduction; i.e., we need to establish the dose/response curve. What then are the remaining problems? These can be discussed under 4 headings: (1) The volume of disruption; (2) appropriate formulations; (3) selection of suitable test sites; (4) economic viability.

UNRESOLVED PROBLEMS

The Nature of Disruption

(a) _The role of chemicals in mating behavior_--Several authors have analyzed the role of pheromones in courtship behavior in Lepidoptera in some detail (e.g. Grant, 1976; Baker and Cardé, 1979; Cardé and Hageman, 1979), identifying in the process several steps in the mating sequence. For the purpose of discussion here, I am going to recognize 3 stages: (1) location of the female; this involves upwind flight by the male along the pheromone plume, followed by walking with fanning wings after landing. (2) recognition of the female, which occurs by physical contact. (3) courtship leading to copulation. These distinctions are somewhat artificial and oversimplified, but they can be clearly differentiated from each other and avoid subjective categories such as long-range and close-range communication.

Most experiments to disrupt mating--and certainly those that I have just described for the spruce budworm--are aimed at preventing males from using the pheromone odor plume to locate the female. It is tacitly accepted that any disruption in the male's ability to recognize a female or in the subsequent courtship behavior (i.e., interplay between male and female once the male has located the female) will be a bonus, adding to the sum total of the disruption effect.

Location of the female, at least in the spruce budworm, is mediated entirely by chemicals. But we know little of the role of chemicals in the recognition of a female by the male and in courtship of the spruce budworm. By observing the mating process, Sanders (1979) concluded that male spruce budworms use their antennae to identify and determine the posture of a female budworm, an observation which is confirmed by the fact that even under crowded conditions where accidental contacts between male and female spruce budworm are frequent, antennated males do not copulate (Sanders, 1976). Attempts to determine if such recognition involves chemicals were unsuccessful (Sanders, 1979) but the possibility is still there. If chemicals are involved, it is possible that they will be different from those involved in location of the female. If we are to exploit mating disruption to the full, we must evaluate the role of chemicals in the recognition of the female by the male, and in the subsequent courtship behavior, as well as in the location of the female by the male.

Even within the process of locating the female, the situation is far from completely resolved. It has been postulated (Kennedy, 1977; Marsh et al., 1978) that plume-following by pheromone-stimulated moths occurs by the male turning diagonally crosswind when he perceives the pheromone. Losing the pheromone then causes him to reverse his turn and travel diagonally crosswind in the opposite direction. Since there is probably no such thing as pheromone-free air in a natural situation but merely pockets or plumes of higher concentration in a matrix of lower concentrations, I would suggest that Kennedy's theory should be modified, by postulating that males turn in response to perceived changes in concentration, without defining how large the differences in concentration have to be for detection. Within this framework, let us examine what we know of the mate-locating behavior of the spruce budworm.

(b) Current Knowledge--The facts we have are as follows:

(1) The female pheromone contains better than 99% of Δ-11-tetradecenal of which 95-97% is the E isomer, and this blend is equivalent in attractiveness to a virgin female moth (Sanders and Weatherston, 1976; Silk et al., 1980; Sanders, this volume).

(2) The number of males captured in traps baited with virgin females is decreased if the trap also contains dispensers containing (a) suboptimum blends of Δ-11-tetradecenal, (b) chemical analogues (the acetate, alcohol, (E)-9-tetradecenal, and (E,E)-9,11-tetradecadienal) commonly referred to as "inhibitors" (Sanders, 1976; Seabrook, unpublished data; Sanders, unpublished data).

(3) The number of males captured in traps baited with virgin females is decreased if the trap is surrounded by dispensers emitting Δ-11-tetradecenal. But the decrease in catch becomes less as the blend of E,Z moves away from the optimum of 97/3 (Figure 1).

(4) The number of males captured in traps baited with virgin females is not decreased if the trap is surrounded by dispensers emitting any of the analogues in 2(b) above (Sanders, 1975; 1976).

(c) A Unifying Theory--How can these facts be incorporated together into a single unifying theory?

The fact that the "inhibitors" have an effect only when they are dispersed from the same trap as the attractant and not when they are released from dispensers a few meters from the trap, suggests that the coincidence or superpositioning of the odor plumes of the 2 chemicals, the inhibitor and attractant, is important (the concept of "aerial mixing" put forward by Nakamura, 1979). Natural pheromone plumes in the field contain considerable structure because of air turbulence. As a result, the male moths encounter eddies or pockets of different chemical concentrations. If 2 chemicals are emitted from the same source, the relative concentrations of the 2 will remain the same regardless of the pattern of turbulence. In other words, the inhibitor will be exactly superimposed upon the attractant. The receptors for the 2 chemicals on the male antennae, therefore, will be firing in synchrony, which presumably is interpreted by the male's central nervous system as a wrong signal. But if the 2 chemicals are released from separated sources, each chemical will be subjected to different patterns of turbulence, and this separation or lack of synchrony apparently does not deter the male. Such a system could have evolved during speciation and the maintenance of species integrity. In the case of the spruce budworm, the related synchronic, allopatric species, the jack pine budworm, utilizes the acetate as a major component of its pheromone system. Thus, the ability to avoid homing-in on the source of acetate production, while still maintaining the ability to locate the source

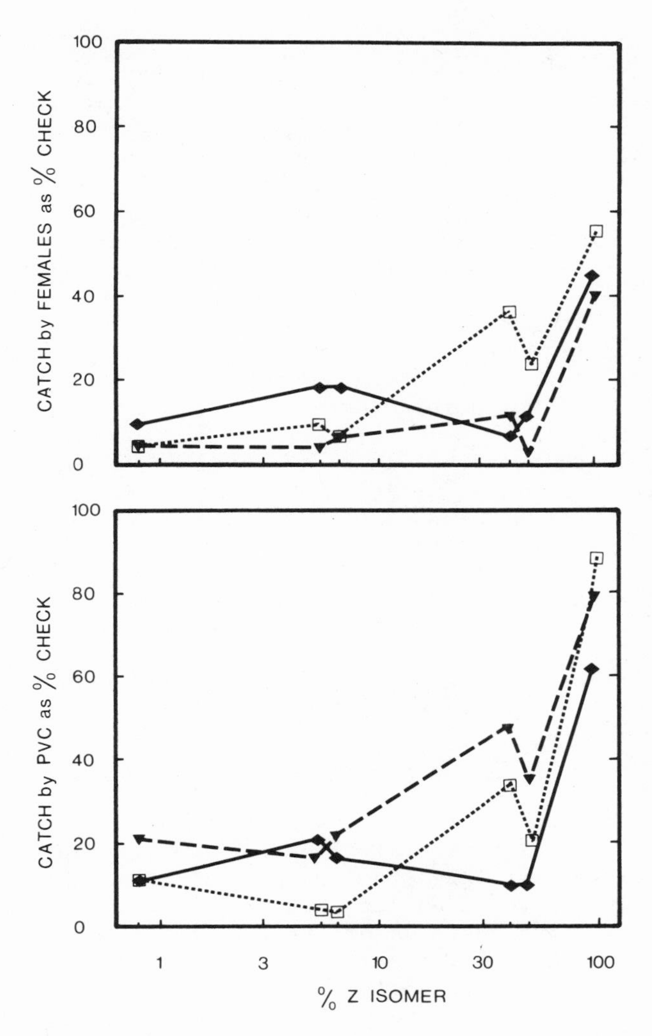

Fig. 1. Disruption of male spruce budworm orientation by blends of (E) and (Z)-11-tetradecenal. Effects assessed by the numbers of male budworm captured in traps baited with virgin females (upper) or synthetic lures (lower). The baited traps were surrounded by 6 dispensers containing the test blend, located at 5 m radius (solid line), 10m radius (dashed line) or 20 m radius (dotted line). Data are presented as % of catch in check traps (i.e. baited traps surrounded by empty dispensers).

of aldehyde in the presence of acetate, would obviously be beneficial to a male spruce budworm.

The effect of the aldehyde itself in attraction, inhibition, and disruption can be explained in a similar fashion. If homing-in on a source is governed by the male continually turning diagonally across wind when he perceives a change in the concentration of the correct chemical blend, hence, zigzagging upwind, then as long as he perceives some pockets of air containing the correct blend, he will proceed upwind and eventually locate the pheromone source. If an incorrect blend is released from the same location as a female, then there will never be any pockets of the correct blend. On the other hand, an incorrect blend is released independently of the female, then there will be some pockets of correct blend, even if they are fewer and further between. Hence, the male will still move ahead toward the source, but perhaps more slowly.

Practical Implications--This, if true, has important implications in the practical application of mating disruption. Let us first consider what happens if the attractant is being released from numerous dispensers, all emitting the correct blend. Now all signals are perceived as correct, the male's turning is still governed by changes in concentration causing him to move upwind. If he is close to a source, then these turns will be likely to bring him to a source, but if he is far away from a source, he is likely to be directed in an erratic series of turns not leading anywhere. The result will be that either many males arrive at a wrong source, or that after long periods of unsuccessful pheromone-mediated flight, the males will become habituated to the pheromone and give up.

Let us first consider the strategy of point source emitters such as Hercon flakes or Conrel fibers. If each point source were equal in potency to a virgin female, then if there were one emitter for each female, we would expect that for every male going to a female one would also go to an emitter. As the number of emitters increased, the number of males going to the females would then decrease. The success of females in attracting males would also be reduced by increasing the potency of the lures. A doubling in potency (i.e., in the number of males attracted) would reduce the number of males going to females by 1/2. But a doubling in pheromone concentration in a lure does not produce a doubling in catch; it would, therefore, be more efficient to increase the number of sources rather than increase individual potency of lures. Clearly, such strategies have to be resolved as a trade-off between the costs of the pheromone, the formulation and its application.

Perhaps the most important inference from this line of reasoning is that success in terms of mating disruption would be affected by population density. In our 1977 trial, the maximum population density of virgin females on any one day was about 4000/ha.

Assuming that each fiber emitted pheromone at a rate equivalent to the female, 76,000 fibers/ha would be required to effect 95% disruption. But each fiber was estimated to be releasing attractant at a rate of 208 ng/fiber/h, about 40X that of a virgin female. If we assume this represents a 3-fold increase in attractiveness then 25,000 fibers would have sufficed. The actual deposit measured was 33,000/ha. In an outbreak situation, however, populations might rise to 500,000 virgin females/ha on one night, which would require 3,125,000 fibers/ha, a formidable figure.

So much for a situation where the pheromone is being released from point sources comparable in potency to that of a female. What if the atmosphere is universally permeated with the correct blend such as should occur with the use of microcapsules? Again, wherever the male turns, he will encounter the correct blend, but regardless of the turbulence, the concentration will remain constant and there will be no edges or changes in concentration to provide the necessary on-off signals to cause him to turn. If a pheromone plume is now superimposed on this, differences in concentration will occur which the male could possibly utilize. To prevent the males from utilizing the plumes in this situation would entail raising the whole background concentration to a level where the change in concentration because of the output of the female was not detectable by the male. This may necessitate the use of very high concentrations of the attractant. One advantage of uniform atmospheric permeation as against the point source emitters is that it would be independent of population density.

Of course, these theories are entirely speculative, and no consideration has been given to the role that adaptation or habituation may play. The objective of the discussion has been to point out the need for us to resolve some of these questions so that synthetic attractants can be used in a more rational and efficient manner.

Appropriate Formulations

Much has been said about the problems of formulating a suitable carrier for pheromones. To a certain extent it can be argued that formulations cannot be designed until the biologists specify the release rates required. But we have very little information on what those concentrations are. As far as I am aware, all field trials to disrupt mating behavior that have so far been reported with any Lepidoptera have been strictly empirical. There is nothing wrong with this if it works and if costs and application techniques were not a problem. But they are and we need to refine systems wherever we can. Therefore, it is essential for optimum efficiency to know how much chemical is required to do the job. At best, there has been very sketchy information on release rates from the candidate formulation tested in the laboratory. Frequently, the only parameter that can be given with any confidence is the amount of material released from the spray plane; reliable estimates of how much of the formulation was actually deposited are usually lacking, and there has seldom been any attempt to monitor release rates under field conditions, let alone measuring the concentrations of material in the air space or adhering to the foliage where the insects will encounter it. Our spruce budworm experiments have been no exception to this rule. In 1975, we made no attempt to measure release rates at all. In 1977, using the Conrel fibers, we had estimates of release rate under laboratory conditions but there were certainly differences between these and the field rates. We also have accurate information on how much material the spray plane released but according to our deposit assessment, only 1/2 of the material released by the plane reached the target plot--where did the rest go?

My own feeling is very strong that we must actually monitor airborne concentrations along the lines developed by Caro et al. (1978). Such work is now being carried out on the spruce budworm by the Research and Productivity Council of New Brunswick (Wiesner et al., 1980). Only when we have this information can we customize our formulations and delivery systems to obtain maximum efficiency. But even with this lack of information much can still be done. Concentrations can be adjusted by applying more or less of a given contents of the formulation (i.e., little, if any, of the active material should remain in the formulation following the required time interval). Little more need be said about this problem here, except to point out that the limiting factors in forestry operations are often very different from those in agriculture. First, there are the areas involved, which may be very extensive lacking the patchwork distribution of agriculture; this makes application over wide, long swaths possible in forestry. Second, there is the problem of accessibility--treatment areas may be many miles from the nearest airport, and the aircraft ferry-time may become the single biggest cost factor. All this points in the direction of bigger payloads, higher application rates, and longer-range aircraft.

Selection of Test Sites

(a) Size of plot--Our attempt to demonstrate the effects of disruption on population densities of the spruce budworm in 1977, was confounded by the possible movement of mated female moths into the treated area from the surrounding untreated areas. The plots were 10 ha in size, but how much bigger would have been big enough? Operationally, we may be faced with treating thousands of hectares, but this would be prohibitively expensive for an experimental trial. The problem is moth mobility. During its search for suitable oviposition sites, a female moth will certainly fly from tree to tree, but just how far such flights may take a female we do not know. It is known that females do not voluntarily fly until their 2nd night, by which time they have already laid 30-50% of their eggs. After 3 more nights of flight and oviposition, they will on the average have laid 80-90% of their egg complement. In 1977, we estimated a maximum movement of 30 m/night and, therefore, allowed a 100 m buffer around a treated plot; it was apparently too small. A 300 m buffer around a plot of 1 ha means a 50-ha treated area, while a 1000 m buffer would mean 440 ha (over 1000 acres), which becomes expensive. Accurate information on the distances that ovipositing females will move in the first few nights of their life could then be of considerable significance.

Another possible problem with small plots is that the use of synthetic attractant in a blend close to that of the natural pheromone will attract male moths from areas surrounding the treated area. In large plots this would not be a problem since adaptation of the male receptors would probably occur long before the insects had reached the center of the plot. But in small plots where the proportion of edge to area is high, sufficient male moths might well be attracted into the plot to increase the probability of the females being successfully inseminated. Perhaps this problem could be alleviated somewhat by the use of blends which are less attractive but still highly disruptive.

(b) Moth dispersal and the location of plots--At high population densities spruce budworm females also make long-range dispersal flights. Since these may carry the moths up to 50 km, it is impractical to even consider an experimental plot large enough to exclude the danger of such invasion. One solution, of course, is to conduct experiments in low density populations where we know dispersal does not occur. But there is sufficient interest in using disruption technique to regulate high density populations to warrant testing at high densities. Also, the cost of accurately sampling low density populations is very high. So, trade-offs have to be made between cost and the danger of invasion. One of the most attractive alternatives is to conduct the trials on isolated populations remote from the chance of dispersal. In Canada at the present time such areas are scarce, but possibly they could be found on the fringes of the current spruce budworm outbreaks in the U.S. or in southern Ontario.

Economic Potential

The yardstick by which the effectiveness of any candidate for regulating spruce budworm has been measured in the past was that of the synthetic insecticide. Up until a few years ago, costs were low, effects were acceptable, and environmental concerns were insignificant. Measured against this, prospects for the use of sex attractants looked dim. But this scene is changing fast. Costs are rising. The average cost (material and application) for operations in New Brunswick in 1979 using Matacil® and fenitrothion were $4.55/

ha; a few years ago they were only $2-3. In Quebec and Ontario, up to $24/ha has been spent on the operational use of the biorational agent, *B.t.* Furthermore, environmental concerns have been sufficient to prevent the use of insecticides to protect forests in Nova Scotia and Newfoundland even when the economics appeared very convincing. It would probably only take one more scare similar to that of Reye's Syndrome, unfounded or not, to cause the use of chemical insecticides to be banned for forest protection. If this happened, then suddenly we might find a big demand for alternative methods of regulating spruce budworm numbers. There is, therefore, strong justification for us to develop the use of such agents as the sex attractant as fast as is practicable.

REFERENCES

Baker, T. C. and Cardé, R. T., 1979, Courtship behavior of the oriental fruit moth (Grapholitha molesta): Experimental analysis and consideration of the role of sexual selection in the evolution of courtship pheromones in the Lepidoptera, Ann. Entomol. Soc. Am., 72: 173.

Cardé, R. T. and Hagaman, T. E., 1979, Behavioral responses of the gypsy moth in a wind tunnel to airborne enantiomers of disparlure, Environ. Entomol., 8: 475.

Caro, J. H., Bierl, B. A., Freeman, H. P., and Sonnet, P. E., 1978, A method for trapping Disparlure from air, and its determination by electron capture gas chromatography, J. Agric. Food Chem., 26: 461.

Grant, G. G., 1976, Courtship behavior of a phycitid moth, Vitula edmansae, Ann. Entomol. Soc. Am., 69: 445.

Kennedy, J. S., 1977, Behaviorally discriminating assays of attractants and repellents, in: Chemical Control of Insect Behavior: Theory and Application, H. H. Shorey and J. J. McKelvey, Jr., eds., John Wiley & Sons, New York.

Marsh, D., Kennedy, J. S., and Ludlow, A. R., 1978, An analysis of anemotactic zigzagging flight in male moths stimulated by pheromone, Phys. Entomol., 3: 221.

Nakamura, K., 1979, Effect of the minor component of the sex pheromone on the male orientation to pheromone source in Spodoptera litura (F.), Chem. Review of Insects (Russia), 4: 153.

Sanders, C. J., 1974, Programs utilizing pheromones: Forest Lepidoptera--the spruce budworm, in: Pheromones, M. C. Birch, ed., [North Holland Res. Monographs] Frontiers of Biol. 32: 435.

Sanders, C. J., 1975, Disruption of mating behavior of the spruce budworm in air permeated by synthetic attractant, Can. Dept. Environ., Bi-Mo. Res. Notes, 31: 10.

Sanders, C. J., 1976, Disruption of sex attraction in the eastern spruce budworm, Environ. Entomol., 5: 868.

Sanders, C. J., 1979, Mate location and mating in eastern spruce budworm, Can. Dept. Environ., Bi-Mo. Res. Notes, 35: 2.
Sanders, C. J. and Weatherston, J., 1976, Sex pheromone of the eastern spruce budworm: Optimum blend of trans and cis-11-tetradecenal, Can. Entomol., 108: 1285.
Schmidt, J. O., and Seabrook, W. D., 1979, Mating of caged spruce budworm moths in pheromone environments, J. Econ. Entomol., 72: 509-511.
Silk, P. J., Tan, S. H., Wiesner, C. J., and Ross, R. J., 1980, Sex pheromone chemistry of the eastern spruce budworm, Choristoneura fumiferana, Environ. Entomol., (In press).
Wiesner, C. J., Silk, P. J., Tan, S. H., and Fullarton, S., 1980, Monitoring of atmospheric concentrations of the sex pheromone of the spruce budworm, Choristoneura fumiferana (Lepidoptera: Tortricidae), Can. Entomol., (In press).

CONTROL OF MOTH PESTS BY MATING DISRUPTION IN FORESTS OF THE WESTERN UNITED STATES

L. L. Sower

G. E. Daterman

C. Sartwell

Pacific Northwest Forest and Range
Experiment Station
USDA Forest Service
Corvallis, Oregon 97331 USA

INTRODUCTION

A specific objective of our project at the Pacific Northwest Forest and Range Experiment Station in Corvallis, OR is to determine whether mating disruption is a promising approach to control selected forest pests and to develop methods and materials for practical use. Disruption efforts have been directed mainly toward control of the Douglas-fir tussock moth, *Orgyia pseudotsugata* (McDunnough), the western pine shoot borer moth, *Eucosma sonomana* Kearfott, and the western spruce budworm, *Choristoneura occidentalis* Freeman. Preliminary work has also been done with the European pine shoot moth, *Rhyacionia buoliana* (Schiffermüller), (Daterman et al., 1975). Results with the spruce budworm are still tentative. Pheromone will substantially reduce reproduction by the Douglas-fir tussock moth and further tests are planned. Western pine shoot borer damage can be controlled with pheromone.

Constraints in Forestry

Insect control techniques for use in forestry have some constraints that are generally uncommon to agriculture. Locations may be remote and topography rough, and applications must usually be by aircraft. Treatment costs cannot be too high and frequent treatment is not usually acceptable. Treatment benefits are somewhat speculative because the economic impact attributable to a particular pest is often poorly understood. Exceptions would be situations where the death of an entire stand is threatened. Vertical spread of foliage in a forest is much greater than in a cotton field, and more material per hectare is required to get equivalent coverage.

Working Hypothesis

We adhere to several working hypotheses, or assumptions, when designing field tests. First, pheromone disruption success is inversely dependent on the insect's population density, other factors being equal. Tests with stored-product insects and observations in the field tend to support this (Sower and Whitmer, 1977; Vick et al., 1979; Sower et al., 1979a; Schwalbe et al., 1979). It is logical that an efficient distance communication is more important when insects are widely separated.

Response curves related to the dose of pheromone are believed to be rather flat (Figure 1) and significant increases in control appear to require at least 3-fold dosage increases. For dose/response data a wide range (100X or more) of dosages are tested. We think uniformity of coverage is much less critical than with contact insecticides.

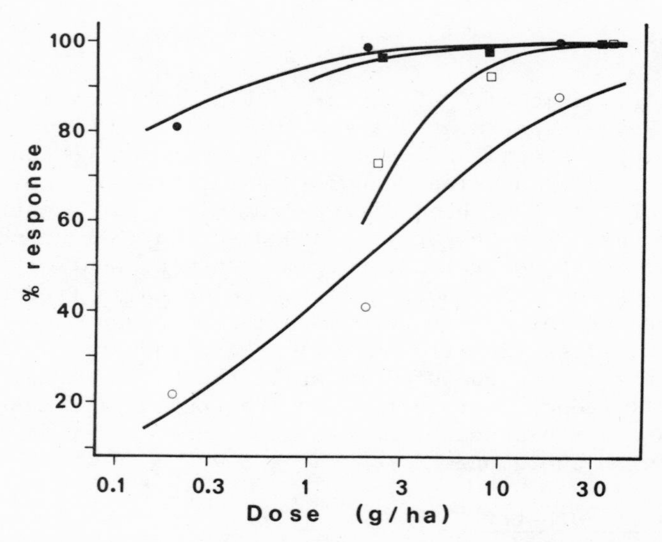

Fig. 1. Dose/response relationships. Criteria were captures of male western pine shoot borers in pheromone-baited traps (solid squares); captures of male Douglas-fir tussock moths in female-baited traps (solid circles); reduction of shoot borer damage (open squares); reduction of tussock moth reproduction (open circles). Curved lines are linear on a probability scale (not shown).

Large, isolated plots are preferred over small plots contiguous with adjacent forests because the impact of females which mate elsewhere and then fly into the treated plots is minimized. Our tests and perceived research needs, as outlined below, have been influenced by these assumptions.

Response Criteria

Criteria we have used to determine disruption effect, in order of importance, are damage reduction, mating reduction, and capture of males in female-baited or synthetic-pheromone-baited traps. Damage reduction, where it can be accurately assessed, is the definitive criterion for efficacy. Mating reduction indicates population control and suggests expected damage reduction. Measurement of mating reduction can be more precise where damage assessment is vague. Captures in female-baited traps indicate the efficiency of disruption of the pheromone communication; however, this does not necessarily indicate a corresponding control of populations or damage. Females often have the potential to attract 50 to 100 males in a lifetime, and mating is still most likely where this is reduced 95% to 2 to 5 males/female. Disruption of female-traps is a good criterion for preliminary tests. Likewise, disruption by synthetic baited traps is convenient for obtaining some kinds of information but should not be used to imply control of the insect.

WESTERN PINE SHOOT BORER

Eucosma sonomana (Kearfott) is a pest of ponderosa pine, *Pinus ponderosa* Lawson, and infests other pines in the Western United States (Stoszek, 1973; Stevens and Jennings, 1977). Larvae mine the pith of the terminal shoot of the tree causing an estimated 25% loss of height growth (Stoszek, 1973). Terminals are occasionally killed outright or break off at the exit hole, which promotes forking and economic degradation. Some lateral shoots, particularly in the first whorl below the terminal, are also mined, but this has little impact on the tree. Damage is most often noted in even-aged pine plantations, but pheromone-baited traps indicate the insect is also ubiquitous in naturally regenerating stands throughout the Western United States (Sartwell et al., 1980b). Trees in the 1.5- to 15-m height range appear most susceptible to damage.

Mating and oviposition occur early in the spring, often shortly after snow melts. Eggs are laid under scales on the pine shoot buds (Koerber, Pacific Southwest Forest and Range Experiment Station, Berkeley, CA, unpublished). Hatching larvae burrow into the shoots and mine the pith. With occasional exception a single larva is present per shoot. Larval growth roughly parallels growth of the pine shoot. In midsummer, larvae bore an exit hole, fall to the ground, and pupate in the forest litter. Pupae have an obligatory diapause and will remain inactive until the following spring.

Population levels of E. sonomana are always low (a few hundred per hectare), perhaps because the numbers of preferred shoots in a pine stand are limited. Population levels are relatively stable from year to year. We have examined damage levels on a number of plots and have yet to find a natural population fluctuation >50% from one year to the next. The insect appears to be a classic "K" strategist (Southwood, 1977); that is, one with stable populations exhibiting arithmetic rather than logarithmic growth.

We were first approached by the Weyerhaeuser Company to consider developing pheromone disruption for control of E. sonomana. They considered the insect damaging to pine plantations, and no control methods existed. They had previously sponsored tests with conventional pesticides, including systemics, but these were found ineffectual, partly because the actively-feeding larval stage is protected by its location inside the shoot.

E. sonomana seemed an excellent target species for control by disruption of mating. It has a single generation each year with stable and predictable populations. Infestations and population levels can be quickly and accurately assessed. Further, its low population density suggested a particular vulnerability to control by disruption of the mating communication.

Pheromone Identification

Screening for potential attractants using field traps was considered a better approach to finding the sex pheromone than isolating and identifying compounds obtained from female extracts. Quantities of adult females are difficult to obtain because of their low field density, obligatory diapause, and related rearing problems. A number of compounds were tested on the basis of availability, known attractiveness to other Olethreutinae, or chemical similarity to such known attractants. Those few females that could be obtained were extracted and examined for the presence of compounds found active by the screening.

Males were strongly attracted to mixtures of (Z)-9- and (E)-9-dodecenyl acetates (Sower et al., 1979b). A ratio of 4:1 Z:E appeared to be present in gas chromatograms of female abdomen extract and approximated the optimum ratio for attraction of males. Minute nanogram quantities of this material formulated in PVC pellets (Daterman, 1974) will attract males about as well as live females. Given the low population densities of the insect, captures of 100 to 200 males/season per trap at 0.1 mg pheromone/trap bait indicate that the mixture is a very effective attractant. Accordingly, we consider a 4:1 mixture of the (Z:E)-9-dodecenyl acetate to be the sex pheromone of E. sonomana.

Disruption Tests

Application of pheromone in the field has repeatedly and consistently reduced the number of terminal shoots damaged (Table 1). Doses of 10 to 20 g of pheromone/ha applied using aircraft or 3.5 to 14 g/ha applied by hand were equally effective. Substantially lower dosages (0.2 to 2 g/ha by air) were not effective.

Table 1. Reduction in number of infested terminal shoots following pheromone disruption trials for western pine shoot borer.

Total area treated (ha)	Dose (g/ha)	Formulation	No. plots	Damage reduction (%)
8	3.5	PVC	3	83
4	14.0	PVC	3	84
20	15.0	Conrel®	3	67
100	20.0	Hercon®	3	88
600	10.0	Conrel	3	76

Controlled-release formulations using PVC (polyvinyl chloride) pellets applied manually and Conrel fibers and Hercon flakes applied from aircraft have been satisfactory for use with *Eucosma* (Overhulser et al., 1980; Sartwell et al., 1980a). All formulations still had 10 to 30% of their original pheromone content present by the end of the season. Treatment was in effect throughout the period during which moths were active (Figure 2). Plastic tarps located on the ground in clearings captured 54 to 63% of the respective intended dosages of Hercon and Conrel formulations, indicating satisfactory coverage of the plots. Stickers kept material in place adequately (Conrel) to totally (Hercon) over the season. Results with the Conrel formulation were consistent in both 1978 and 1979, the Hercon material was tested in 1979 only.

Pheromone release particles applied at various intervals, but which maintain similar total dosages, were about as effective. Only 100 to 400 PVC pellets were required/ha to get 3.5 to 14 g of pheromone/ha in contrast to 30,000 Conrel fibers/ha or 63,000 Hercon flakes/ha with respective pheromone doses of 10 and 20 g/ha. Consistent results (Table 1) with such diverse application patterns, but similar quantities of pheromone is remarkable. We plan further evaluation of effects of interval between evaporators as a variable separate from pheromone dosage *per se*.

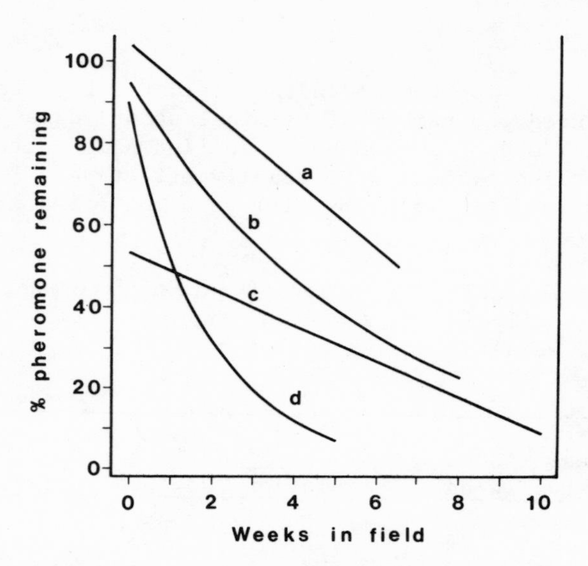

Fig. 2. Pheromone decline in controlled-release dispensers in the field. a) Z-6 Heneicosen-11-one for Douglas-fir tussock moth in Conrel fibers. b) 9-Dodecenyl acetates for western pine shoot borer in Hercon flakes. c) 9-Dodecenyl acetates for shoot borer in Conrel fibers. d) E-11 Tetradecenal for western spruce budworm in Hercon flakes. The end of each line approximates the end of the respective adult flight season which began during or after week 0. Weather was cool to cold for all data except (d) where temperatures were moderate.

Clearly, pheromone disruption can be used to reduce damage caused by *E. sonomana*. The pheromone appears to be registerable with the EPA in existing formulations. Toxicity of the pheromone appears very low based on various acute toxicity tests. Whether to use pheromone to control *E. sonomana* will now become an economic decision rather than a technical one. The specific question is: Will application on a given site cost less than predicted damage? This becomes an interesting point with *E. sonomana* since no previous control measures exist and, consequently, no previous guidelines exist on when to treat.

DOUGLAS-FIR TUSSOCK MOTH

Orgyia pseudotsugata is a serious defoliator of Douglas-fir, *Pseudotsuga menziesii* Franco, and true firs, *Abies* spp., in Western North America (Brookes et al., 1978). Small larvae prefer the more succulent new growth but late instar larvae will eat older needles. Tops of trees are usually consumed first, but in severe outbreaks complete defoliation of a stand can occur. Defoliation can result in top kill or death of the entire tree. Trees have to be almost totally defoliated before they are killed, and a stand may recover rapidly where defoliation is less severe (Wickman, 1979).

The Douglas-fir tussock moth is a sporadic pest. It is continually present throughout its range but normally at low to very low levels. Often the insect can only be detected by use of pheromone-baited traps. Sparse populations cause no damage; however, populations occasionally build rapidly to levels where damage occurs (Mason, 1979). Natural checks then catch up and the populations crash because of disease, parasitism, or starvation.

Tussock moth dispersal is passive -- the small larvae drift around on silk threads (Mitchell, 1979). Adult female tussock moths are wingless and mate and lay eggs on the cocoon from which they emerged. Adults emerge, mate, and lay eggs in the fall. The egg mass is the diapausing/overwintering stage. Hatch occurs in the spring about simultaneously with the growth of new foliage.

Development of pheromone disruption for control of the moth has been supported by the Douglas-fir Tussock Moth Research and Development Program. This program was formed in response to widespread outbreaks in Washington and Oregon in 1971 and 1972 which occurred coincidentally with the ban of DDT. The R&D Program supported development of a number of management methods including pheromone disruption. The final report of this program is a book which summarizes and updates available information about the Douglas-fir tussock moth (Brookes et al., 1978).

Pheromone Identification

The sex attractant pheromone of the Douglas-fir tussock moth was isolated and identified as (Z)-6-heneicosen-11-one (Smith et al., 1975). Daterman et al. (1976) suggested that additional unidentified co-attractants may also be released by the females, but none is nearly as effective as (Z)-6-heneicosen-11-one which is extremely attractive. Less than 0.5 μg of pheromone per 35 mg PVC formulated bait is sufficient for a survey trap and will last throughout the flight season with a maximum loss of only 10%. This kind of weak bait is used for insect survey traps where the survey objective is to capture numbers of males representative of the local population density. This survey system is now semi-operational, and is being deployed to provide early warning of sudden population increases concurrently with its final development (Daterman, 1978).

Disruption Tests

Pheromone application for mating disruption sharply reduces tussock moth reproduction. Pheromone applied at 9 to 36 g/ha in

110 to 450 g of Conrel fibers/ha reduced tussock moth reproduction by nearly 100% in a low density population in southern Oregon (Table 2). Dosages of 36 g/ha in Conrel fibers decreased reproduction by almost 80% in an outbreak population in New Mexico (Table 2). Formulations suitable for hand application, developed by Bend Research, Inc. (Sower and Daterman, 1977), had previously yielded similar results (Table 2). Dosages are not directly comparable because the manual application was confined to a single level 1.5 m above ground whereas trees, and presumably the aerially applied fibers, were up to 30 m. Pheromone applied at 1.5 m has little effect at 20 m (Sower and Daterman, 1977).

Table 2. Mating reduction following pheromone disruption of Douglas-fir tussock moth.

Treated area (ha)	Dose (g/ha)	Number plots	Population density	Mating reduction (%)
		Bend Research Formulations		
3.0	<1.0	3	low[1]	~100
0.3	<1.0	1	high[2]	81
		Conrel Formulations		
6.0	2.3	3	low[1]	73
6.0	9.0	3	low[1]	92
6.0	36.0	3	low[1]	~100
4.0	36.0	1	high[2]	77

[1]Test conducted in Oregon.

[2]Test conducted in New Mexico.

(Z)-6-Heneicosen-11-one is an unusual pheromone. It is relatively non-volatile and appears to be quite stable. We have recovered traps, lost the previous season and originally baited with 0.1 μg pheromone, that were still capturing fresh males 1 year later. Both Conrel fibers (Sower et al., 1979a) and the Bend Research formulations (Sower and Daterman, 1977) have easily lasted the field season with 40 to 50% of the original pheromone still present at the end.

When populations are sufficient to cause partial defoliation, we expect application of pheromone to result in about an 80% reduction in numbers of larvae the next season. This should hold damage to acceptable levels the next year and still allow natural control agents time to overtake and collapse the population before severe damage occurs. A stand that is not totally defoliated is expected to show substantial recovery (Wickman, 1979). If the population buildup was anticipated a year earlier, perhaps upwards of 90% reduction in reproduction would occur and the outbreak would be prevented. We do not know if such preventive treatment can be made cost-effective, but the ability to detect building moth populations before significant damage occurs appears feasible now (Daterman, 1978).

Low-level populations could presumably be eliminated from isolated areas with a multiple-year treatment program. A 90+% reduction each year should drive an already low population to zero. Since tussock moth dispersal is passive and probably slow reinvasion of larger areas from which the moth has been eliminated might take a long time. We do not know if this approach is economically feasible but it appears technically possible.

WESTERN SPRUCE BUDWORM

Western spruce budworm is a destructive defoliator of true firs, Douglas-fir, and spruce, *Picea* spp., (Furniss and Carolin, 1977). Periodic outbreaks persist for 4 to 5 years. Larvae feed on the current year's growth. Tree growth is decreased, top kill and other deformities occur, and trees are occasionally killed. Adults are active and oviposit in July and August with one generation/year.

We have conducted preliminary tests for disruption of mating of the western spruce budworm. Because results of these studies have not yet been published or concurrently submitted, collated data discussed can be made available on request. These tests are supported by the PNW Station and the Canada/U.S. Spruce Budworm Program-West.

Pheromone Identification and Disruption Tests

The pheromone is principally (*E*)-11-tetradecenal (Weatherston et al., 1971). Several additional materials, including the alcohol and acetate forms and the *Z* isomer of all of these, are also thought to be present in the females. Except for small quantities of the *Z* isomer of the aldehyde, however, none of these additional compounds have yet been shown to enhance attraction into traps. (*E*)-11-Tetradecenal was tested on small plots (20 x 20-m, 3 reps.) monitored with female-baited traps in the presence and absence of the other materials. Disruption with (*E*)-11-tetradecenal alone was 91% relative to checks, about equal to disruption of 84% with all of the other materials also present.

Disruption was tested on a larger scale in north central Oregon in 1979. Population densities were moderately high. Plots of 10 to 145 ha were treated (3 reps.) with 2 or 20 g of (E)-11-tetradecenal carried in 950 g of Hercon flakes/ha. Application was by aircraft.

Disruption appeared insufficient to substantially inhibit reproduction. Posttreatment egg mass samples were attempted but sample results were inadequate to document moderate amounts of disruption and no massive disruption was indicated. Female-baited traps in untreated plots captured 165 ± 91 ♂/trap. [Traps showed an 80% (33 ± 25 ♂/trap)]. Traps showed an 80% (33 ± 25 ♂/trap) reduction in males captured at 2 g/ha and 87% (22 ± 15 ♂/trap) reduction at 8 g/ha. If the average female can attract 22 to 33 males, mating is still likely even though substantial reduction occurred.

Low disruption effect was not due to any qualitative failure of materials or application. Pheromone content of flakes declined progressively until 5 to 10% of the original amount remained at the end of the adult flight (Figure 2). Of flakes found on foliage and marked just after treatment, 95% remained in place throughout the season. Disruption effect on male captures in traps was about as good at the end of the season as at the beginning. Further, trap captures at 20 m above ground were disrupted about equally to those 2 m above ground indicating a vertical treatment effect. Plastic sheets collected 59% of the intended treatment (8 ± 3 flakes/m^2) indicating good coverage. The pheromone content of individual flakes (0.2-2.0%) was lower than we have used with other species.

Disruption effects on western spruce budworm were quite low relative to results with western pine shoot borer on Douglas-fir tussock moth. High population densities were a likely contributing factor, and the insects may have been less dependent on long-range pheromone communication because of their close proximity. Higher pheromone dosages may be required, and further exploratory tests are anticipated.

SAFETY AND ENVIRONMENTAL IMPACTS

Several acute mammalian toxicity tests of pheromone compounds are done under contract before they are used in substantive field disruption tests. Costs for these tests are low with purchase of the 500 g of required pheromone being a major share of the cost. Results are well worth having during the research phase and may eventually be applicable toward registration requirements. The types of tests conducted were as follows: oral LD_{50}--rat, dermal LD_{50}--rabbit, primary skin irritation--rabbit, eye irritation--rabbit, and acute inhalation--rat. (Western pine shoot borer pheromone apparently irritated the eye of one (of several) rabbits with no other effects noted on tested animals.) Douglas-fir tussock moth pheromone and western spruce budworm pheromone had no effects on test animals.

We also were careful to watch for unexpected adverse environmental impacts which might occur. For example, host finding could involve kairomone-mediated behavior on the part of parasites. Accordingly, we are interested in any possible effect of the synthetic pheromones on parasites. Parasitization of tussock moth eggs was not reduced by pheromone application (Sower and Torgersen, 1979). Emergence of *Telenomus californicus* Ashmead and *Tetrastichus* spp. from tussock moth eggs was the same in a New Mexico plot treated with pheromone and in an adjacent untreated check plot.

RESEARCH AND DEVELOPMENT NEEDS

Interaction between effects of disruption treatments and insect population dynamics needs further study. For example, we do not know whether 95% control in 1 year is likely to be more effective than an 80% control for 3 consecutive years. Can the rise and fall of cyclic pests be monitored and predicted accurately enough to determine when prophylactic treatment is warranted? Is population density always a critical factor? Some other technical generalities require further clarification. Do dose response curves follow the log scale plot as shown in Figure 1?

The cost of pheromones can be a limiting factor in research and operational use. Prices of $3/g and upward prevent potential use of dosages much beyond 10 to 20 g of active pheromones/ha for many insects. Dosages of 100 g to 1 kg/ha (only 1.5 to 15 oz/acre!) are not really high except for price. These dosages are not getting proper research attention for field and forest pests due to cost. When females are good fliers and large plots are required to prevent inundation, high pheromone costs prevent research applications. Cheaper pheromones would remove many constraints on pheromone use.

Pheromone treatment should be considered for integrated use with other methods such as biological insecticides. Such methods alone are often not 100% effective but neither do they have severe environmental impact. We have discussed integrated uses often enough but know of no actual investigation.

SUMMARY

Control of damage by western pine shoot borer with disruption treatment has been accomplished with dosages that appear economically feasible. Larger-scale applications for control are planned, and registration appears likely.

Disruption was highly to moderately effective in reducing Douglas-fir tussock moth reproduction. Further tests stressing impact on damage and population levels are anticipated. Test results with western spruce budworm are still tentative.

We have yet to find evidence of any safety problems or significant adverse environmental impacts associated with pheromone disruption of forest Lepidoptera.

Research on disruption of appropriate forest pests should continue. Focal points could include improved methods for applying and using materials and cost reduction. Relevant studies of insect population dynamics are welcomed. Possibilities for integration with other methods of insect control should be explored.

REFERENCES

Brookes, M. H., Stark, R. W., and Campbell, R. W., 1978, The Douglas-fir tussock moth: A synthesis, USDA Forest Service Tech. Bull., 1585, 331 pp.

Daterman, G. E., 1974, Synthetic sex pheromones for detection survey of European pine shoot moth. USDA Forest Service Res. Paper PNW-180, 12 pp.

Daterman, G. E., 1978, Monitoring and early detection in the Douglas-fir tussock moth, in: "The Douglas-fir Tussock Moth: A Synthesis," M. H. Brookes, R. W. Stark, and R. W. Campbell, eds., pp. 99-102, USDA Forest Service Tech. Bull., 1585.

Daterman, G. E., Daves, G. D., Jr., and Smith, R. G., 1975, Comparison of sex pheromone versus an inhibitor for disruption of pheromone communication in *Rhyacionia buoliana*, Environ. Entomol., 4: 944.

Daterman, G. E., Peterson, L. J., Robbins, R. G., Sower, L. L., Daves, G. D., Jr., and Smith, R. G., 1976, Laboratory and field bioassay of the Douglas-fir tussock moth pheromone, (*Z*-6-heneicosen-11-one). Environ. Entomol., 5: 1187.

Furniss, R. L., and Carolin, V. M., 1977, Western Forest Insects, USDA Forest Service Misc. Publ. No. 1339, 654 pp.

Mason, R. R., 1979, Synchronous patterns in an outbreak of the Douglas-fir tussock moth, Environ. Entomol., 7: 672.

Mitchell, R. G., 1979, Dispersal of early instars of the Douglas-fir tussock moth, Ann. Entomol. Soc. Am., 72: 291.

Overhulser, D. L., Daterman, G. E., Sower, L. L., Sartwell, C., and Koerber, T. W., 1980, Mating disruption with synthetic sex attractant controls damage by *Eucosma sonomana* (Lepidoptera: Tortricidae, Olethreutinae) in *Pinus ponderosa* plantations. II. Aerially applied hollow fiber formulations, Can. Entomol., 112: 163.

Sartwell, C., Daterman, G. E., Sower, L. L., Overhulser, D. L., and Koerber, T. W., 1980a, Mating disruption with synthetic sex attractants controls damage by Eucosma sonomana (Lepidoptera: Tortricidae, Olethreutinae) in Pinus ponderosa plantations. I. Manually applied polyvinyl chloride formulation, Can. Entomol., 112: 159.

Sartwell, C., Daterman, G. E., Koerber, T. W., Stevens, R. E., and Sower, L. L., 1980b, Distribution and hosts of Eucosma sonomana in the western United States as determined by trapping with synthetic sex attractants, Ann. Entomol. Soc. Am., (In press).

Schwalbe, C. P., Pascek, E. C., Webb, R. E., Bierl-Leonhardt, B. A., Plimmer, J. R., McComb, C. W., and Dull, C. W., 1979, Field evaluation of controlled release formulations of disparlure for gypsy moth mating disruption, J. Econ. Entomol., 72: 322.

Smith, R. G., Daterman, G. E., and Daves, G. D., Jr., 1975, Douglas-fir tussock moth: Sex pheromone identification and synthesis, Science, 188: 63.

Southwood, T. R. E., 1977, Entomology and mankind, Proc. of the International Congress of Entomol., Washington, D.C., (1976), 15: 36.

Sower, L. L., Daterman, G. E., 1977, Evaluation of synthetic pheromone as a control agent for Douglas-fir tussock moths, Environ. Entomol., 6: 889.

Sower, L. L., and Torgersen, T. R., 1979, Field application of synthetic Douglas-fir tussock moth sex pheromone did not reduce egg parasitism by two Hymenoptera, Can. Entomol., June: 751.

Sower, L. L., and Whitmer, G. P., 1977, Population growth and mating success of Indian meal moth and almond moths in the presence of synthetic sex pheromone, Environ. Entomol., 6: 17.

Sower, L. L., Daterman, G. E., Orchard, R. D., and Sartwell, C., 1979a, Reduction of Douglas-fir tussock moth reproduction with synthetic sex pheromone, J. Econ. Entomol., 72: 739.

Sower, L. L., Daterman, G. E., Sartwell, C., and Cory, H. T., 1979b, Attractants for the western pine shoot borer, Eucosma sonomana, and Rhyacionia zozana determined by field screening, Environ. Entomol., 8: 265.

Stevens, R. E., and Jennings, D. T., 1977, Western pine-shoot borer: A threat to intensive management of pondoresa pine in the Rocky Mountain area and southwest, USDA Forest Service General Tech. Report RM-45, 8 pp.

Stoszek, K. J., 1973, Damage to ponderosa pine plantations by the western pine-shoot borer, J. Forest., 71: 701.

Vick, K. W., Coffelt, J. A., and Sullivan, M. A., 1979, Disruption of pheromone communication in the Angoumois grain moth with synthetic female sex pheromone, Environ. Entomol., 7: 528.

Weatherston, J., Roelofs, W., Comeau, A., and Sanders, C. J., 1971, Studies of physiologically active arthropod secretions. X. Sex pheromone of the eastern spruce budworm, _Choristoneura fumiferana_, _Can_. _Entomol_., 103: 1741.

Wickman, B. E., 1979, A case study of a Douglas-fir tussock moth outbreak and stand condition 10 years later. _USDA Forest Service Res_. _Paper PNW-244_, 22 pp.

DISRUPTION OF SOUTHERN PINE BEETLE INFESTATIONS WITH ATTRACTANTS AND INHIBITORS

Thomas L. Payne

Department of Entomology
Texas Agricultural Experiment Station
Texas A&M University
College Station, TX 77843 USA

INTRODUCTION

The southern pine beetle, *Dendroctonus frontalis* Zimmerman (Coleoptera: Scolytidae), is the most destructive insect pest of pine forests in the southern and southeastern United States and in parts of Mexico and Central America. Historically, efforts to control the beetle have been remedial in nature and primarily included the use of insecticides and salvage. In general, these methods have not been greatly successful, as evidenced by the continuing epidemics of the pest. Salvage is still practiced, but in the last decade cost and government restrictions have all but eliminated the operational use of insecticides in the forest. As a result, the forest manager is greatly limited in the choice of methods with which to attempt to deal with the beetle.

Remedial control techniques are still urgently needed, and through efforts over the past 15 years, behavioral chemicals have been investigated for their potential use in filling that need. Several behavioral chemicals, including attractants and inhibitors, have been found to play a role in the landing and attack behavior of the southern pine beetle (Borden, 1974; Vité and Francke, 1976; Payne et al., 1978a).

ATTRACTANTS

In nature, attractants function to orient flying beetles to a common host tree so that the beetles arrive in sufficient numbers over a relatively short period of time in order to overcome the resistance of the tree and successfully colonize it.

Frontalin (1,5-dimethyl-6,8-dioxabicyclo [3.2.1] octane [Kinzer et al., 1969]), produced by the female and believed to be released when she makes contact with a suitable host tree (Renwick and Vité, 1969), is considered the primary aggregation pheromone of the southern pine beetle (Kinzer et al., 1969; Payne et al., 1978a). By itself, frontalin attracts flying beetles of both sexes, but in the presence of a host odor such as _alpha_-pinene, its effect can be greatly enhanced (Kinzer et al., 1969; Payne et al., 1978a). _Alpha_-pinene proposedly functions as an arrestant in combination with frontalin (Renwick, 1970); that is, the pheromone attracts beetles to the tree and the host odor arrests their flight such that they land.

The attractant mixture, frontalure (frontalin plus _alpha_-pinene), has been evaluated in a trap tree application incorporating the herbicide, cacodylic acid (Vité, 1970). The method resulted in limited success and was influenced by several variables (Coulson et al., 1973ab, 1975). In an aerial application, frontalure was also evaluated for its effectiveness in suppressing an infestation through disrupting communication of flying beetles and the process by which they colonize new host trees (Vité et al., 1976). However, the treatment did not reduce attack by the beetles and resulted in increased aggregation on host trees.

INHIBITORS

Inhibitors are believed to function in directing populations of flying beetles from one tree to the next. As an attacked tree becomes colonized by a sufficient number of beetles, the population of flying beetles is steered away from that tree by the level of inhibitor being released.

Endo-brevicomin (7-ethyl-5-methyl-6,8-dioxabicyclo [3.2.1] [Silverstein et al., 1968]), produced in the hindgut of the male (Pitman et al., 1969; Silverstein and West, personal communication), functions to inhibit the response of both male and female beetles to attractive host trees and thus facilitates proliferation of an attack on new trees (Vité and Renwick, 1971; Payne et al., 1978a). The pheromone may also function in male competitive behavior on the host tree since it has been shown to elicit "rivalry chirps" (Rudinsky et al., 1974).

Verbenone, another inhibitor and produced essentially by the male beetle, is believed to be "multifunctional" (Rudinsky, 1973). At one level it affects the population of beetles attracted to the host tree by reducing the number of males and thereby balancing the sex ratio more toward 1:1 (Renwick and Vité, 1969; Payne et al., 1978a). In this regard it tends to differentially inhibit the response of males and females to the host tree. In contrast, when released in much smaller amounts by the female, verbenone is believed to synergize the attractant pheromone mixture (frontalin, trans-verbenol and host odor) in close-range orientation of males to the female entrance hole on the bark of the host tree (Rudinsky, 1973).

Results from earlier field bioassays showed that the aggregation of the southern pine beetle on attractant-baited traps could be significantly reduced with the addition of one or more inhibitors. Vitê and Renwick (1971) found that endo-brevicomin and its exo-isomer greatly reduced response to frontalure-baited traps. Later Payne et al. (1978a) found the inhibitory effect to be attributed to only the endo-isomer and that the addition of verbenone increased the effect further.

The earlier findings on both attractants and inhibitors prompted the following field studies, the results of which suggest the potential for development of behavioral chemicals for use in management of pest populations of the southern pine beetle.

TECHNIQUES AND FINDINGS

Attractant Studies

In an attempt to determine the effects of frontalure on southern pine beetle infestations in a combination of interruption and bait tree methods, Richerson et al. (1980) conducted a series of field tests in 2 active infestations of ca. 75 loblolly pines (*Pinus taeda* L.) in Montgomery County, TX, from September to October 1978.

Frontalure, used as a bait, was a 1:2 mixture of racemic frontalin and alpha-pinene eluted from polyethylene caps (Glass et al., 1970) containing 200 µl of the attractant (Fig. 1). Each cap released ca. 18 mg frontalure/day at a temperature range of 27-35°C based upon gravimetric analysis. Four caps were spaced equidistantly from one another at 3 m on selected host and non-host trees (hardwoods) within each infestation. All host trees containing either emerging brood adults, callow adults, pupae, or 1st to 4th instar larvae and all non-host trees in the area of those infested host trees were baited. The synthetic attractant was not placed adjacent tc naturally attractive sources at the front of the infestation.

Trap catch data were analyzed based upon the zone of the infestation as determined by the stage of attack and predominant life stages of beetles at the diameter at breast height level in infested trees (Payne et al., 1978b). Morisita's index of dispersion was used to measure the dispersion pattern of the southern pine beetle and a major predator, the clerid beetle *Thanasimus dubius* .(F.) (Morisita, 1962).

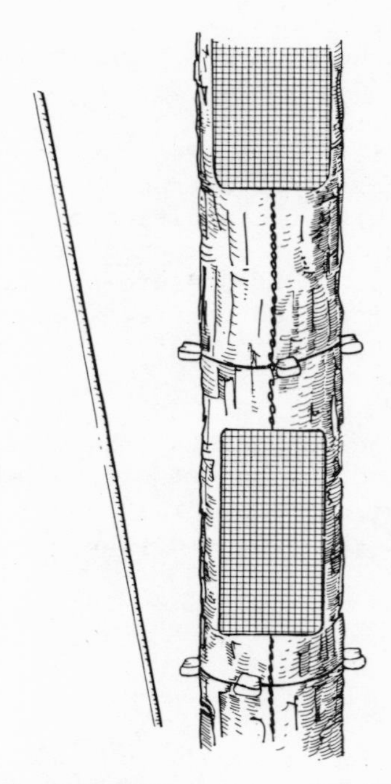

Fig. 1. Schematic of pine bole with landing trap and behavioral chemical elution caps (Payne and Richerson, 1979).

In all tests, no previously uninfested trees came under succesful mass attack during the frontalure treatment periods as opposed to during the pre- and post-treatment periods when southern pine beetles were successfully attacking host trees (Table 1).

Table 1. Number of loblolly pines mass attacked by southern pine beetles during pre-treatment, treatment, and post-treatment periods -- September 14-October 8, 1978 (after Richerson et al., 1980).

	No. of pines mass attacked					
Test no.	Pre-treatment	No. days	Treatment	No. days	Post-treatment	No. days
1	8	(6)	0	(6)	7	(1)
2	3	(3)	0	(3)	12	(1)
3	6	(3)	0	(5)	12	(2)

The placement of frontalure within the infestation interrupted the spot growth phenomenon of beetle activity by containing emerging broods and randomly redistributing the number of beetles throughout the infestation (Table 2). Consequently, insufficient numbers of beetles were present at the active front to successfully overcome host resistance. Beetle activity outside the treated area did not occur. The normal infestation spot growth/beetle attack behavior was reestablished after the frontalure treatment was removed from the infestation.

Table 2. Trap catch of southern pine beetles within infestations during pre-treatment and treatment periods in frontalure tests -- September 14-October 8, 1978 (after Richerson et al., 1980).*

Predominant life stage in test tree	No. trees	Pre-treatment	Treatment
		Test 1	
Brood adults and pupae	15	5 (46)	43 (355)**
Larvae (1st-4th instar)	11	2 (13)	9 (77)
Parent adults	33	93 (800)	48 (401)**
		Test 2	
Brood adults and pupae	7	3 (20)	54 (735)**
Larvae (1st-4th instar)	25	4 (22)	12 (167)**
Parent adults	16	93 (541)	34 (454)

*Percent total catch (n = total catch).
**T-probability <5%.

Although clerid distribution in the infestations was modified similarly to that of the southern pine beetle, the predators were not adversely affected by the treatment in that they remained in close association with their beetle host.

Payne et al. (unpublished data, 1980) conducted a subsequent test near Athens, GA from late August through October 1979 in which frontalure was used to keep emerging beetles in those areas of the infestation which were not conducive to population growth and were thus disruptive to spot growth (Fig. 2).

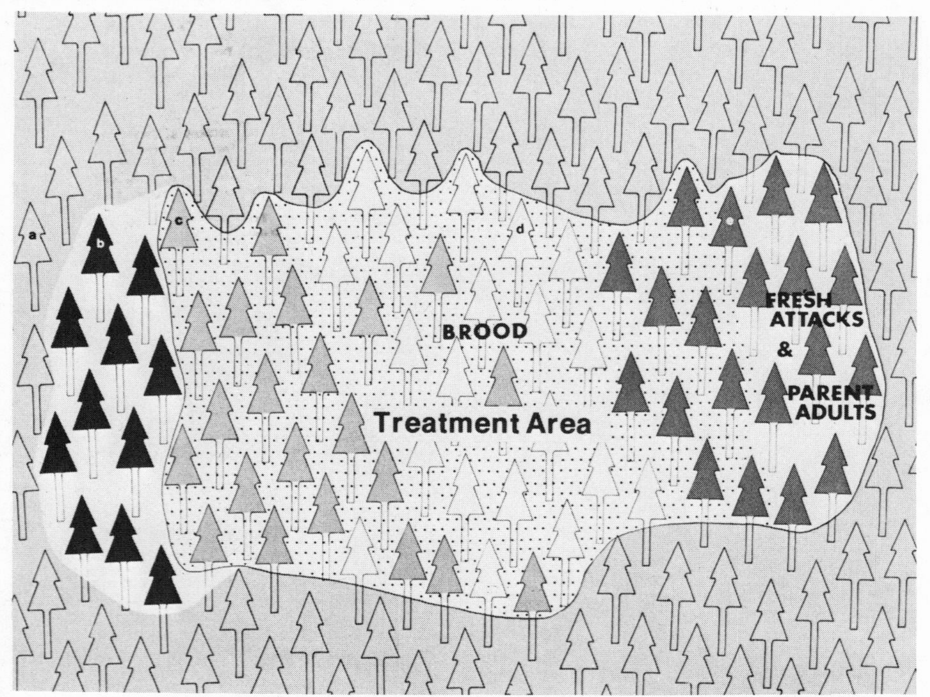

Fig. 2. Schematic of the attractant-treated area (frontalure) within an active southern pine beetle infestation. The clear area represents the infestation. The stippled area represents the treated trees and developing brood. a) Treated trees. b) Old attacked trees, beetles no longer present. c) Primarily pupae, callow adult and brood adult trees. d) Primarily late egg larval brood trees. e) Freshly attacked parent adult trees.

Frontalure, eluted from cellulose filters in glass vials (2/ tree at 3 m) at a rate of ca. 40 mg/tree/day for 50 days, was placed on 95 trees, nearly all of which were host pines. All host trees in the infestation containing developing brood, dead trees from which all brood beetles had emerged, and non-host trees in the area of these trees were treated. One modified "window trap" (Chapman and Kinghorn, 1955) was placed at each of 3 different locations in the spot as an indicator of beetle activity concentration.

The application of frontalure resulted in an abrupt halt of beetle activity. In a 3-week pre-treatment period, 13 pines were attacked, whereas within 7 weeks after frontalure was applied, only one pine was attacked (Fig. 3). Trap catch indicated that during the pre-treatment period, beetle activity focused on the area of the infestation containing predominantly parent adults but shifted to the brood area after treatment (Fig. 4). These data agree with results from the earlier frontalure test (Richerson et al., 1980).

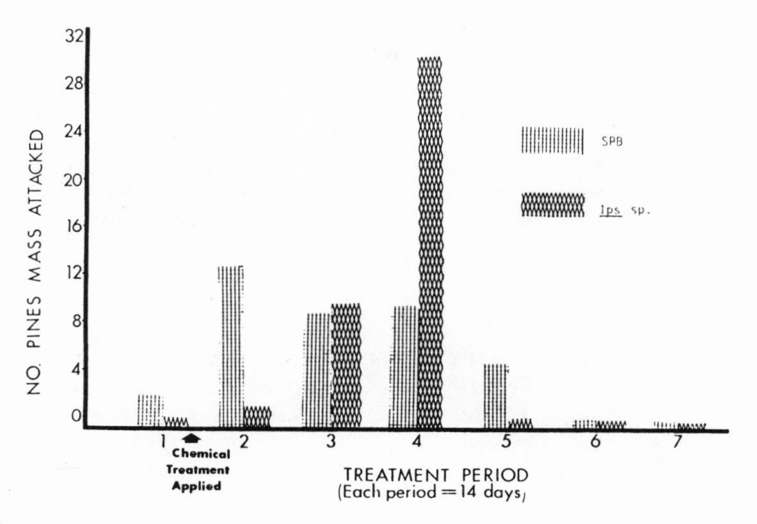

Fig. 3. Number of pines attacked during frontalure tests (Payne et al., 1980, unpublished).

Results from these 2 attractant studies supplement earlier data indicating the feasibility of using attractants in southern pine beetle control. By keeping beetles in an area of the infestation not conducive to population growth, spot dynamics may be disrupted to the extent that expansion is slowed greatly or stopped altogether. It is hypothesized that by keeping beetles from aggregating on suitable hosts for a sufficient number of days (as yet unknown), they will succumb to the influence of abiotic and biotic factors.

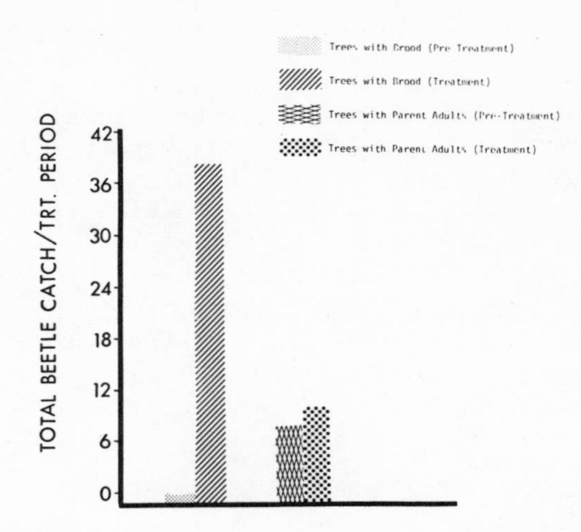

Fig. 4. Trap catch of southern pine beetles during pre-treatment and treatment periods on trees containing predominantly brood or parent adults during frontalure test (Payne et al., 1980, unpublished).

Inhibitor Studies

Payne et al. (1976) carried out tests monitoring the effects of inhibitors on both landing and flight activity of the southern pine beetle over a 24-day period from October 9 to November 2, 1975 on a study site located in a 500-tree southern pine beetle infestation (loblolly pine/mixed hardwood forest) at Bald Hill, south of Lufkin, TX.

Southern pine beetle activity in the infestation was monitored with the use of unbaited flight and landing traps coated with Stickem Special®. The inhibitors, endo- and exo-brevicomin, were released from Conrel® hollow fiber, controlled-release dispensers, (approximately 600 mg/h/day) which were attached to host trees at selected release points in the plot.

Check and treatment tests were conducted sequentially for 3 replicates of each in the same infestation. Consisting of a 3 x 3 grid, the plot was limited to that area containing primarily freshly attacked and unattacked trees at the head of the infestation.

As the result of these methods, the slow-release formulation of endo- and exo-brevicomin significantly reduced (70%) landing activity of beetles on host trees, but did not affect their flight activity (Table 3). This test was the first demonstration of an area-wide influence of synthetic inhibitors on the landing behavior of the southern pine beetle in an infestation.

Table 3. Total and mean daily landing trap catches of southern pine beetles in synthetic endo- and exo-brevicomin treated vs. check tests for each 4-day replication (Payne et al., 1976).

	Numbers		
	Total	Mean*	Significance
Replicate 1			
Treatment	114	29 ± 5.20	$\alpha \leq 0.005$
Check	458	115 ± 30.34	
Replicate 2			
Treatment	438	110 ± 30.14	$\alpha \leq 0.05$
Check	980	245 ± 78.33	
Replicate 3			
Treatment	199	50 ± 22.55	$\alpha \leq 0.05$
Check	1088	272 ± 110.22	
Totals			
Treatment	751	63 ± 15.43	$\alpha \leq 0.005$
Check	2526	211 ± 46.46	

*Means ± $SE_{\overline{X}}$

Richerson and Payne (1979) conducted subsequent field bioassays which revealed that the addition of verbenone to endo-brevicomin resulted in an even greater inhibitory effect.

Trees in 3 active southern pine beetle infestations (loblolly pine/mixed hardwood forests) in San Jacinto County, TX were treated with 1 of 4 inhibitory treatments. Two of the treatments were mixtures of endo- and exo-brevicomin (50:50 at 16 and 32 caps/tree and 85:15 at 32 caps/tree); one was a mixture of endo- and exo-brevicomin (50:50) and verbenone (16 caps of each/tree); and one was of verbenone alone at 16 caps/tree. Polyethylene caps were used as elution devices. The brevicomin isomer mixture eluted at ca. 10 mg/day/cap, and verbenone at ca. 5 mg/day/cap. Sufficient compound (500 µl) was placed in each cap to provide up to 50 days of release of inhibitors.

Southern pine beetle activity was monitored in the 3 study areas by (1) daily counts of the number and sex of beetles caught on unbaited landing traps on test trees, (2) daily inspection of the infestation for evidence of fresh beetle activity, and (3) an estimate of beetle attack density and brood production from a series of 100 cm^2 bark samples taken from each tree at the conclusion of each test, after procedures of Coulson et al. (1975). *Ips avulsus* activity was similarly monitored in that the incidence of the beetle was noted to increase with the treatments.

The brevicomin isomers alone and with verbenone did affect southern pine beetle activity; however, verbenone by itself did not. The effective inhibitors reduced landing by southern pine beetles on treated trees by ca. 84% (Table 4). The treatments effected an increase in *I. avulsus* landing by ca. 70% (Table 5). None of the inhibitor treatments prevented the treated trees from succumbing to southern pine beetle and/or possible *Ips* attack, indicating that the use of only inhibitors to protect individual uninfested trees will be ineffectual; however, there were significantly fewer total numbers of southern pine beetle galleries in treated trees (89% reduction), indicating that the inhibitors could cause significant reduction in brood production (Table 6). Foltz et al. (1976) reported an average of 1.59 eggs/cm of gallery length. Using this estimation of egg production, the inhibitor treatment effected an 88% reduction in the estimated number of eggs laid in the treated trees. These results, therefore, suggest the potential use of the compounds in ultimately reducing tree mortality by reducing brood production.

Watterson et al. (unpublished data, 1980) later conducted an experiment to determine if the inhibitor treatment used by Richerson and Payne (1979) would (1) significantly reduce the number of southern pine beetles attacking host trees, (2) increase egg laying by attacking southern pine beetles as the result of reduced competition, and (3) significantly reduce the within-tree populations and subsequent emerging populations of the southern pine beetle.

The study was conducted in 2 infestations of ca. 90 trees/ha (host and non-host trees) in loblolly pine/mixed hardwood forests near Cut'n'Shoot, Texas. The dosage and placement of the inhibitor pheromones were the same as those used by Richerson and Payne (1979). Techniques for sampling the within-tree, re-emerging and emerging southern pine beetle populations followed the guidelines set forth by Coulson et al. (1975). A topological mapping procedure (Pulley et al., 1976) was used to estimate southern pine beetle and *I. avulsus* population densities.

Table 4. Landing trap catches of southern pine beetles on inhibitor-treated and control trees (after Richerson and Payne, 1979).

	$\bar{X} \pm$ SE beetles caught/tree[a,b]		
Trap height (m)	Treated (n=3)	Control (n=3)	F prob.[c]
3	45 ± 16 a	181 ± 40 ab	*
5	46 ± 2 a	236 ± 78 a	*
7	15 ± 3 b	201 ± 74 a	*
9	8 ± 1 c	85 ± 34 b	*
Total catch	343	2108	*

[a]Means and SE are rounded to nearest whole number.

[b]Means followed by same letter within columns are not significantly different at the 5% probability level based on pairwise comparison of means using a one-way analysis of variance.

[c]One-way analysis of variance based upon pooled trap catch data/treatment. * = 5% Probability.

Table 5. Landing trap catches of *Ips avulsus* on inhibitor-treated and control trees (after Richerson and Payne, 1979).

	$\bar{X} \pm$ SE Beetles caught/tree[a,b]		
Trap height (m)	Treated	Control	F prob.[c]
3	139 ± 7 a	14 ± 1 a	*
5	111 ± 5 a	17 ± 1 a	*
7	114 ± 5 a	42 ± 2 b	*
9	97 ± 3 a	67 ± 2 c	NS
Total catch	1390	402	

[a]Means and SE are rounded to nearest whole number.

[b]Means followed by same letter within columns are not significantly different at the 5% probability level based on pairwise comparison of means using a one-way analysis of variance.

[c]One-way analysis of variance based upon pooled trap catch data/treatment. * = 5% Probability, NS = Non-significant.

Population densities of the re-emerging southern pine beetle, a measure of attacking adult density, were reduced 73% on the treated trees. Emergence population density was reduced 63% as compared to control (untreated) trees; however, the population density of *I. avulsus* was significantly higher (2550%) in treated trees than in control trees.

The inverse correlation between population densities of the southern pine beetle and *I. avulsus* was repeated in gallery data. Southern pine beetle gallery density was significantly lower in the treated trees (80% reduction) while that for *I. avulsus* was higher (1625%).

Table 6. Gallery lengths of southern pine beetles from 100 cm^2 bark samples from inhibitor-treated and control trees (after Richerson and Payne, 1979).

	$\bar{X} \pm$ SE Gallery length (mm)[a,b]	
Sample height (m)	Treated (n)	Control (n)
3	66 ± 10 (5)	65 ± 5 (41) NS
5	73 ± 12 (5)	82 ± 7 (28) NS
7	120 ± 0 (1)	63 ± 6 (34) *
9	89 ± 27 (3)	89 ± 27 (25) NS

[a]Gallery lengths based upon one sample/height. Means and SE rounded to nearest whole number. n = Number of galleries.

[b]One-way analysis of variance calculated on total number of galley lengths/height/treatment. F probability level: * = 5%, NS = Non-significant. Number of galleries (n) significantly different between treated and control for all heights.

These findings supported the earlier findings of Richerson and Payne (1979) and were somewhat disconcerting in that it appeared as if a secondary bark beetle pest was taking the place of the primary pest we were attempting to manipulate. However, *I. avulsus* is generally not aggressive and in the absence of the southern pine beetle it might not continue an infestation. Such a possibility led to continued tests to develop inhibitors for potential use in pest management.

Payne et al. (unpublished data) conducted field studies involving ground and aerial applications of inhibitors in an attempt to disrupt population growth by inhibiting beetle activity on trees (Figure 5).

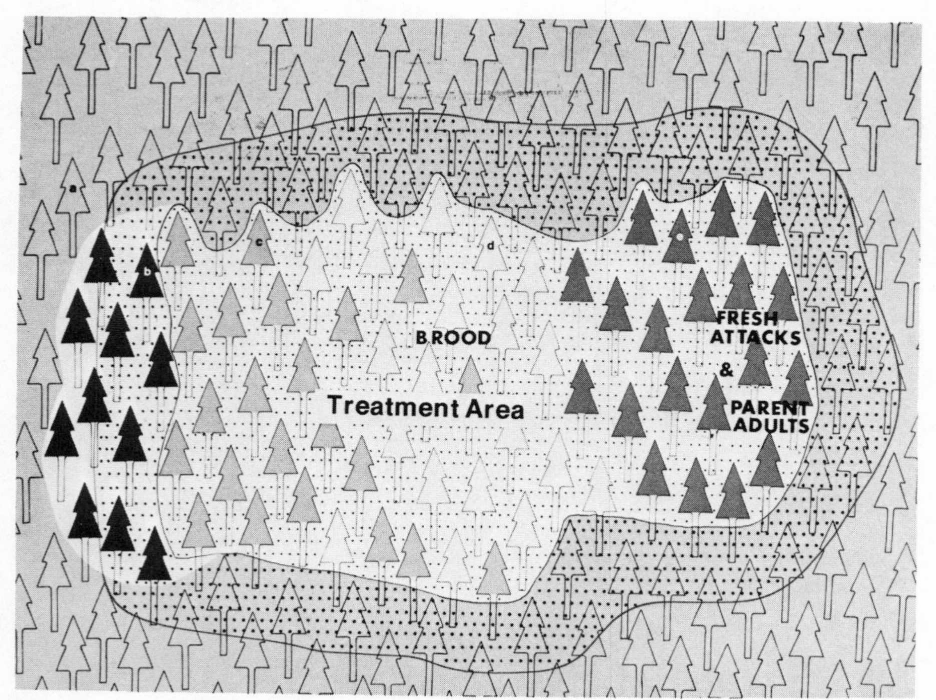

Fig. 5. Schematic of the inhibitor-treated area (endo-brevicomin and verbenone) within an active southern pine beetle infestation. The clear area represents the infestation. The inner stippling represents the treated area within the infestation. The outer stippling represents a 20 m treated buffer zone. (a) Treated trees. (b) Old attacked trees, beetles no longer present. (c) Primarily pupae, callow adult and brood adult trees. (d) Primarily late egg larval brood trees. (e) Freshly attacked parent adult trees.

A ground-applied inhibitor test was conducted near Macon, GA from mid-July through October 1979. Each compound of a 1:2 combination of the inhibitors, verbenone and endo-brevicomin, was released from polyethylene caps placed at 3 m on cane poles. Poles were placed on 2 m centers in a 0.2 ha area which included a 4 m buffer zone around a 0.15 ha infestation. The combined release rate of the inhibitors was ca. 11.5 gm/ha/day. The test was divided into a pre-treatment and treatment period, each including population estimates (Foltz et al., 1977) which were made twice at 2-week intervals. The infestation was cruised for new attacks ca. 3 times/week.

Results from the ground-applied inhibitor test are presented in Figure 6. During the 2-week pre-treatment period, the first population estimate indicated approximately 78,000 beetles were present within the infestation. At the time of the 2nd population estimate, beetle numbers were found to have increased to 188,000. (Such an indication of population expansion during the pre-treatment period also occurred for the aerially-applied inhibitor test). Another population estimate made during pre-treatment indicated an increase in the number of pupae present in the infestation (from 28,500 to over 46,300). During the treatment period, southern pine beetle dynamics were disrupted to the extent that trees meeting the sampling criteria were not available.

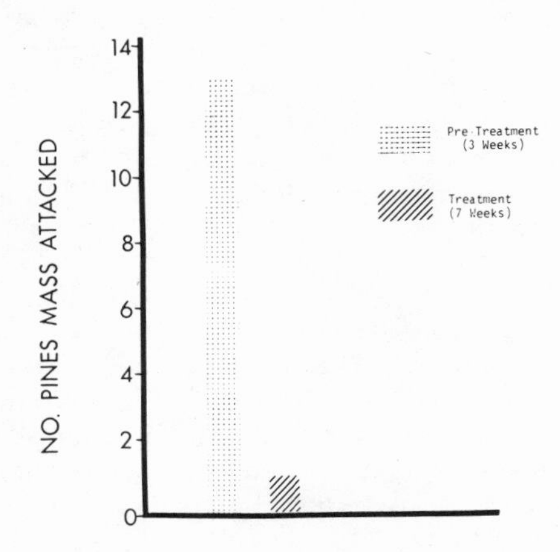

Fig. 6. Number of pines attacked during ground-applied inhibitor tests (Payne et al., unpublished).

In the pre-treatment period, only 2 pines were attacked by the southern pine beetle and none by *Ips* species; however, within 2 weeks following inhibitor application, a total of 13 trees was attacked by the beetle, as opposed to only one by *Ips* species. Throughout the remainder of the test the number of southern pine beetle-attacked trees declined concurrent with an increase in attacks by *Ips* beetles, predominantly *I. avulsus* and *I. calligraphus*. This trend continued until the infestation reached a point where no new trees were attacked by any beetles.

The phenomenon of few attacks, prior to treatment and many soon after treatment, possibly is the result of a coincidental burst in beetle activity. Such a spurt in activity during the period immediately following treatment is indicated by the large buildup of beetles in the pupal stage during the pre-treatment period.

Another factor possibly contributing to both the relatively high number of trees attacked subsequent to treatment and the length of time required to halt spot growth was the narrowness of the buffer zone around the infestation. This possible shortcoming was remedied in subsequent tests.

It is hypothesized that the inhibitor treatment disrupted southern pine beetle spot dynamics to the extent that beetles did not mass attack in numbers sufficient to monopolize tree resources but rather were able only to attack and weaken trees, making them susceptible to *Ips* colonization. Where *Ips* generally colonize those areas of host trees not infested by southern pine beetles, they were able to successfully compete with the beetle for resources. Most trees attacked by the beetle subsequently were found to be monopolized by *Ips* species. This is commensurate with previous work (Richerson and Payne, 1979; Watterson, 1979), indicating increased activity of *Ips* in infestations in which verbenone and *endo*-brevicomin were used. Once southern pine beetle populations were too low to initiate any further attack, *Ips* populations declined in the absence of susceptible trees.

In early October 1979, an aerial application of inhibitors was applied to an infestation near Juliette, GA. A 1:2 ratio of verbenone and *endo*-brevicomin was formulated into Hercon® plastic flakes at the ratio of 1 gm pheromone/5.2 gm formulated flakes which were mixed with a sticker and applied by helicopter to an area of ca. 0.56 ha. This area was comprised of an infestation of ca. 0.2 ha plus a 20 m buffer zone around the infestation. As previous calibration indicated, only 19% of aerially-applied flakes reached the understory of a pine forest; the concentration of pheromone applied was increased from that of the ground-applied treatment to a total of 759 gm/ha for a 50-day test, leaving an "effective application" of 3 g/ha/day. Two other beetle infestations were monitored as controls with methods identical to those used in the ground-applied test.

Population estimates made twice at a 2-week interval prior to treatment indicated southern pine beetle numbers increased from approximately 39,000 (first estimate) to approximately 162,000 (2nd estimate). That is, the infestations were not declining.

Estimates following the treatment revealed a 48% decrease in the beetle population. However, these results were inconclusive. Inclement weather had postponed application of the inhibitors until later than anticipated, and rain and cool weather subsequent to treatment severely limited beetle activity in the treated spots as well as in both control spots. Although some within-tree brood development did occur, no new trees came under attack in the treatment or control spots. Unquantified beetle activity was found, however, in other infestations near the treated area.

Outlook

For the future, tests are planned for the spring, summer and fall of 1980 for evaluation of both the attractants and the aerial applications of inhibitors to provide sufficient replication to quantify effective treatments.

As indicated by the results reported here, both attractant and inhibitor techniques are now at a point where they should be considered for their potential as direct control tools in suppressing southern pine beetle infestations. The attractants will likely be most practical for use in smaller infestations and will afford the forester and/or pest management specialist the potential for reducing or halting spot growth of infestations that cannot be salvaged or otherwise eliminated for economic or other reasons. Inhibitors, on the other hand, may offer aerial application to larger, unmanageable infestations.

REFERENCES

Borden, J. H., 1974, Aggregation pheromones in the Scolytidae, *in*: "Pheromones," M. C. Birch, ed., North-Holland Publishing Co., Amsterdam.

Chapman, J. A., and Kinghorn, J. M., 1955, Window-trap for flying insects, *Can. Entomol.*, 82: 46.

Coulson, R. N., Oliveria, F. L., Payne, T. L., and Houseweart, M. W., 1973a, Variables associated with use of frontalure and cacodylic acid in suppression of the southern pine beetle. 1. Factors influencing manipulation to prescribed trap trees, *J. Econ. Entomol.*, 66: 893.

Coulson, R. N., Oliveria, F. L., Payne, T. L., and Houseweart, M. W., 1973b, Variables associated with use of frontalure and cacodylic acid in suppression of the southern pine beetle. 2. Brood reduction in trees treated with cacodylic acid, *J. Econ. Entomol.*, 66: 897.

Coulson, R. N., Foltz, J. L., Mayyasi, A. M., and Hain, F. P., 1975, Quantitative evaluation of frontalure and cacodylic acid treatment effects on within-tree populations of the southern pine beetle, *Dendroctonus frontalis* Zimmerman, *J. Econ. Entomol.*, 68: 671.

Foltz, J. L., Mayyasi, A. M., Hain, F. P., Coulson, R. N., and Martin, W. C., 1976, Egg-gallery length relationship and within-tree analyses for the southern pine beetle, *Dendroctonus frontalis* (Coleoptera: Scolytidae), *Can. Entomol.*, 108: 341.

Foltz, J. L., Pulley, P. E., Coulson, R. N., and Martin, W. C., 1977, Procedural guide for estimating within-spot populations of *Dendroctonus frontalis*. Texas Agric. Exp. Stn. Misc. Publ., 1316, 27 pp.

Glass, E. H., Roelofs, W. L., Arn, H., and Comeau, A., 1970, Sex pheromone trapping red banded leaf roller moths and development of a long-lasting polyethylene wick, J. Econ. Entomol., 63: 370.

Kinzer, G. W., Fentiman, A. F., Page, T. F., Foltz, R. L., Vité,, J. P., and Pitman, G. B., 1969, Bark beetle attractants: Identification, synthesis and field bioassay of a new compound isolated from Dendroctonus. Nature, 221: 477.

Morisita, M., 1962, Iδ index, a measure of dispersion of individuals. Res. Pop. Ecol., 4: 1.

Payne, T. L., and Richerson, J. V., 1979, Management implications of inhibitors for Dendroctonus frontalis (Coleoptera: Scolytidae). Bull. de la Societe Entomol. Suisse, 52: 323.

Payne, T. L., Hart, E. R., Edson, L. J., McCarty, F. A., Billings, P. M., and Coster, J. E., 1976, Olfactometer for assay of behavioral chemicals for the southern pine beetle, Dendroctonus frontalis (Coleoptera: Scolytidae)., J. Chem. Ecol., 2: 411.

Payne, T. L., Coster, J. E., Richerson, J. V., Hart, E. R., Hedden, R. L., and Edson, L. J., 1978a, Reducing variation in field tests of behavioral chemicals for the southern pine beetle, J. Ga. Entomol. Soc., 13: 85.

Payne, T. L., Coster, J. E., Richerson, J. V., Edson, L. J., and Hart, E. R., 1978b, Field response of the southern pine beetle to behavioral chemicals, Environ. Entomol., 7: 578.

Pitman, G. B., Vité, J. P., Kinzer, G. W., and Fentiman, Jr., A. F., 1969, Specificity of population-aggregating pheromones in Dendroctonus, J. Insect Physiol., 15: 363.

Pulley, P. E., Foltz, J. L., Mayyasi, A. M., and Coulson, R. C., 1976, Topological mapping to estimate numbers of bark-inhabiting insects, Environ. Entomol., 4: 640.

Renwick, J. A. A., 1970, Chemical aspects of bark beetle aggregation, Symp. on Population Attractants, Contrib. Boyce Thompson Inst., 24: 337.

Renwick, J. A. A., and Vité, J. P., 1969, Bark beetle attractants: Mechanism of colonization by Dendroctonus frontalis, Nature, 224: 1222.

Richerson, J. V., and Payne, T. L., 1979, Effects of bark beetle inhibitors on landing and attack behavior of the southern pine beetle and beetle associates, Environ. Entomol., 8: 260.

Richerson, J. V., McCarty, F. A., and Payne, T. L., 1980, Disruption of southern pine beetle infestations with frontalure, Environ. Entomol., 8: 360.

Rudinsky, J. A., 1973, Multiple functions of the southern pine beetle pheromone verbenone, Environ. Entomol., 2: 511.

Rudinsky, J. A., Morgan, M. E., Libbey, L. J., and Putnam, T. B., 1974, Antiaggregation rivalry pheromone of the mountain pine beetle and a new arrestant of the southern pine beetle, Environ. Entomol., 3: 89.

Silverstein, R. M., Brownlee, R. G., Bellas, T. E., Wood, D. L., and Browne, L. E., 1968. Brevicomin: Principal sex attractant in the frass of the female western pine beetle, Science, 159: 889.

Vité, J. P., 1970, Pest management systems using synthetic pheromones. Symp. on Population Attractants. Contrib. Boyce Thompson Inst., 24: 343.

Vité, J. P., and Francke, W., 1976, The aggregation pheromones of bark beetles: progress and problems, Naturwiss., 63: 550.

Vité, J. P., and Renwick, J. A. A., 1971, Inhibition of *Dendroctonus frontalis* response to frontalin by isomers of brevicomin, Naturwiss., 58: 418.

Vité, J. P., Hughes, P. R., and Renwick, J. A. A., 1976, Southern pine beetle: Effect of aerial pheromone saturation on orientation, Naturwiss., 63: 44.

Watterson, G. P., 1979, Effects of verbenone and brevicomin on within-tree populations of *Dendroctonus frontalis* and *Ips avulsus* (Coleoptera: Scolytidae), M.S. Thesis, Texas A&M Univ., 40 pp.

DISRUPTION OF LONG-DISTANCE PHEROMONE COMMUNICATION IN THE ORIENTAL FRUIT MOTH: CAMOUFLAGING THE NATURAL AERIAL TRAILS FROM FEMALES?

Ring T. Cardé

Department of Entomology and
Pesticide Research Center
Michigan State University
East Lansing, MI 48824 USA

INTRODUCTION

The feasibility of disrupting "long-distance" pheromone communication by atmospheric permeation with synthetic pheromone has now been demonstrated in the field for numerous species of Lepidoptera (Roelofs, 1979), including the Oriental fruit moth, *Grapholitha molesta* (Gentry et al., 1975; Rothschild, 1975, 1979; Cardé et al., 1977a; Charlton and Cardé, 1980). Notwithstanding the substantial progress achieved within the last several years in the application of this technique, the mechanisms by which disruption of communication is effected remain speculative. This paper summarizes, for the Oriental fruit moth, the orientation and mating behaviors of the male as mediated by the female-emitted pheromone and suggests how these reactions may be modified by atmospheric permeation with synthetic disruptant. Understanding these processes requires characterization of the chemical components comprising the natural chemical communication system, description of how the pheromone disperses in a wind field, and definition of the threshold concentrations of pheromone eliciting the natural mating behaviors.

THE NATURAL PHEROMONE COMMUNICATION SYSTEM

The existence of a female-produced sex pheromone in *G. molesta* was first demonstrated by George (1965). One component of the pheromone was identified from female abdominal tips as (*Z*)-8-dodecenyl acetate (Z8-12:Ac) by Roelofs et al. (1969). In subsequent empirical field screening trials, Z8-12:Ac alone was not found to be an effective attractant and several congeners of Z8-12:Ac were shown to augment considerably the catch elicited by Z8-12:Ac. When its geometrical isomer, (*E*)-8-dodecenyl acetate (E8-12:Ac), was admixed, maximal trap catch was effected with about 5-7% (E) (Beroza et al., 1973a,b; Roelofs and Cardé, 1974; Gentry et al., 1974, 1975; Cardé, et al., 1975a,b; Rothschild and Minks, 1977; Baker and Cardé, 1979a). Using an optimal (Z) to (E) ratio of these 2 acetates, trap catch was found to be enchanced further by the addition of either n-dodecanol (12:OH) (Roelofs et al., 1973; Beroza et al., 1973a,b; Gentry et al., 1974, 1975; Cardé et al., 1975a,b; Rothschild and Minks, 1977) or (Z)-8-dodecen-1-ol (Z8-12:OH) (Cardé et al., 1975a; Rothschild and Minks, 1977; Baker and Cardé et al., 1979a). These 4 compounds are now known to be emitted by the calling female (Cardé et al., 1979) and hence can be considered as components of the pheromone bouquet.

Although assuredly not all moth species use identical mechanisms for orientation to and recognition of the female, the available evidence strongly suggests many common features in the behavior of those species that have been studied (Table 1). First, upwind flight to the female generally appears to be under guidance of a pheromone-induced anemotactic mechanism in which upwind progress is gauged by optomotor cues (Kennedy and Marsh, 1974). Second, very near the chemical source, ground speed of flight may be reduced, probably because of an increase in pheromone concentration (orthokinesis), and appropriate visual cues (Farkas et al., 1974; Cardé, and Hagaman, 1979). The next behavioral event is likely to be landing, mediated by similar sensory inputs, followed by walking, probably while wing fanning, toward the female. Actual recognition of the female and elicitation of the copulatory attempt may involve interplay of pheromonal, visual and tactile sensory inputs. Each behavior may be viewed as occurring either comparatively "early" and at "long-range" or "late" and at "close-range." These terms may be useful for categorizing behavior, but the distinctions between such terms are probably arbitrary inasmuch as the order, timing and frequency of these events is in part dependent upon the observational milieu (Baker et al., 1979a).

The role in G. molesta of each of the chemical components of the female's pheromone and visual, tactile and anemotactic cues in evoking the foregoing male behaviors has been investigated in considerable detail (Cardé et al., 1975a,b; Baker and Cardé, 1979a,b; Baker and Roelofs, 1980). The female pheromonal components Z8-12:Ac, E8-12:Ac and Z8-12:OH act largely as a unit to elicit both "early" and "late" behavioral responses. The behavioral effects of each of these components are described most precisely when one component is added to the other 2, rather than alone or in a binary combination. Upwind flight in a wind tunnel bioassay requires a combination of Z8-12:Ac and E8-12:Ac, whereas postflight walking and the hairpencil courtship display requires Z8-12:OH along with the natural binary acetate blend. Z8-12:OH in this combination also elevates over the acetates alone the proportion of males flying upwind. The 4th pheromonal component, 12:OH, discernibly affects the courtship hairpenciling only when the other alcohol is present in suboptimal amounts. Other measures of male response, such as trap catch and orientation to synthetic lures, appear in maximal frequencies when the ratios and dosages of Z8-12:Ac, E8-12:Ac and Z8-12:OH are optimized (Baker and Cardé, 1979a).

Table 1. Generalized organization of male location of females and mating behavior in moths.

Time	Distance	Behavior	Proximal stimuli
Early	Long-range	Flight initiation	Environmental cues regulating circadian rhythm, pheromone
		Upwind flight	Pheromone, visual, anemotactile
Late	Close-range	Orthokinetic reduction in ground speed of flight	Pheromone concentration, visual, anemotactile
		Landing	Pheromone concentration, visual, tactile
		Walking to source	Pheromone, visual, tactile
		Recognition and mating attempt	Pheromone, visual, tactile

DISRUPTION OF "LONG-DISTANCE" PHEROMONE COMMUNICATION IN *GRAPHOLITHA MOLESTA*

Efficacy of disruption of communication can be evaluated in several ways; orientation to synthetic- or virgin female-baited traps, mating, and crop damage can be contrasted in disruptant-treated and experimental check areas. Many of the early tests with *G. molesta* evaluated disruption with synthetic lures that lacked the Z8-12:OH component and also provided no estimate of the airborne concentration of disruptant. Rothschild (1975) established that ca. 1.2×10^{-1} g/ha/day of Z8-12:Ac [2-3% (*E*)] was effective in disrupting orientation to virgin females, even when the point source disruptant dispensers were deployed at a density of as few as 25/ha. In field trials in which the season's end damage was the criterion of efficacy, Rothschild (1975) found both peach fruit and shoot infestations were below damage levels of the comparable check (conventional insecticide treatment) plot: Similarly, Cardé et al. (1977b) showed complete disruption of males to female-baited traps at estimated emission rates of 1.5 x 10^{-1} g/ha/day from hollow fibers with a relatively even atmospheric permeation of a 93:7 mix of Z8-12:Ac and E8-12:Ac (ca. 1700 dispenser points/ha). Rothschild (1979) estimated a lower threshold for disruption using the 2-acetate blend at 2.1 x 10^{-4} g/ha/day.

In view of the major behavioral effects mediated in part by Z8-12:OH, the potential role of this compound in conjunction with Δ8-dodecenyl acetates in communication disruption required assessment. Thus, in 1979 we conducted field tests in replicated, small plots (0.065 ha) to evaluate the disruptive effects of the Z8-12:Ac [6.5% (E)] blend alone at dosages of 2.5 x 10^{-2}, 2.5 x 10^{-3} and 2.5 x 10^{-4} g/ha/day and the same rates of acetates plus Z8-12:OH (Charlton and Cardé, 1980). The levels of disruption were measured in 2 ways. The percent disruption of communication to the synthetic 4-component pheromone and to a cage containing 3 virgin females in sticky traps gave a relative evaluation of efficacy between treatments. A 2nd method, the index of source location, assessed the likelihood of a single male locating a female or synthetic pheromone source during any sample interval. This latter measure equates location of a pheromone source by even a single male as a breakdown in disruption and hence an indicator of potential mating. The index of source location is more sensitive to the sampling interval employed and to changes in the population density than percent disruption of communication.

These indices suggested that at the highest levels tested, the acetates and the 3-component mixture were equally efficacous (Table 2). As the levels of disruptant were decreased by decade steps, the 2-acetate blend appeared less effective than the alcohol-containing mixture, especially when efficacy was evaluated by the index of source location. Therefore, the addition of the Z8-12:OH component to the disruptant mixture reduces the quantity of disruptant necessary to achieve disruption of "long-distance" orientation. Because of the substantial levels of disruption at the highest dose of acetates alone, we could not measure if the degree of disruption at higher levels of emission was also increased by the addition of Z8-12:OH, but this is an obvious possibility.

DISRUPTION OF COMMUNICATION MECHANISMS

The actual means by which the airborne disruptant modifies "normal" male behavior are little understood. Three principal hypotheses seem most appealing when the disruptant is the synthetic of the natural pheromone. These mechanisms are not mutually exclusive and may act in concert in the disruption of some species.

Table 2. Disruption of Grapholitha molesta attraction to baited traps using different atmospheric concentrations of Z8-12:Ac (6.5% E) alone and in combination with Z8-12:OH dispensed from hollow fiber sources (Charlton and Cardé, 1980).

Disruptant treatment (fibers/ .065 ha)	Trap bait[a]					
	Pheromone[b]			3 Virgin ♀		
	Trap catch[c]	% Disr.[d]	Index[e]	Trap catch[c]	% Disr.[d]	Index[e]
			Z8-12:Ac (6.5% E)			
4000 (2.5×10^{-2})[f]	1 a	100	3 ab	0 a	100	0 a
400 (2.5×10^{-3})	7 a	98	18 cd	0 a	100	0 a
40 (2.5×10^{-4})	39 b	90	57 e	3 a	98	12 bcd
			Z8-12:Ac (6.5% E) + Z8-12:OH			
4400 (2.5×10^{-2})	1 a	100	3 ab	2 a	99	3 ab
480 (2.6×10^{-3})	2 a	99	6 abc	1 a	99	3 ab
80 (3.0×10^{-4})	18 a	95	24 d	8 a	94	6 abc
Check (no disruptant)	398 c	-	88 f	139 c	-	57 e

[a]Catch in unbaited traps is not included due to the negligible (<1%) male capture.

[b]Septa loaded with 100 μg Z8-12Ac (5% E), 1 μg Z8-12:OH and 300 μg n-12:OH.

[c]Values represented by the same letter are not significantly different ($P<0.05$) according to Student-Newman-Keul's test of means transformed to $\sqrt{X + 0.5}$.

[d]Percent disruption for a specific treatment was calculated as:

$$\frac{\text{(catch in untreated areas)} - \text{(catch in disruptant area)}}{\text{catch in untreated area}} \times 100\%.$$

[e]Index of source location = Percentage of traps catching ≥ 1 male/trapping interval. Percentages in same column having no letters in common are significantly different according to a χ^2 2 x 2 test of independence ($P<0.05$).

[f]Figures in ( ) = release rate in g/ha/day.

First, continuous or even brief exposure to pheromone may require that the concentration of pheromone requisite to elicit behavior be elevated or pheromone responsiveness may be eliminated altogether. Both of these modifications could result from either sensory adaptation of peripheral receptors or habituation at a more central integrative level. Although there are innumerable laboratory trials demonstrating this phenomenon (Bartell, 1977), the relevance of this mechanism to field trials of disruption remains to be established. The laboratory assays generally did not assess "long-distance" reactions and the molecular concentrations of pheromone were not measured. Indeed, at least in G. molesta, caged males exposed for 24 h in the field to an atmospheric release of 2.5×10^{-2} g/ha/day of the 3-component pheromone nonetheless engaged in typical courtship displays and mated readily (Charlton and Cardé, 1980).

A 2nd hypothesis proposes that males "search" normally in disruptant-treated areas and are attracted to the numerous sources of synthetic pheromone. This mechanism is particularly appealing when synthetic pheromone is released from a point source matrix such as hollow fibers, which emit pheromone at approximately the same rate as a female emits. The success of this mechanism, which may be termed the "competition" effect, clearly is dependent upon the relative attractiveness of the point sources and females as well as their ratio. For Pectinophora gossypiella (Saunders), the pink bollworm, the rate of pheromone emission from a female and hence its direct comparability to hollow fibers is unknown. However, direct behavioral observations in field trials have established that some P. gossypiella males locate hollow fiber synthetic sources (applied at a rate of $1/m^2$) and may attempt to copulate with the fiber (Staten, unpublished). As yet we do not know what proportion of wild males engage in such behavior or what proportion of their normal activity time is spent "searching" for fibers. In G. molesta, when the disruptant employed is ca. a 95:5 mix of Z8-12:Ac and E8-12:Ac plus Z8-12:OH in fibers, a single 8 mil fiber would be emitting pheromone at a rate well above the optimal emission rate for "close-range" behaviors (Cardé et al., 1977b; Baker and Cardé, 1979a; Baker et al., 1980). Moreover, the 2-acetate blend alone, even if it was emitted at an optimal rate from numerous point sources, would elicit substantially less upwind flight and close-range behaviors than the Z8-12:OH-containing mixture. G. molesta may be somewhat atypical in that males respond optimally in the "late" behaviors to a rather narrow range of pheromone concentration (Cardé et al.,

1975b; Baker and Cardé, 1979a; Baker and Roelofs, 1980).

A 3rd mechanism supposes that camouflaging the natural aerial trials by raising the concentration of synthetic pheromone sufficiently above the density emanating from the female, thereby rendering the boundaries of the natural plume indiscernible. A male in this milieu would lack sufficient information to detect the boundaries or even the presence of natural plumes and thus to negotiate a typical zigzag course upwind to the females. Verification of this mechanism for a given species requires information on the rate of pheromone emission from a female, an adequate model to describe its dispersal in a wind field, the atmospheric concentration of emitted disruptant, and the concentration of pheromone requisite for the various male behaviors. In *P. gossypiella*, there is some behavioral evidence that a camouflaging mechanism contributes to disruption. Doane and Brooks (1980) found that a "typical" application of pink bollworm disruptant (ca. 1 fiber/m^2) was sufficient to disrupt communication to 10 fiber sources of pheromone used as bait in traps. However, when the number of sources in a bait trap was increased to 20 or more fibers, the disruptive treatment was no longer as effective in reducing trap catch. These findings with *P. gossypiella* strongly implicate a camouflage effect which can be overcome by increasing the rate of pheromone emission from a monitoring trap.

For *G. molesta* in those field tests of disruption where the atmospheric concentration has been estimated by loss from the release matrix, a camouflaging mechanism seems the most plausible explanation. The female's rate of pheromone emission is ca. 3.2 ng/h of the acetate components (Baker et al., 1980). Downwind concentrations of pheromone from a point source generally have been estimated by Sutton's equation (Bossert and Wilson, 1963). The maximum concentration of pheromone from a calling *G. molesta* female in a 100 cm/sec wind can be estimated for various distances (Fig. 1). This procedure for estimating pheromone dispersal in a wind field is based upon empirical measurements which yield a 3-min average of concentration (Sutton, 1953). This method underestimates peak ("nearly instantaneous") concentrations, especially close to the female, and assuredly it is peak concentration rather than a long interval average that determines the likelihood of eliciting behavior and thus defines the active space. Moreover, flux of pheromone rather than concentration may determine the threshold for responses.

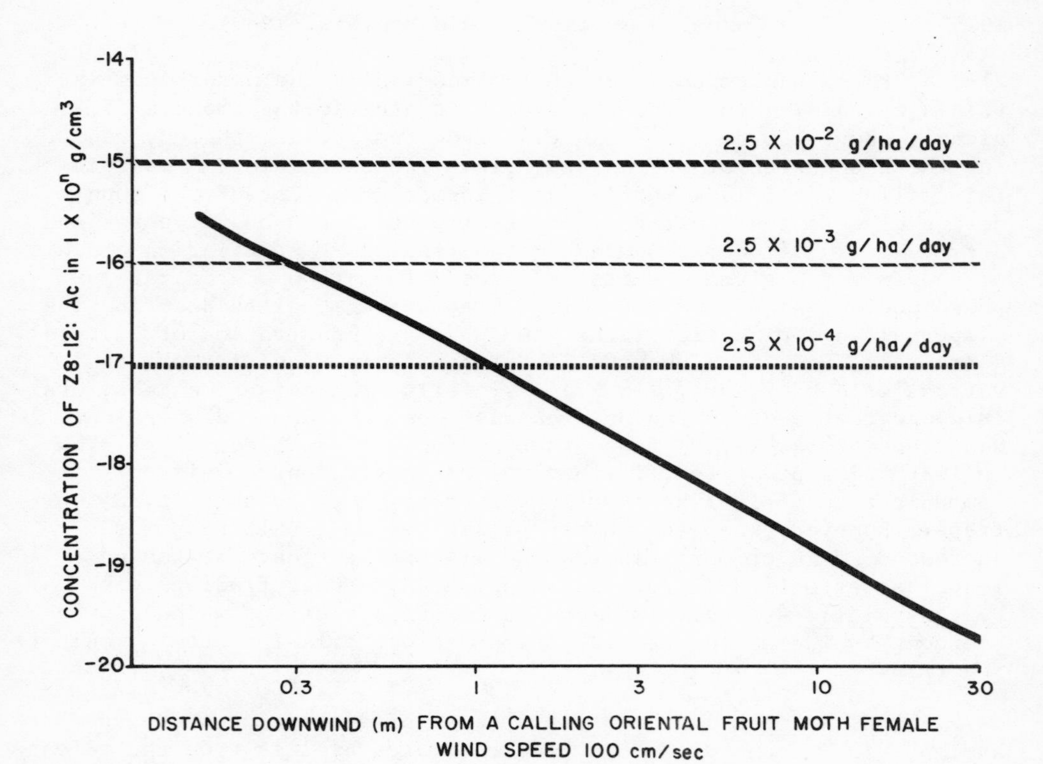

Fig. 1 Relationships between estimated concentrations of disruptant in 1 ha plots and the maximal downwind concentrations of pheromone from a calling G. molesta female as calculated by Sutton's formula.

Airborne concentration of disruptant in the field could be measured directly, although to date this has been accomplished only for one moth disruption system (Plimmer et al., 1978). In the absence of direct quantification, an upper estimate of disruptant concentration can be derived from assuming dispersal only into 3 m of orchard canopy, and no loss by absorption into the leaf or other surfaces. In large-scale field applications, neglecting such losses, disruptant concentration would be progressively elevated downwind.

Given the above assumptions, the concentration of synthetic pheromone in disruptant plots (Charlton and Cardé, 1980) can be estimated for a 1 ha source in a 100 cm/sec wind. The comparative levels suggest that, at a disruptant emission rate of 2.5 x 10^{-2} g/ha/day, the concentration of synthetic pheromone is well above the maximal concentration from the female, even within decimeters. When the camouflage is viewed in this way, its effect must be considered in terms of the distance to a female. Estimates of concentration of airborne disruptant and female pheromone plumes would be similar at other typical wind velocities or if the evaluations are of flux rather than concentration.

These estimates are tentative. Understanding the camouflaging mechanism in G. molesta and other species first necessitates a model of point source pheromone dispersal in a wind field. This model should provide concentrations or fluxes of pheromone over time intervals corresponding to the response latency. A 2nd requirement is actual measurements of disruptant concentration in the field, thereby accounting for the possibly substantial loss of synthetic by adsorption, especially onto leaf surfaces, and loss above the canopy. However, in the preliminary analysis presented here, a camouflage of the female's aerial plume seems likely.

If the formulated disruptant is not the pheromone but is only a part of the blend of an "incorrect" ratio, then the mechanisms responsible for disrupting communication may differ (Roelofs, 1978). If a camouflage is involved, then an atmospheric concentration of disruptant, for example, 10-fold above the concentration several decimeters from the female, should render the natural plume indiscernible and alter the ratio of natural components markedly. But if the mechanism achieving disruption is due to competing sources, then an "inferior" (less attractive) blend of chemicals in the release matrix should be relatively ineffective. Considering the same factor in a multispecies pest complex where competition contributes to efficacy, it would be important to formulate and disperse the sources for each species separately, because it is rarely possible to maintain attractancy for several species in a single source containing a mixture of pheromonal components from several species.

If the disruptant is not the natural system but a parapheromone comprised of one or more analogues and eliciting the same behaviors, then disruption may be caused by any of the 3 principal mechanisms enumerated previously. However, if an antogonist (variously termed an antipheromone, inhibitor, masking agent, etc.) is employed, then a competition effect seems unlikely, and yet additional modifications of normal behavior are possible (Cardé, 1980).

EVOLUTION OF RESISTANCE

Few direct chemical control methods for insects have failed in time to generate partial or complete "resistance" in at least some target (and indeed non-target) species (Brown and Croft, 1975). Selection clearly would favor individuals which would mate despite the omnipresence of disruptant. Now that disruption of pheromone communication is an operational technique for some moth pests and under development for a host of others, some consideration of the potential evolution of resistance is warranted.

The potential modifications to "overcome" the disruptant system are manifold (Cardé, 1976) and obviously are related to the mechanisms by which disruption is achieved. When a camouflage is the effective mechanism, simply raising the actual emission rate from the female, in turn, could raise the discernibility of her plume. Also, males which are capable of perceiving slight changes in atmospheric concentration of pheromone may be more efficient at discerning natural plumes above the background of disruptant.

In species such as G. molesta, where the pheromone is a blend of compounds, a shift in their ratio could provide a plume sufficiently distinct over the background concentration. This path toward resistance would be especially advantageous where disruption was achieved wholly or in part by the competition effect. Many moth species appear to employ relatively precise blends in which the optimal behavioral responses are elicited by a narrow range of component ratios (Baker et al., 1976; Baker and Cardé, 1979a). In such cases, the pheromone components generally are structurally similar and can be expected to have quite similar vapor pressures at typical environmental temperatures. Hence, their ratio would remain constant over this range of temperatures. In other pheromone blends where the components differ by chain length or moiety, the component volatilities may differ sufficiently at typical environmental temperatures to prevent emission of a consistent ratio (Cardé et al., 1977a). Thus, in these systems males do not respond optimally to as narrow a ratio of components; hence, evolution of resistance by alteration of component ratios seems less likely.

Another possible modification would add a new component to the pheromone. In some situations components could be chemicals already emitted by the females but not now sensed by the responding males. In the first case, evolution of resistance would involve release of new compounds and concomitant evolution of appropriate peripheral receptors, a CNS integrative mechanism, or both. In the second example, only the male sensing system would be altered.

Resistance also could be conferred by placing greater emphasis on other communication modalities, such as visual or tactile cues, or the use of favored habitat locations for calling and mating, such as the top of the vegetative canopy. Males that continue the "search" for females, especially in the walking mode, may be successful in locating them, despite the omnipresence of disruptant.

The eventual prospects for any of these mechanisms evolving are obviously speculative, but it is clear that "long-distance" communication via attractant pheromones is the dominant means of mate location in moths, as well as many other insects, and it undoubtedly is a major factor in the evolutionary success of this group. Other communication modalities, such as visual or tactile orientation, are considerably less efficient at low population densities. Thus, it seems improbable that resistance would involve elimination of attractant pheromone communication, although modification of this system is possible.

CONCLUSIONS

In G. molesta, the available evidence shows that a combination of Z8-12:Ac and E8-12:Ac and a blend of these acetates plus Z8-12:OH are effective in disrupting long-range pheromone communication (Charlton and Cardé, 1980). Estimates of airborne disruptant concentrations suggest that their levels are well above those emanating from the female; hence, they support a camouflage mechanism for interruption of attraction. The 3-component blend would appear to be efficacious at lower concentrations than the 2-acetate mixture; thus, the 3-component blend would have the most potential for direct population manipulation. However, further field studies where damage is the criterion of efficacy are warranted, because at some atmospheric concentrations the alcohol-containing disruptant mix may elicit proportionately more late or close-range behavior than the 2-acetate blend. Males in the 3-component disruptant area might be more apt to "search" for females and possibly be more successful in locating them. Certainly Rothschild's (1975) successful test of damage reduction with the 2-acetate blend indicates great potential for mating reduction in the Grapholitha species.

In a more general view, the development of operational systems effective in interruption of moth mating is obviously a complex process, particularly if we are to understand the sensory inputs and behavioral reactions governing normal mate location and recognition, as well as to how the disruptant modifies these processes. We have measurements of the natural rates of female pheromone emission or male response thresholds for only a few species, and the process of pheromone dispersal in a wind field is poorly understood. If the mechanism effecting interruption of mating is indeed a camouflage, then such parameters of the natural communication system can be used to design the requisite disruptant system. If the mechanism is competition, then the number of sources and their competitiveness with feral females can be designed for various pest densities. Without knowledge of these parameters, testing of disruptant systems must proceed largely on an empirical basis.

Finally, the development of the use of semiochemicals for pest management must be undertaken in the context of an integrated approach accounting for aspects of moth behavior other than mating. In G. molesta, for example, Steiner and Yetter (1932) in extensive mark-release recapture studies (using "bait pails") found that females dispersed a mean distance of over 400 m before recapture. Such detailed observations will have much to say about the degree of isolation required in successful control efforts. Many moth species routinely disperse even longer distances and the interplay

(Cardé, 1979) of such "vegetative" long-distance movement, with the potential influence of semiochemicals upon the induction of both male and female emigration, remains uncharted. The efficacy of some disruptant systems could be based in part on their mediation of vegetative-type dispersal behavior.

REFERENCES

Baker, T. C., and Cardé, R. T., 1979a, Analysis of pheromone-mediated behaviors in male Grapholitha molesta, the Oriental fruit moth (Lepidoptera: Tortricidae), Environ. Entomol., 8: 956.

Baker, T. C., and Cardé, R. T., 1979b, Courtship behavior of the Oriental fruit moth (Grapholitha molesta): Experimental analysis and consideration of the role of sexual selection in the evolution of courtship pheromones in the Lepidoptera, Ann. Entomol. Soc. Am., 72: 173.

Baker, T. C., and Roelofs, W. L., 1980, Factors affecting sex pheromone communication distance in the Oriental fruit moth, Grapholitha molesta, Environ. Entomol., 9: (In press).

Baker, T. C., Cardé, R. T., and Miller, J. R., 1980, Oriental fruit moth pheromone component release rates measured after collection by glass surface adsorption, J. Chem. Ecol., (In press).

Baker, T. C., Cardé, R. T., and Roelofs, W. L., 1976, Behavioral responses of male Argyrotaenia velutinana (Lepidoptera: Tortricidae) to components of its sex pheromone, J. Chem. Ecol., 2: 333.

Bartell, R. J., 1977, Behavioral responses of Lepidoptera to pheromones, in: "Chemical Control of Insect Behavior," H. H. Shorey and J. J. McKelvey, Jr., eds., Academic Press, New York.

Beroza, M., Muschik, G. M., and Gentry, C. R., 1973a, Isomer content and other factors influencing captures of Oriental fruit moth by synthetic pheromone traps, J. Econ. Entomol., 66: 1307.

Beroza, M., Muschik, G. M., and Gentry, C. R., 1973b, Small proportion of opposite geometric isomer increases potency of synthetic pheromone of Oriental fruit moth, Nature, 244: 149.

Bossert, W. H., and Wilson, E. O., 1963, The analysis of olfactory communication among animals, J. Theoret. Biol., 5: 443.

Brown, A. W. A., and Croft, B. A., 1975, Responses of arthropod natural enemies to insecticides, Ann. Rev. Entomol., 20: 285.

Cardé, A. M., Baker, T. C., and Cardé, R. T., 1979, Identification of a four component sex pheromone of the Oriental fruit moth, Grapholitha molesta (Lepidoptera: Tortricidae), J. Chem. Ecol., 5: 423.

Cardé, R. T., 1976, Utilization of pheromones in the population management of moth pests, Environ. Health Persp., 14: 133.

Cardé, R. T., 1979, Behavioral responses of moths to female-produced pheromones and the utilization of attractant-baited traps for population monitoring, in: "Movement of Highly Mobile Insects:

Concepts and Methodology in Research," R. L. Rabb and G. G. Kennedy, eds., North Carolina State Univ. Press, Raleigh.

Cardé, R. T., 1980, Precopulatory sexual behavior of the adult gypsy moth, in: "The Gypsy Moth: Research Toward Integrated Pest Management," C. C. Doane, ed., USDA Tech. Bull., 1584 (in press).

Cardé, R. T., Cardé, A. M., Hill, A. S., and Roelofs, W. L., 1977a, Sex pheromone specificity as a reproductive isolating mechanism among the sibling species Archips argyrospilus and A. mortuanus and other sympatric tortricine moths (Lepidoptera: Tortricidae), J. Chem. Ecol., 3: 71.

Cardé, R. T., Baker, T. C., and Castrovillo, P. J., 1977b, Disruption of sexual communication in Laspeyresia pomonella (codling moth), Grapholitha molesta (Oriental fruit moth) and G. prunivora (lesser appleworm) with hollow fiber attractant sources, Entomol. Exp. Appl., 22: 280.

Cardé, R. T., Baker, T. C., and Roelofs, W. L., 1975a, Ethological function of components of a sex attractant system for Oriental fruit moth males, Grapholitha molesta (Lepidoptera: Tortricidae), J. Chem. Ecol., 1: 475.

Cardé, R. T., Baker, T. C., and Roelofs, W. L., 1975b, Behavioural role of individual components of a multichemical sex attractant system in the Oriental fruit moth, Nature, 253: 348.

Cardé, R. T., and Hagaman, T. E., 1979, Behavioral responses of the gypsy moth in a wind tunnel to air-borne enantiomers of disparlure, Environ. Entomol., 8: 475.

Charlton, R. E., and Cardé, R. T., 1980, Comparing the effectiveness of sexual communication disruption in the Oriental fruit moth (Grapholitha molesta) using different combinations and dosages of its pheromone, J. Chem. Ecol. (In press).

Doane, C. C., and Brooks, T. W., 1980, Research and Development of pheromones for insect control with emphasis on the pink bollworm, Pectinophora gossypiella, in: "Management of Insect Pests with Semiochemicals," E. R. Mitchell, ed., Plenum Press, New York.

Farkas, S. R., Shorey, H. H., and Gaston, L. K., 1974, Sex pheromones of Lepidoptera. Influence of pheromone concentration and visual cues on aerial odor-trail following by males of Pectinophora gossypiella, Ann. Entomol. Soc. Am., 67: 633.

Gentry, C. R., Beroza, M., Blythe, J. L., and Bierl, B. A., 1974 Efficacy trials with the pheromone of the Oriental fruit moth and data on the lesser appleworm, J. Econ. Entomol., 67: 607.

Gentry, C. R., Beroza, M., Blythe, J. L., and Bierl, B. A., 1975 Captures of the Oriental fruit moth, the pecan bud moth, and the lesser appleworm in Georgia field trials with isomeric blends of 8-dodecenyl acetate and air-permeation trials with the Oriental fruit moth pheromone, Environ. Entomol., 4: 822.

George, G. A., 1965, Sex pheromone of the Oriental fruit moth, Grapholitha molesta (Busch) Lepidoptera: Tortricidae) Can. Entomol., Vol. 97, 1002-1007.

Kennedy, J. S., and Marsh, D., 1974, Pheromone-regulated anemotaxis in flying moths, Science, 184: 999.
Plimmer, J. R., Caro, J. H., and Freeman, H. P., 1978, Distribution and dissipation of aerially-applied disparlure under a woodland canopy, J. Econ. Entomol., 71: 155.
Roelofs, W. L., 1978, Threshold hypotheses for pheromone perception, J. Chem. Ecol., 4: 665.
Roelofs, W. L., ed., 1979, "Establishing Efficacy of Sex Attractants and Disruptants for Insect Control," Entomol. Soc. Am., 97 pp.
Roelofs, W. L., and Cardé, R. T., 1974, Oriental fruit moth and lesser appleworm attractant mixtures refined, Environ. Entomol. 3: 586.
Roelofs, W. L., Cardé, R. T., and Tette, J., 1973, Oriental fruit moth attractants and synergists, Environ. Entomol., 2: 252.
Roelofs, W. L., Comeau, A., and Selle, R., 1969, Sex pheromone of the Oriental fruit moth, Nature, 224: 723.
Rothschild, G. H. L., 1975, Control of Oriental fruit moth (Cydia molesta (Busck) (Lepidoptera: Tortricidae)) with synthetic female pheromone, Bull. Entomol. Res., 65: 473.
Rothschild, G. H. L., 1979, A comparison of methods of dispensing synthetic sex pheromone for the control of Oriental fruit moth, Cydia molesta (Busck) (Lepidoptera: Tortricidae), in Australia, Bull. Entomol. Res., 69: 115.
Rothschild, G. H. L., and Minks, A. K., 1977, Some factors influencing the performance of pheromone traps for Oriental fruit moth in Australia, Entomol. Exp. Appl., 22: 171.
Steiner, L. F., and Yetter, W. P., 1932, Second report on the efficacy of bait traps for Oriental fruit moth as indicated by the release and capture of marked adults, J. Econ. Entomol., 26: 776.
Sutton, O. G., 1953, "Micrometeorology," McGraw Hill, New York.

FUTURE THRUSTS ON MATING DISRUPTION

OF PEST INSECTS WITH SEMIOCHEMICALS

The members of the panel on mating disruption are greatly encouraged by the many positive developments in this field during the past decade. Problem areas that are common to most projects and that should be given high priority if progress in mating disruption control is to continue are summarized below.

Application Technology

A major obstacle to successful implementation of mating disruption control is the development of suitable formulations and application systems. There must be a continued application of new slow-release technology to meet the specific needs of mating disruption projects. Improvements are needed in existing formulations to produce a more predictable release of semiochemicals.

The ideal delivery or application system would utilize existing agricultural equipment. For this reason microcapsules have great appeal and efforts should continue to develop suitable microcapsular formulations. Specialized equipment for broadcast application of existing formulations must be developed or improved. Suitable ground equipment and formulations must be developed for use in field crops during the early season when little foliar surface is available or for use in crops where broadcast material must not contact fruit or edible foliage.

These objectives can be met only through collaboration between the appropriate governmental agencies and industry, and may require modification of existing contractual arrangements, particularly with regard to patenting new developments.

Semiochemicals Used for Disruption

Identification of sex pheromones should continue on a rational basis, with particular emphasis on the full identification of pheromone blends of important species known to be incompletely identified at present, and with greater emphasis on the identification of the pheromones of the entire pest complex of selected orchard, vegetable, and field crops. Research into disruptant compounds other than pheromones should be intensified, particularly where the latter are unstable or difficult to synthesize. New procedures for the large-scale production of disruptant compounds must be developed, as the present cost structure is based on processes for synthesizing smaller experimental or semi-commercial quantities of these materials. Production centers should be established for custom synthesis of smaller quantities of high quality compounds for experimental use. These activities should be conducted by industry under approrpiate contracts with government agencies.

Ecological and Behavioral Characteristics of the Target Pests

The scientific basis for disruption control should be further strengthened by a significant intensification of work; the behavioral and sensory mechanisms involved in disruption and male orientation behavior should receive particular attention. Attention should be given to the effects on male behavior of numerous point source emitters of the disruptant, each less potent than a calling female vs. fewer, but more potent emitters. Investigations are required to determine the possible effects of large concentations of sex attractant on female behavior, particularly dispersal and ovi-positional behavior.

Relevant population studies are required, particularly research designed to demonstrate the extent to which mating has to be suppressed in order to achieve an acceptable pest population level. The influence of initial population density on the outcome of disruption control should be determined. These studies should include preliminary attempts to develop meaningful, but not necessarily complex, models using relevant biological and ecological data where these are available.

Evaluation of Disruption in Large-Scale Trials

Current sampling and data analysis techniques, as well as criteria for assessing disruption, are considered to be inappropriate for the very large control programs now being undertaken in some countries. Suitable techniques for obtaining and handling data from such programs need to be developed.

Strategies for Adopting Disruption Control

There should be an intensification of operational research to demonstrate that disruption can be incorporated into integrated pest management (IPM) programs. This would help to emphasize that disruption with semiochemicals is not usually regarded as a single control strategy developed to replace pesticides, but that it can be an integral component of IPM, compatible with other biological control agencies as well as with pesticides.

Recognition of Disruption with Semiochemicals as an Important Component of IPM

The potential of disruption as a part of IPM is at present not recognized by those involved in the planning and implementation of current major IPM programs in the USA. This has led to inadequate government support for both basic and operational research in this area. The potential for disruption with semiochemicals, and its compatibility with other components of IPM, has now been clearly demonstrated for a number of major pests of forests, field and orchard crops both in the USA and elsewhere. These include the pink bollowrm, western pine shoot borer, and the codling moth. If this potential is to be fully exploited it is imperative that disruption with semiochemicals receive the same attention as other components of IPM.

PANEL: G. H. L. Rothschild, Australia; J. R. McLaughlin, USDA; D. G. Campion, England; T. Henneberry, USDA; C. C. Doane, Conrel; P. S. Beevor, England; S. Tatsuki, Japan; W. L. Roelofs, Cornell University; R. T. Cardé, Michigan State University; H. Arn, Switzerland; C. J. Sanders, Canada; L. L. Sower, USFS; T. Payne, Texas A&M University; E. R. Mitchell, USDA.

FORMULATION AND REGULATION: CONSTRAINTS ON THE DEVELOPMENT OF SEMIOCHEMICALS FOR INSECT PEST MANAGEMENT

Jack R. Plimmer

Organic Chemical Synthesis Laboratory
AR-SEA, USDA
Beltsville, MD 20705 USA

INTRODUCTION

Formulation, toxicology, and registration fundamentally influence the successful use of semiochemicals. In this context, "successful" implies that the farmer and the manufacturer will benefit economically and that the potential hazards to man and the environment will not be greater than those from current pest management techniques.

Formulation is a technical problem, and we can readily assess progress toward its solution. Toxicology and registration are closely linked. Regulatory policies have profoundly affected the development of insecticides by emphasizing the importance of toxicological data, and the resultant toxicological studies of insecticides have revealed potential effects that have led manufacturers to modify or abandon many structural types. As a result, the demonstrable harmful effects of insecticide use have been reduced, and greater attention has been focused on alternative methods of insect control and on combined strategies such as integrated pest management (Office of Technology Assessment, 1979). However, the regulatory requirements appear to be creating a situation in which returns from investment in the development of new pesticides are limited. Much greater effort is now required to discover a new insecticide and to demonstrate that it shows no potential for adverse effects. Because of the high costs involved, industry has found it generally unprofitable to develop pesticides to protect crops other than those that represent the maximum potential market. Thus, many needs for crop protection chemicals remain unfilled.

Likewise, uncertainty concerning regulatory policies has clouded the future of semiochemicals. Although the consensus of opinion is that these materials will rank low in the hierarchy of environmental contaminants, such unproven assumptions are usually treated with skepticism until evidence to the contrary is obtained. Existing evidence based on consideration of many known pheromone structures suggests that anticipated use patterns will not result in levels that can be detected in the environment. Problems may arise because data concerning environmental fate and toxicology must be obtained at considerable cost to industry; the proprietary nature of these data, as well as that of data relating to synthesis and formulation, is important for economic success in a competitive market. If semiochemicals are to be developed for practical use, the symbiotic relationship between private and public efforts must continue; at the practical level this is not difficult to ensure but many issues are still unresolved. The momentum of research efforts and the scale of field programs have increased considerably in recent years. The present situation is in direct contrast to some earlier predictions and we should examine present limitations on continued development.

In 1976, a gloomy assessment appeared in a report on biological substitutes for chemical pesticides (Lawless et al., 1976). The major problems appeared as follows: The difficulty of identification and isolation of pheromones; the fact that most work was conducted in government, university, and other public laboratories and the results published so they did not receive patent protection; the difficulties of ensuring their effective use in the field, given the sensitivity of materials to environmental conditions; and the fact that pheromones were less specific in relation to other species than had been previously suggested so that their use would have to be monitored for possible adverse effects. Only 2 commercial companies had taken a major interest in the field at that time (Zoecon and 3M). The conclusion was as follows:

> "...in spite of the substantial current research interest in pheromones, great obstacles remain to be overcome before pheromones can be used on a major scale in the management of insect pests. Progress toward solving registration problems, and toward commercial production of pheromones at feasible costs has been slow. Industrial participation in the development and implementation of this technology is very limited. The results of the extensive tax-supported research efforts remain to be reduced to practice in an effective economically feasible manner.

In the light of this rather slow progress in the practical application of pheromones thus far, we do not anticipate that pheromones will become a major factor in insect control before the 1980's. The situation is not likely to change substantially in the next decade unless and until the major problems identified above are resolved. Pheromones will probably be used more widely for diagnostic purposes, but their use for insect reduction may remain underdeveloped indefinitely."

This report summarized major obstacles perceived in 1976. Unfortunately, there is often a considerable time lag between the actual experiments and their publication. Thus, information in reviews such as this may be misleading in relation to the actual state of the art. In fact, considerable progress has been made toward the solution of the problems discussed above. Great strides have been made in the processes of identification and synthesis; and the problem of defining efficacy has been opened to critical examination.

The future of pheromones in pest management now plainly depends upon economics. For example, the introduction of a semiochemical for protection of crops by air permeation involves great commercial risks. In 1978, synthesis expenditures for formulation and registration of such a product were estimated to be at least $1,335,000 over a period of 4 to 5 years (Siddall, 1979).

However, future prospects for pheromones are rather different from those for conventional insecticides, and many more questions must be raised concerning their use. How will pheromones be used? Their limited spectrum of activity probably restricts their use, and their application will affect only a limited number of insect species. How much pheromone will be used? We can anticipate that the amount of active ingredient applied will be relatively small, but it is becoming clear that specialized formulation techniques will be required and will probably be the key to successful application and commercialization of semiochemicals. Therefore, major questions today relate to formulation, toxicology and registration.

FORMULATIONS

General Properties of Controlled-Release Formulations

The use of insect semiochemicals for trapping or permeating the air requires an effective delivery system, but one must distinguish between trap use and use in mating disruption techniques. Traps will require a delivery system (including application and formulation) somewhat different from that for air permeation.

Ease of application is important because it will profoundly affect economic success.

Formulation usually implies that the semiochemical is combined with other ingredients to provide an effective dose rate and to insure that biological activity will be maintained. The chemical must be protected from atmospheric oxidation, light, and other environmental agents that may rapidly destroy it.

Because semiochemicals show extremely high biological activity, the amount of chemical used can be quite small; this is compatible with a relatively expensive controlled-release formulation.

Some desirable features of controlled-release systems for air permeation are:

(1) Convenience in use;

(2) A constant and reproducible rate of emission for active material throughout a predetermined period;

(3) Effective use of an expensive chemical;

(4) Ease of application;

(5) Absence of adverse effects from the chemical or the formulation; and

(6) Low cost of production and application.

A single formulation is unlikely to meet all these criteria. Formulations must be designed to suit a forest-insect or crop-insect pest system, and, if possible, their emission characteristics should synchronize with the biological properties of the target insect.

Concerning air permeation, controlled-release devices are fabricated to act as discrete sources of pheromone suitable for distribution over a large area. Such formulations include microcapsules, flakes, hollow fibers, or particulate materials. If thickeners or other additives are to be combined with the formulation, they may substantially affect the emission rate.

Although efficacy must finally be determined in the field, many parameters in the formulation's design can be altered, and it is common practice to select the most promising candidates by laboratory screening. A number of physicochemical tests are needed for this which include measuring the loss of semiochemical content and determining the release rate by trapping and quantitative analysis (Plimmer and Inscoe, 1979). For example, laboratory measurements of emission rates over a range of environmental conditions will indicate whether the formulation is compatible with specific semiochemicals and whether it is likely to perform effectively.

After these measurements have been obtained under controlled laboratory conditions, it is useful to conduct the same determinations in a typical field situation. The question of material balance should also be investigated to determine the fate of the chemical that has been released.

Multicomponent pheromones present a special problem if all components are to be included in the same formulation. For instance, (Hedin et al., 1976), grandlure, the pheromone of the boll weevil (_Anthonomus grandis_ Boheman), is a mixture of 4 compounds. Several components of grandlure are costly, and it is desirable to use small quantities as efficiently as possible for large-scale application. The compounds were originally impregnated on firebrick for use in traps. Later, a variety of formulations were compared, which included polyethylene caps treated with a methylene dichloride solution of the components; impregnated firebrick, nylon resin or polyethylene-glycol 1000 tablets; and cotton dental wicks impregnated with solutions in glycerol, water, methanol, or polyethylene 1000 at various concentrations. All these formulations were only active for a short period. In other experimental formulations, grandlure was mixed with antioxidants and then combined with resins, clays or melted paraffin. It was found that water acted in stabilizing and releasing the alcohol components of grandlure, and a water-containing formulation of alcohols and polyhydroxy compounds was prepared. The formulation was effective when applied to a cotton dental roll but lost its activity after 4 days. A commercial gel-type formulation extended activity to 7 days and was as active as males in the field.

Further modification of grandlure formulations led to the widespread adoption of a cigarette filter roll contained in a glass vial with an open end. This device gave a linear loss of pheromone over 2- to 3-week periods and it could be mass-produced mechanically at a low cost. Grandlure was later incorporated into a 3-layer laminated polymeric dispenser, which had proved useful for controlled release of several pheromones. This dispenser and a further modification of the cigarette filter dispenser in which an outer plastic cylinder was used to reduce the emission rate have been used on a large scale in programs to control the boll weevil.

Many types of material have been used for dispensing other pheromones. Often the choice was determined by materials that were most readily available in the laboratory. One convenient device that has been widely used is the rubber septum, which can be permeated with pheromone. It provides a useful and reproducible technique, and its release characteristics have been studied in detail (Butler and McDonough, 1979).

A major difficulty today in predicting the utility of individual formulations is the paucity of data obtained in large-scale experiments. However, the following commercially available types of controlled-release formulations proved to be compatible with spray application.

Microencapsulated Pheromones

Microencapsulation, which is valuable for stabilization of oxygen-sensitive compounds, involves the coating of small droplets of liquid with a film of polymeric material. The polymer forms a protective barrier through which the volatile liquid may diffuse and volatilize from the surface of the microcapsule.

Microcapsules are prepared by dispersing or emulsifying a solution of a pheromone in a suitable medium--frequently, a solution of the encapsulating material. The coating is then deposited onto the core material by changing the solvent characteristics so that the wall material separates. Finally, the wall solidifies. Polymers used include such materials as gelatin, methylcellulose, polyvinyl alcohol, polyamides, and polyethylene oxide. The choice of the wall depends upon the characteristics of the encapsulated material and the end use of the product.

Microcapsular formulations of pheromones have several desirable features. One prominent feature is their compatibility with conventional equipment used for aerial spray application of pesticides. They can be applied at measured rates over forests and rugged terrain, and the microencapsulation protects the semiochemicals from oxidation and allows release over a prolonged period. Measurements of emission rates and of residual pheromone content in the laboratory and in the field indicate that pheromone from microcapsules is not efficiently utilized, but insufficient data are available to permit valid comparisons of microcapsule field performance with the performance of other formulations.

Gelatin microcapsules have formed the basis for a number of disparlure formulations used to disrupt mating of the gypsy moth, *Lymantria dispar* (L.), that infests forest areas in the northeastern United States. Since several factors (wall thickness, pheromone loading, pheromone solvent, capsular coating, and capsular size) can be varied during the manufacture of the microcapsules, a number of formulations were made available; laboratory measurements of emission rate were used to identify eligible candidates for development (Plimmer et al., 1977). Two factors, emission rate and capsular size, have an important influence on biological activity. The former affects the longevity of the formulation and the latter determines the number of point sources of pheromone. In practice, although the optimization of release rate was given prior consideration, a maximum limit on capsular size was imposed by the size of the delivery nozzle (400 microns) in the aircraft spray booms.

Hollow Fiber Dispensers

Volatile liquids vaporize from open capillary tubes by a diffusion-controlled process. The evaporation of liquid at the liquid-vapor interface is followed by diffusion of vapor to the open end of the tube, and at the open end of the tube the vapor is removed by convection. Thus, the rate of evaporation is independent of the amount of liquid within the tube, and the rate of loss at constant temperature can be predicted from the diffusivity of the liquid and its vapor pressure. The Conrel® hollow fiber fabricated from Celcon®, a commercial polyacetal resin, embodies these principles and has been used for controlled-release formulations for a number of semiochemicals (Ashare et al., 1976; Iyer et al., 1978). Because the vapor pressure is affected by the curved meniscus in the fiber, the rate of evaporation is affected by the nature of the material of which the fiber is made.

The hollow fiber has been used to formulate semiochemicals both for traps and for mating disruption. Gossyplure H.F. is a formulation of gossyplure, the pheromone of the pink bollworm, *Pectinophora gossypiella* (Saunders), a major pest of cotton. In 1977, this formulation was used to treat 9300 ha of growing cotton to suppress pink bollworm populations, and evidence of its efficacy was provided by a concomitant reduction in the need for conventional insecticide treatment of 50 to 80% (Brooks et al., 1979).

Laminated Polymeric Dispensers

The laminated plastic or multi-layered polymeric dispenser has been used for sustained release of a number of pheromones. The 3-layer Hercon® polymeric dispenser was first used as a pheromone dispenser in 1974. It consists of an inner layer that acts as a reservoir to contain the pheromone and outer membranes above and below the inner layer. The reservoir may contain from 0.5 to 40% of the active substance. The material passes through the nonporous outer membranes by absorption, solution, diffusion, and desorption at the surface, where it evaporates (Kydonieus, 1977). Its rate of release is affected by the construction of the dispenser; i.e., reservoir concentration and wall thickness. The chemical structure of the molecule also affects the release rate, as does the chemistry of the polymer. In addition, substances that actively diffuse from the reservoir and interact with the polymer matrix influence the rate of release and may be added to the formulation. The effects of chemical functionality and molecular weight have been studied over a range of molecular types. As would be anticipated, in a homologous series of acetates there was a dramatic increase in half-life as the molecular weight increased (Kydonieus et al., 1976).

The rate of emission is affected by the rate of air flow over the dispenser and by temperature. The log of the emission rate was found to be inversely proportional to the absolute temperature (Bierl-Leonhardt et al., 1979). In tests to obtain optimized formulations of racemic disparlure, a dispenser of 6 mil thickness containing 6.2 mg disparlure/in^2 was found to emit 0.24 μg/hr at 27°C. At 32°C the rate increased to 0.45 μg/hr. The over-all rate will remain relatively constant for a long period because only a small percentage of the attractant is lost daily.

The laminated plastic dispenser has proved very useful for formulating many semiochemicals. The optically active form of disparlure is formulated commercially in a dispenser 1" x 1/8" that contains approximately 1 mg of material. Dispensers of 1" x 2" containing ca. 22 mg racemic disparlure/in^2 have been registered by the Environmental Protection Agency for suppression of gypsy moth populations. They are placed on a grid measuring 10 m x 10 m over the infested area. Dispensers are useful for treatment of orchards and areas where spray application is undesirable.

More recently, the laminated dispenser principle has been used to fabricate "flakes" by cutting or cryogenically grinding sheets of laminated polymer. These flakes possess similar sustained-release characteristics but can be used in spray applications.

TOXICOLOGY

Webster's Dictionary defines toxicology as "that branch of medicine which treats of poisons and their antidotes." Here "toxicology" is used in its broadest sense, which includes potential harmful effects for man and the environment. Pesticides are toxic substances that are intentionally introduced into the environment. We now know a great deal about the hazards associated with their use because there have been many detailed studies concerning their long- and short-term toxicity, metabolism and environmental fate.

Because there is increasing concern that long-term effects of pesticides at low levels may present potential risks to the population at large, emphasis has been placed on identifying teratogens, mutagens and carcinogens. There is reason to be concerned with pesticides that are not readily degraded in the environment, either chemically or biologically. Important factors that determine its persistence are the breakdown rate of a pesticide by air and sunlight and the rate of its transformation by plants or microorganisms.

Many so-called "hard" pesticides possess structural features that make them very resistant to breakdown by microorganisms or other biological systems. Lipid-soluble compounds such as DDT may become sequestered within fatty tissues and remain unavailable to many enzymes responsible for the processes of detoxification. Pesticides and their alteration products do not usually resemble biological molecules, and sophisticated techniques are used to trace their path through biological systems and determine their fate in the environment. They generally cause the death of the target organism by interfering with metabolic processes.

Pheromones tend to be quite different. Naturally occurring semiochemicals act on very specific biological systems. They do not kill the target organism, and their acute mammalian toxicity is extremely low. Most pheromones are closely related to commonly occurring molecules of biological origin, and the enzymes capable of degrading similar substrates are widely distributed in nature. Finally, the anticipated patterns of use will restrict the amounts entering the environment. For example, when insect population suppression is to be achieved by air permeation, there will be no more than the saturated vapor concentration of the pheromone present in the air, and this amount is a maximum that will rarely be encountered. Very low concentrations of pheromone will enter the environment if present use patterns of a few grams per acre are maintained. In addition, available data (Knipling, 1976; Beroza et al., 1975) suggest that pheromones show a very low order of toxicity (disparlure, looplure, western pine beetle pheromone [*exo*-brevicomin:frontalin: myrcene, 1:1:1], muscalure, etc.). Recently, we found that several long-chain unsaturated aldehydes are not very toxic to rats. Tests of sex attractants for acute dermal toxicity and aerosol inhalation toxicity in rats, eye irritation and primary skin irritation in rabbits, and for toxicity to rainbow trout and bluegill sunfish suggest a very low potential for hazard.

Nevertheless, if relatively large amounts of naturally occurring or totally synthetic chemicals are to be manufactured and released into the environment, some attention must be given to their potential for harm. The fact that pheromones are different from conventional insecticides does not reduce the obligation to ensure their safe use. The Environmental Protection Agency commissioned a study on which guidelines could be based. In this study (prepared under contract by the Zoecon Corporation) pheromones are said to differ from other chemical pesticides mainly because they are designed to volatilize as completely as possible and their use rates will be very low. The product chemistry guidelines written for pesticides are thus applicable to pheromones with minor modifications. A further conclusion is that the types of naturally occurring products that will probably be used should be readily degraded by most organisms in normal fatty acid catabolic pathways. Assessment of hazards to fish and wildlife should be based on the amount of active ingredient, the area to which it is to be applied and its frequency of application; toxicity assessment should be based on a mouse oral toxicity study in which the pheromone is administered in a single dose at a single dose level. Studies of inhalation and dermal toxicity would also be appropriate because pheromones will be volatilized into the atmosphere over an extended period. The judgment was as follows: "Chronic studies would only be required if a use pattern generated residues in agricultural commodities, food, or feed significantly (100 fold) above the background of residue from natural sources.... Likewise toxicity to wildlife need be studied only in the event that environmental levels will be significantly (100 fold) increased" (Anonymous, 1975).

There are only a few studies on the environmental fate of pheromones. In contrast, there is a wealth of studies on the environmental fate of pesticides, detergents, and other man-made chemicals that have been introduced into the environment, and the data thus obtained make it easy to recognize structural features that might enhance the resistance of molecules to biological or nonbiological breakdown. Environmental agents principally responsible for non-biological degradation of materials are sunlight, air and water. Pheromones, in general, are so susceptible to degradation that they must be protected if they are to retain their activity. Biological degradation may occur through plant, animal or microbial metabolism; oxidative degradation of naturally occurring compounds closely related to long-chain lipids may take place by slight adaptation of well-established biochemical pathways.

The complex of volatile compounds (grandlure) emitted by the male boll weevil acts as a sex-aggregation pheromone. The structures of the pheromone components, (+)<u>cis</u>-2-isopropenyl-1-methylcyclobutaneethanol, (<u>Z</u>)-3,3-dimethyl-$\Delta^{1,\beta}$-cyclohexaneethanol, (<u>Z</u>)-3,3-dimethyl-$\Delta^{1,\alpha}$-cyclohexane acetaldehyde, and (<u>E</u>)-3,3-dimethyl-$\Delta^{1,\alpha}$-cyclohexaneacetaldehyde, are sufficiently novel to recommend a study of their environmental breakdown. Such a study (Henson et al., 1976) elucidated the fate of grandlure in soil and water and identified the oxidative decomposition products. None of the compounds were found in soil or water at 32°C within 24 h after application to the soil surface or adding them to water. Aldehydic components gave acids or esters as oxidation products. Volatilization of the compounds was a significant factor in their removal from soil or water.

In future investigations of pheromones, attention should be directed to their fate and movement in the atmosphere. Since the techniques of mating disruption require the deliberate release of pheromone vapor into the air, a study of change in concentration with time and distance from the source may be important in assessing biological effects and potential hazards. In one study, the quantity of disparlure in air was measured at selected time intervals after application of several formulations. These determinations were conducted in an open field and under a forest canopy. The results provide a valuable indication of potential exposure levels and very useful information concerning the potential value of the formulation as a controlled-release vehicle. This information can also be correlated with biological data to give a better understanding of efficacy requirements. The formulations used were shown in field tests to be capable of disrupting mating for the duration of male gypsy moth flight. The problem is that aerial concentrations could only be measured when the application rate was at least an order of magnitude greater than that used for disruption. Although the method of analysis is capable of detecting small quantities of pheromone, it is not feasible at the present time to obtain satisfactory analyses of pheromones in the air at practical use levels.

Thus, the current knowledge of the toxicology of semiochemicals is broadening. There can be little concern that most of those in current use today will present any appreciable hazard to man and his environment if present use patterns are maintained. Methods of analysis present a challenge and are expensive. Therefore, rigorous studies of the fate of many semiochemicals may be wasteful of resources and may also be repetitious, since the various molecules have many common features. Extension of the use of semiochemicals may take unforeseen directions in the future, and it is conceivable that behavior-modifying chemicals of novel structures may be introduced. In that case, it would be wise to consider the quantitative aspects of their use and the potential human exposure. Then, if necessary, fuller toxicological studies would be conducted following the guidelines established for pesticides when appropriate.

PHEROMONE REGISTRATION

Registration of pheromones in the United States has been closely linked with registration of pesticides. Although the Departments of Agriculture and Health, Education and Welfare, the Food and Drug Administration, and the Environmental Protection Agency (EPA) have all participated in pesticide regulation, major responsibility now rests with the Administrator of the EPA. The transfer of regulatory authority was accomplished by the Federal Environment Pest Control Act of 1972, basically an amendment to the Federal Insecticide, Fungicide and Rodenticide Act (FIFRA), which was again amended in 1975 and 1978.

To obtain registration, the manufacturer of a pesticide must submit basic data concerning the chemistry of the pesticide, directions for use, and the claims made for the pesticide. The label must give full information, including the directions for use, warnings and antidotes, registration number, and procedures for storage and disposal. Also required is an assessment of risks to man and the environment, "taking into account the economic, social and environmental costs and benefits of the use of any pesticide." If the pesticide is to be used in such a manner that residues might enter raw agricultural commodities, the registrant is also required to obtain a "tolerance" or an exemption therefrom.

In 1975 at the American Chemical Society meeting in Chicago, EPA registration requirements for insect behavior-controlling chemicals were discussed by W. G. Phillips of the Criteria and Evaluation Division, U. S. Environmental Protection Agency (Phillips, 1976). This discussion of philosophy and mandates closed on an optimistic note: "'The Guidance and Criteria' package will be developed in a separate document paralleling the general registration guidelines in format, but applicable specifically to the development and testing of these type compounds.... The anticipated completion date for the pheromone guidance and criteria will be the end of the current calendar year, as we move forward in the area of new pesticide products." The technical and scientific data to be used for preparation of these guidelines were provided in the document already discussed that was prepared under contract by the Zoecon Corporation (Anonymous, 1975).

In 1977, the EPA, through the American Institute of Biological Sciences, commissioned a number of studies designed to provide a compendium of methods for testing the efficacy and hazards of chemical pesticides. A study group on insect behavior-modifying chemicals prepared a report; some of the difficulties encountered in transmitting the report to the EPA have been recounted (Tucker, 1978). The committee report (Roelofs, 1979) provided critiques of a number of tests performed to determine the efficacy of pheromones as insect control agents, where efficacy is defined as "the prevention or reduction of (a) crop damage or (b) the pest population where (1) there has been a clear relationship established between the density of the damaging stage and yield (quantity or quality) or (2) a lowered population density is the only objective of the treatment, e.g. quarantine maintenance, elimination of nuisance or noxious populations, etc. Efficacy must be demonstrated under field conditions."

Questions of efficacy are indeed important and in the marketplace efficacy will ultimately determine the impact of pheromones in pest management systems. The economics of production and development will, however, be fundamentally influenced by regulatory constraints, and an adverse regulatory climate has been held responsible for the slow pace of development of pheromones in pest management. Considerable efforts were required to demonstrate the safety of gossyplure, the first pheromone to be registered for protection of a field crop (Johnson, 1978). Potential residues in cottonseed oil were made the subject of extensive tests and cost of registration was close to $1 million (Tucker, 1978). If comparable costs must be incurred for every registration, the application of semiochemicals will necessarily be limited to applications that affect major pest insects in major crops or forests. Development for minor uses will not be economically feasible, except for limited use in detection.

On the other hand, EPA has recently indicated that it intends to consider biorational agents such as pheromone on a basis other than that used with conventional pesticides, and it issued a statement in May, 1979 on the regulation of "biorational" pesticides (Jellinek, 1979). This stated that "biorational pesticides include biological pest control agents and certain naturally occurring biochemicals which differ in their mode of action from most organic and inorganic pesticide compounds currently registered with EPA." The statement and the background paper (Rogoff, 1979) were prepared for public comment and the concerned public, therefore, is awaiting the outcome of the resultant discussion and the agency response. EPA noted that a very small percentage of the active ingredients registered for pest control do differ in their mode of action from the majority of organic and inorganic pesticides, which possess innate toxicity. Examples of agents that function by different modes of action were living or replicable biological entities such as viruses, bacteria, fungi, and protozoans. "Naturally occurring biochemicals, such as plant growth regulators and insect pheromones and hormones, also function by modes of action other than innate toxicity." The policy statement addressed the following questions. (1) Over what classes of biorational pesticides should the EPA exert regulatory authority? (2) What data will be required to assess hazards to man and the environment? (3) What research and development efforts are needed to fill gaps in testing methodology? (4) How should EPA be concerned in testing and developing biorational pesticides?

The policies set forth in relation to biorational pesticides were that (1) EPA will take into account the fundamental difference between biorational pesticides and conventional pesticides and the lower risks of adverse effects in its data requirements; (2) EPA will develop programs to resolve safety issues and monitor the effects of biorational pesticides on man and the environment; (3) EPA will develop guidelines for testing human and environmental safety within the next 24 months while seeking the participation of other Federal agencies and the public; (4) the registration of environmentally acceptable biorational pesticides that present alternatives to conventional pesticides will be facilitated by ensuring that data requirements are "appropriate to their nature and not unduly burdensome;" (5) data will be reviewed expeditiously and allotted some priority in the registration process; (6) EPA will encourage demonstration of the practical value and safety of biorational pesticides; and (7) EPA will rely heavily on the expertise of other Federal agencies concerned with biological pesticides for development and implementation of EPA's biorational pesticides regulations and programs.

The background paper elaborates on some aspects of the policy document and provides the rationale for the statements therein. It deals largely with biological pest control agents, presumably because there is much greater experience in their practical application. The paper cites 2 recent developments that might improve the position in relation to patent protection. The first was the grant of patent protection to a developer of a new life form, a bacterium. The second was that FIFRA, as amended in 1978, permits the registrant of a new pesticide exclusive use of the data submitted in support of registration for a period of 10 years. Thus, an original registrant might recover some part of development cost through data compensation.

The renewed commitment of the Federal government to integrated pest management (IPM) is also viewed as a factor that will encourage the development of biorational pesticides. EPA feels that it is necessary to offset the negative impacts of cancellation, suspension, and reregistration on the availability of conventional pesticides by positive support for development of environmentally integrated alternatives. Because the document deals very specifically with pheromones and hormones at this point, it is appropriate to quote fully:

> "Pheromones and Hormones - Of all the biorational pesticides, the naturally occurring biochemicals, namely insect pheromones and hormones, present the fewest problems to the Agency in hazard assessment. Their chemical nature, patterns of use and the extremely small amounts of biochemical, which must be applied for effect in the field (at least for pheromones), may, in fact, indicate risk levels of so low a magnitude that much formal safety testing for mammalian effects prediction may not be necessary. For example, when pheromones currently registered are used under actual agricultural (or silvicultural) field conditions, no significant changes in the environmental background of these naturally occurring biochemicals has been observed. This indicates little or no potential for ecosystem effects since levels of the compounds normally found in nature are not significantly altered."

The issue of toxicological testing is then addressed in the document. Demands for data, particularly environmental fate data, have been considered excessive by registrants, especially when use patterns are considered. The EPA argument is that although pheromones are natural products, this does not preclude adverse health effects, and a tier-testing approach is proposed as an alternative to comprehensive data requirements. There is a suggestion that the possibility of unconventional approaches to testing should not be excluded, although the policy statement does not proceed beyond this statement.

After this discussion of the type of data required, the lack of guidelines for the registration of biological pesticides and the resulting specter of continued change in requirements for data are addressed, and the comment is made that "the potential for change in data requirements for biorational pesticides may be even higher than in the case of the more traditional chemicals." It is to be anticipated that some respite will be provided by the approach in Section 3 of the amended FIFRA, which permits registration on the condition that if new data requirements are imposed during the development of a new product, the new material may be registered subject to the condition that the new data are obtained.

EPA does not propose to use the chemical guidelines as a starting point but will prepare *de novo* guidelines that take into account the unique properties of biorational chemicals. Such an effort may require the recruitment of new staff who can complement existing expertise.

The other relevant issue addressed in the document is the role of EPA in encouraging, promoting, subsidizing, or facilitating the development and registration of biorational pesticides. The Agency considers that there are two routes by which it can implement the Congressional mandate to EPA to take a direct and active role in the development of biologically integrated alternatives for pest control. The first is to take actions that reduce registration costs to the developer and the second is to give priority to reviews of data and registration actions to expedite passage through the regulatory procedures.

These statements represent EPA policy in May, 1979. All those concerned with pheromone development now await the appearance of guidelines. A concern that may be expressed at this stage is that the policy document emphasizes only "naturally occurring biochemicals, namely pheromones and hormones." This may raise some questions. Naturally occurring juvenile hormones, for example, are unlikely to be used in insect control because less expensive synthetic analogues have been developed. Also, the use of semiochemicals not occurring in nature appears promising. Therefore, guidelines must address the question of the use of totally synthetic as well as naturally occurring chemicals, and it is to be hoped that a comprehensive document will soon appear before the current impetus of experimental programs is lost.

NATO Pilot Study on Pheromones

The North Atlantic Treaty Organization (NATO), under the auspices of its Committee on the Challenges of Modern Society (CCMS), accepted a draft for a NATO/CCMS Pilot Study on Integrated Pest Management Systems. Among the objectives listed was the following:

"Pest monitoring - important advances are being made in the use of sex attractants (pheromones) in monitoring pest populations. Work underway in member countries would benefit by more effective exchange of information and cooperative studies."

The question of facilitation of registration of new chemical pesticides by an exchange of testing data was also raised in the draft.

The Science Committee of NATO decided to refer the recommendations of the Advanced Research Institute held at Leeuwenhorst in 1978 to the CCMS. These recommendations (Ritter, 1979) are as follows:

"I. Regulatory, managerial and financial affairs concerning the application and (commercial) production of behavior modifying chemicals.

1. Registration by regulatory agencies

1.1 Regulatory agencies of all countries should publish special guidelines during 1979 for registration of Behavior Modifying Chemicals for pest control. Such chemicals which must provide insectistasis without insecticidal action should be clearly distinguished from insecticides. Suitable terminology should be adopted internationally.

1.2 The measurement of the efficacy of a pheromone treatment should conform to internationally recognized methods and standards, as befits the particular situation, similar to those stated in the AIBS Study Team report."

It was suggested that a pilot study be undertaken to determine the scale of the problem of legislation and regulation as it may exist in the countries of the NATO alliance. Dr. F. J. Ritter of the Department of Chemistry, Netherlands Organization of Applied Scientific Research (TNO), heads the pilot study. The use of pheromones in pest management is an international concern, and their value as alternatives to potentially more hazardous insecticides indicates that their possible use for insect control will have many applications throughout the world. However, their potential will remain unexploited, and their development will be hindered unless conditions for safe and effective use can be simply defined. International channels should be created to develop and share the information necessary to ensure their safety and efficacy, and in this endeavor, the understanding and cooperation of regulatory agencies is critically important.

REFERENCES

Anonymous, 1975, Scientific Data and Technical Information to Enable the Environmental Protection Agency to Develop Registration Guidelines for Insect Growth Regulators and Pheromones. Final Report, EPA Contract 68-01-2444 (July, 1975), Zoecon Corp., 975 California Ave., Palo Alto, CA 94304.

Ashare, E., Brooks, T. W., and Swenson, D. W., 1976, Controlled release from hollow fibers, in "Controlled Release Polymeric Formulations" (D. R. Paul and F. W. Harris, eds.), ACS Symp. Ser. 33, pp. 273-282, American Chemical Society, Washington, D.C.

Beroza, M., Inscoe, M. N., Schwartz, P. H., Jr., Keplinger, M. L., and Mastri, C. W., 1975, Acute toxicity studies with insect attractants, Toxicol. Appl. Pharmacol., 31: 421.

Bierl-Leonhardt, B. A., DeVilbiss, E. D., and Plimmer, J. R., 1979, Rate of release of disparlure from laminated plastic dispensers, J. Econ. Entomol., 73: 319.

Brooks, T. W., Doane, C. C., and Staten, R. T., 1979, Experience with the first commercial pheromone communication disruptive for suppression of an agricultural insect pest, in "Chemical Ecology: Odour Communication in Animals" (F. J. Ritter, ed.), pp. 375-388, Elsevier/North-Holland Biomedical Press, Amsterdam.

Butler, L. I., and McDonough, L. M., 1979, Insect sex pheromones: Evaporation rates of acetates from natural rubber septa, J. Chem. Ecol., 5: 825.

Hedin, P. A., Gueldner, R. C., and Thompson, A. C., 1976, Utilization of the boll weevil pheromone for insect control, in "Pest Management with Insect Sex Attractants and Other Behavior-Controlling Chemicals" (M. Beroza, ed.), ACS Symp. Ser. 23, pp. 30-52, American Chemical Society, Washington, D.C.

Henson, R. D., Bull, D. L., Ridgway, R. L., and Ivie, G. W., 1976, Identification of the oxidative decomposition products of the boll weevil pheromone, grandlure, and the determination of the fate of grandlure in soil and water, J. Agr. Food Chem., 24: 228.

Iyer, P. O., Yates, W. E., Akesson, N. B., and Horgan, P. M., Jr., 1978, Cotton pink bollworm control using the sex pheromone gossyplure--I. Controlled release of gossyplure, Paper 78-1502 presented at the American Society of Agricultural Engineers, 1978 Winter Meeting (Dec. 1978), Chicago, IL.

Jellinek, S. D., 1979, Regulation of "biorational" pesticides; policy statement and notice of availability of background document, Fed. Reg., 44(94): 28093.

Johnson, E. L., 1978, Approval of application to register pesticide product containing a new active ingredient, Fed. Reg., 43(82): 18018.

Knipling, E. F., 1976, Role of pheromones and kairomones for insect

suppression systems and their possible health and environmental impacts, Environ. Health Perspectives, 14: 145.
Kydonieus, A. F., 1977, The effect of some variables on the controlled release of chemicals from polymeric membranes, in "Controlled Release Pesticides" (H. B. Scher, ed.), ACS Symp. Ser. 53, pp. 152-167, American Chemical Society, Washington, D.C.
Kydonieus, A. F., Smith, I. K., and Beroza, M., 1976, Controlled release of pheromones through multi-layered polymeric dispensers, in "Controlled Release Polymeric Formulations" (D. R. Paul and F. W. Harris, eds.), ACS Symp. Ser. 33, pp. 283-294, American Chemical Society, Washington, D.C.
Lawless, E. W., von Rumker, R., Kelso, G. L., Lawrence, K. A., Maloney, J. P., and Thomson, R. R., 1976, A technology assessment of biological substitutes for chemical pesticides. MRI report, p. 209. National Science Foundation, Washington, D.C.
Office of Technology Assessment, Congress of the United States, 1979, "Pest Management Strategies in Crop Protection," Vols. I and II. OTA-F-98 and OTA-F-99. U. S. Government Printing Office, Washington, D.C.
Phillips, W. G., 1976, EPA's registration requirements for insect behavior controlling chemicals--philosophy and mandates, in "Pest Management with Insect Sex Attractants and Other Behavior-Controlling Chemicals" (M. Beroza, ed.), ACS Symp. Ser. 23, pp. 135-144, American Chemical Society, Washington, D.C.
Plimmer, J. R., Bierl, B. A., Webb, R. E., and Schwalbe, C. P., 1977, Controlled release of pheromone in the gypsy moth program, in "Controlled Release Pesticides" (H. B. Scher, ed.), ACS Symp. Ser. 53., pp. 168-183, American Chemical Society, Washington, D.C.
Plimmer, J. R., and Inscoe, M. N., 1979, Insect pheromones: Some Chemical problems involved in their use and development, in "Chemical Ecology: Odour Communication in Animals" (F. J. Ritter, ed.), pp. 249-260, Elsevier/North-Holland Biomedical Press, Amsterdam.
Ritter, F. J., ed., 1979, "Chemical Ecology: Odour Communication in Animals", p. 404, Elsevier/North-Holland Biomedical Press, Amsterdam.
Roelofs, W. L., ed., 1979, "Establishing Efficacy of Sex Attractants and Disruptants for Insect Control," Entomological Society of America, College Park, Md.
Rogoff, M., 1979, Background paper: Regulation of biorational pesticides. Appendix to EPA Federal Register statement on biorational pesticides. Office of Pesticide Programs, Environmental Protection Agency, Washington, D.C.
Siddall, J. B., 1979, Commercial production of insect pheromones--Problems and prospects, in "Chemical Ecology: Odour Communication in Animals" (F. J. Ritter, ed.), pp. 389-402, Elsevier/North-Holland Biomedical Press, Amsterdam.
Tucker, W., 1978, Of mites and men, Harper's 257(1539): 43.

POSITION OF THE U. S. ENVIRONMENTAL PROTECTION AGENCY ON REGULATION OF BIORATIONAL PESTICIDES

Michael J. Dover

Benefits and Field Studies Division
Office of Pesticide Programs
U. S. Environmental Protection Agency
Washington, D. C. 20460 USA

I would like to address directly the feeling that many may have that the registration process is a serious constraint to full utilization of the potential of semiochemicals. Stated as simply as possible, I submit to you that it is not.

In May of 1979, the Environmental Protection Agency published a proposed policy on regulation of biorational pesticides (including microbial pesticides, synthetically produced biochemicals that occur naturally, and analogs of such biochemicals). The proposed policy includes recognition of the fact that these materials are different from conventional pesticides and that because of different modes of action their use is expected to result in lower risks of adverse effects. It also states that EPA will:

- develop and implement programs to resolve outstanding safety issues and monitor probable effects on the environment and man;

- develop guidelines for human and environmentally acceptable biorational pesticides;

- review data submitted in support of registration in an expeditious manner; and

- encourage the demonstration of the practical value and safety of biorational pesticides.

That is the stated policy. Let us have a look at what has been done.

- On an interim basis, in 1978 the Integrated Pest Management Unit, of which I am a member, was asked by the Deputy Assistant Administrator for Pesticide Programs to "shepherd and expedite" the registration of biorationals. That responsibility has now shifted to its proper place in the organization, namely, to the Registration Division in the person of Dr. Alvin Chock. He is now the focal point for information regarding registration of these materials.

- With regard to facilitating registration of biorationals, interim criteria are in place while the full set of guidelines are developed.

- To date, 15 biorational pesticides have been registered, 7 of them in 1978 and 1979. Of those 7, 4 were semiochemicals.

- In 1979, the average elapsed time, that is, from date of original submission to registration, was 11.5 months for biorationals as compared to an average of 21.5 months for conventional pesticides.

As far as EPA is concerned, registration is no longer an impediment to widespread use of semiochemicals. I believe that a much more serious constraint is that of economics. Economics is in part a technological issue involving formulation, synthesis, and the like. Another major element in the economics of these materials is not subject to technological breakthroughs, and that is marketing. Without a clearly defined market and a way to reach that market, even materials that appear economically feasible simply will not be adopted.

That market is the expanding community of IPM practitioners, as I am sure you will agree. I strongly believe that it is in the best interests of researchers and industry representatives to support every effort to develop and expand that market, regardless of whether existing programs currently utilize semiochemicals or not. These programs will eventually be your principal supporters -- in fact, you may soon find them pressing you to do even more than you are now able to handle.

As I have said, my position does not directly involve registration of biorational pesticides. Our unit's responsibilities are much broader: To support the development and implementation of integrated pest management. We see IPM as the benchmark for evaluating <u>any</u> pest control practice, existing or proposed, and we believe that EPA's regulatory posture over the next few years will come to reflect that point of view. The policy, and the

the record over the last two years with respect to biorational pesticides, is one example of this new posture. You can help bring it about. I hope that you will take every opportunity to do so. My short experience with EPA has led me to the belief that visibility and expectation help the process. I urge you to participate in the process: Ask questions, provide information, and make suggestions. Keep your eye on the system and keep your expectations high. I am confident that your effort will be rewarded.

METHODOLOGY FOR DETERMINING THE RELEASE RATES OF PHEROMONES FROM HOLLOW FIBERS

J. Weatherston

M. A. Golub

T. W. Brooks

Albany International
Controlled Release Division
110 A Street
Needham Heights, MA 02194 USA

Y. Y. Huang

M. H. Benn

Department of Chemistry
The University of Calgary
Calgary, Alberta T2N 1N4
Canada

INTRODUCTION

In the quest for efficacious, environmentally acceptable agents for the control of pest insects, strategies involving the use of pheromones, attractants and other behavior modifying chemicals have been receiving much attention and publicity. To ensure viability of behavior modifying chemicals as an alternative to conventional insecticides, a necessary feature of their formulation is a controlled release system, in which the 2 vital criteria are rate of release of the semiochemical and the longevity of the formulation.

During the last 10 years or so in the evolution of the practical use of behavior modifying chemicals in pest control it is of interest to note that in the strategies employing trapping techniques, at least a dozen different release substrates have been used, including rubber septa, polyethylene tubing, vials and vial caps, glass vials, cigarette filters, dental wicks, waxed cardboard, filter paper, PVC pellets, plastic laminates and hollow fibers. In the strategies aimed at manipulating the pest populations by disrupting their natural orientation and mating behavior, the variety of substrates has been equally varied and included cork fragments, corn cobs, string, metal planchets, petri dishes, rubber tubing, microcapsules, polyamide beads, impregnated polyethylene fragments, polymer laminates and hollow fibers. By far, the greater number of the field tests carried out using the above listed substrates were completed without knowledge of the rate of release of the active ingredient of the formulation (Roelofs, 1979).

Studies of methodology for the determination of emission rates of pheromones have been reported over the last 8 years notably by Bierl et al., (1975, 1976), Bierl-Leonhardt et al., (1979), Cross (1980a) and Cross et al. (1980). The methods used in the measurement of the release rate have been characterized by Roelofs (1979) as (a) weight loss measurements, (b) extraction of residual semiochemical, (c) trapping of effluent vapors, and (d) direct observation of volume changes in capillary reservoirs, all determined as a function of time. The analytical techniques used in the quantification include gravimetric (Fitzgerald et al., 1973; Bierl and DeVilbiss, 1975; Beroza et al., 1975; and Rothschild, 1979); gas chromatographic (Beroza et al., 1975; Bierl and DeVilbiss, 1975; Campion et al., 1978; Bierl-Leonhardt et al., 1979; Butler and McDonough, 1979; Baker et al., 1980; Cross, 1980a,b; and Cross et al., 1980); high performance liquid chromatographic (Look, 1976); radiolabel scintillation counting (Kuhr et al., 1972); cathetometric and meniscus regression measurements (Ashare et al., 1975; 1976; and Cross, 1980b). Before examining the methodology in more detail it is appropriate, at this juncture, to simply state for future reference some of the external factors which influence pheromone emission from controlled release systems, these are temperature, humidity, wind speed, number and type of pheromonal components, decomposition rate of the pheromone, and additives such as stickers and extenders.

METHODOLOGY

Weight Loss Measurements

The most extensive report of the weight loss technique is that of Rothschild (1979) who, in relation to disruption trials against the Oriental fruit moth, *Cydia molesta* (Busck), in peach orchards, studied the emission rate of (*Z*)-8-dodecen-1-yl acetate from 5 types of dispensers at 3 temperatures and 1 wind speed in the laboratory and also in the field. He noted, as did Bierl and DeVilbiss (1975) that the volatility of the pheromone and its release rate are directly related to the accuracy of the method. If the rate is such that weighing intervals are short, then accurate daily release rates can be obtained. On the other hand, if the release rate is slow, volatiles from a substrate may be released at a faster rate than the active ingredient. If the interval between weighings is large, water and other materials may be absorbed into the substrate. Such effects preclude accurate measurement; however, such errors may be reduced by recording the weight loss of a control dispenser and applying a correction factor.

The release rate of (Z)-8-dodecen-1-yl acetate from 70 fiber dispensers of polyester terephthalate has been determined at 15, 20, and 29°C (Rothschild, 1979). The average daily rate of these temperatures over 70 days was respectively, 1.04, 1.25, and 1.63 mg/dispenser. Calculations based on Rothschild's data show that at 15°C the average daily release rate over the 1st 10 days was 1.5 mg/dispenser; then the rate stabilized to 0.95 mg/dispenser over the next 60 days. Ashare et al. (1975, 1976) on theoretical grounds predicted that the release rate of pheromones will double with a 17°C rise in temperature which is in reasonably good agreement with the 1.57-fold rise observed by Rothschild.

The most serious disadvantages of this method are that the identity of the released material is not evident; any pheromone lost by decomposition or oxidation cannot be accounted for and with multicomponent pheromones the rate of release of the individual components cannot be determined.

Extraction and Quantification of Residual Semiochemical

In this method which is also appropriate to both laboratory and field application the amount of pheromone remaining in the formulation is usually quantified by solvent extraction followed by gas chromatographic analysis. The variables in the extraction process which will affect the measurement are choice of solvent, extraction time, and temperature. Few authors have given details especially with regard to the efficiency of the extraction. Maitlen et al. (1976) reporting on baits for trapping studies with codling moth, *Laspeyresia pomonella* (L.), presented procedures and data indicating the recovery efficiency of (E,E)-8,10-dodecadien-1-ol from rubber bands (88-105%), septa (92-109%), and polyethylene vial caps (92-112%). Butler and McDonough (1979) and Flint et al. (1978) working with rubber septa, simply state that the extraction is carried out in 50 ml hexane:dichloromethane (1:1) over 1 h with mechanical shaking. Similarly, Campion et al. (1978), working with microcapsules and polyethylene vials, tersely report that the release rate "...was measured either by g.c. analysis of residual pheromones after chloroform extraction or by...."

This method identifies the pheromone as evidenced by the retention time of the peak on the chromatogram, and the amount present determined by accurate and reliable procedures. However, like the weight loss method, it characterizies the remaining compound but offers no insight into any oxidation, reaction with the substrate (Steck et al., 1979), or decomposition, the products of which are neither extractable, nor eluted from the gas chromatograph under the conditions used. Alternatively, if a decomposition product is as volatile as or more volatile than the pheromone under investigation, the determined rate would again be erroneous.

Kuhr et al. (1972) using radiolabeled compounds measured the release rate of pheromone analogues from polyethylene vial caps, determining the residual radioactivity by scintillation counting.

Trapping of Effluent Vapors

This technique should give the most accurate measurement of the release rate (cf. Plimmer et al., 1977); however, it is fraught with many difficulties. Vapor collection may be achieved both in a static atmosphere and in a dynamic environment; the former leads to "static air collection" while the latter will be referred to as "air flow collection."

As the name implies, in the air flow method the substrate containing the pheromone is held in a vessel while a current of air is passed over it; volatile materials thus eluted may then be captured in cold traps (Browne et al., 1974) or solvent traps (Beroza et al., 1975; Bierl and DeVilbiss, 1975; Bierl-Leonhardt et al., 1979) or on adsorbents such as charcoal (Campion et al., 1978), Porapak Q® (Look, 1976; Wiesner and Silk, 1979; Cross, 1980b; Cross et al., 1980) and Tenax® (Cross, 1980a,b; Cross et al., 1980). The use of polymeric adsorbents in release rate studies was as a result of the pioneering work of Byrne et al. (1975) and Peacock et al. (1975) who used them in pheromone isolation and identification studies. The semiochemical is then isolated from the trap by solvent extraction and concentration, or desorption and concentration, which ever is appropriate, and then analyzed by gas chromatography. The analytical method of choice was high performance liquid chromatography in the case of 3-methyl-2-cyclohexen-1-one, an aggregation pheromone of the Douglas fir beetle (Look, 1976). For a description of the several apparatus used, the reader is referred to the reports of Cross (1980a) and Cross et al. (1980) and the references contained therein.

The "static air" technique developed by Baker et al. (1980) involves the use of a glass surface to collect semiochemicals; in studies with both septa and live moths, 250 ml round-bottomed flasks have served as both the aeration chamber and the collecting trap. Removal of the pheromone source from the flask is followed by solvent rinsing, concentration and quantification by gas chromatography.

Direct Observation of the Volume Changes in Capillary Reservoirs

This method is specific to hollow fibers and microcapillaries. Ashare et al. (1975; 1976) report the use of a cathetometer to follow the meniscus regression, while Cross (1980b) preferred an optical micrometer mounted in the eyepiece of a dissecting microscope. The cathetometer, which has an accuracy which translated into ±1 μg, has been superceded by a Wilder Varibeam® optical comparator fitted with IKL digital positioners connected to IKL Microcode® digital readout systems. With this system release rates can be determined to ±0.015 μg. Fibers (25 or 50), affixed to microscope slides are aged in environators usually at 22.2 and 37.7°C and examined at regular intervals. The differences in the length of the liquid column with time, taken in conjunction with the internal diameter of the capillary and the density of the active ingredient, are translated into a release rate. The disadvantages of this method are that the identity of the active material is not evident, and when used with multicomponent pheromones formulated in the same fiber, the release rate of the components cannot be determined.

RELEASE RATE OF GOSSYPLURE FOR HOLLOW FIBERS

Perusal of the pertinent literature indicates a multiplicity of release rates for a pheromone emitting from the same substrate, under the same conditions, when measured by different methods irrespective of whether these determinations are carried out in one laboratory or several.

In general, methods which determine the release rate by residue analysis techniques (e.g. weight loss measurements, residue extraction, meniscus regression, etc.) yield a greater release rate than methods which rely on quantifying trapped volatiles emitted from the substrates. This phenomenon has been evident in most of our studies, on the release rate of gossyplure from hollow fibers, as indicated by Table 1 which details the mean release rates in μg per fiber/day for fibers without Bio-Tac® over at least 17 days.

Table 1. Release rate of gossyplure from hollow fibers.

Method	Mean rate* (μg/fiber/day)	% Standard deviation
	Weighing	
Weight loss	1.50	3.60
	Optical Comparison	
Meniscus regression**	2.86	27.30
	Gas Chromatography	
Static air	0.04	11.30
Airflow**	0.04	1.9
	Scintillation Counting	
Static air	0.34	2.3
Airflow	0.74	0.7
Residue analysis	2.02	2.6
Residue analysis	2.25	7.97
Airflow***	3.11	----

*Mean rate of at least 3 replicates.

**Mean rate from only 2 replicates.

***The result from a single experiment.

Data given in Table 1 as it refers to the meniscus regression method is for 2 individual fibers which have been designated fibers #2 and #3 in Figure 1. The standard deviation of 27.3% is misleading and can be greatly reduced with an increased sample size. As stated above, routine release rate measurements are made on lots of 25 or 50 fibers. This method also allows us to recognize defective fibers such as fiber #1 whose increased release rate is symptomatic of an improperly sealed fiber.

The static air methods used are based on that elaborated by Baker et al. (1980). When scintillation counting was the quantification technique, the apparatus used as an aeration chamber and collecting device is shown in Figure 2, and consists of an inverted round-bottomed flask and a "through tube" sealed a short distance from the end so as to form a support for a petti-cup. When the petti-cup was charged with 5 fibers containing gossyplure labeled

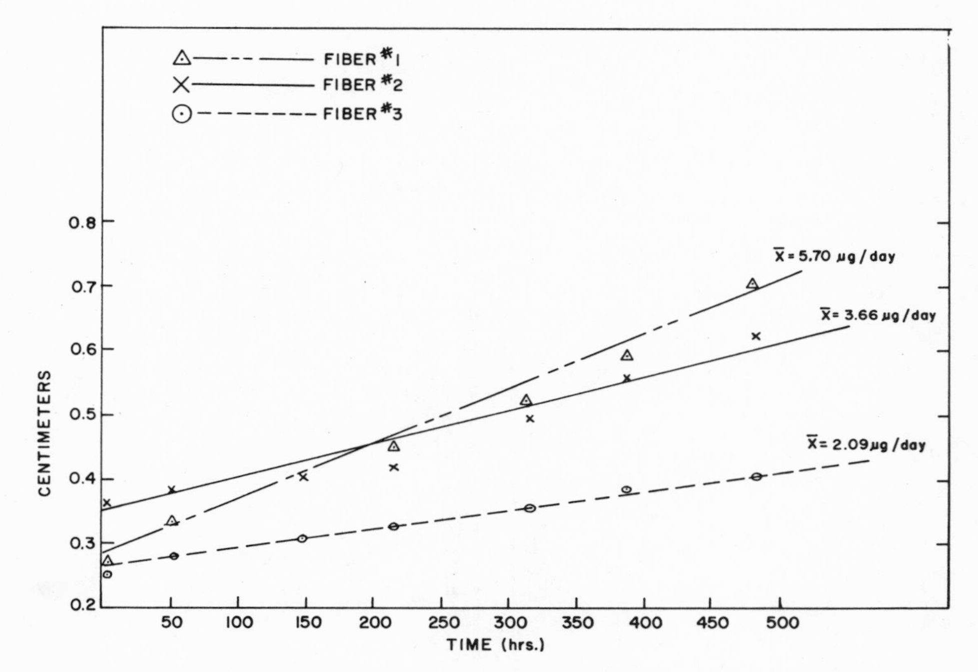

Fig. 1 Release rate of gossyplure from 3 individual fibers measured by the meniscus regression method at 22°C with the Wilder Varibeam Optical Comparitor.

in the acetate carbonyl group with ^{14}C, held at 21 $\pm$ 1°C, and sampled everyday, the release rate profiles obtained are exemplified by Figure 3. Over 21 days the average release rate was 0.323 μg per fiber/day with individual daily rates varying from a minimum of 0.25 μg/fiber to a maximum of 0.37 μg/fiber. The data may also be exhibited as a graph of the cumulative amount released versus time, giving a straight line plot. In this series of experiments the effect of varying the size of the chamber flask, and of coating the fibers with sticker were also observed. The release rates determined using 250, 500 and 1000 ml flasks were not significantly different from each other. As can be seen from Figure 4, which illustrates the mean profile of 3 runs in which the fibers were coated with Bio-Tac, and 3 runs without Bio-Tac, the Bio-Tac causes a reduction of 23% in the release of gossyplure from fibers as determined by this method. At the conclusion of the static air experiments which

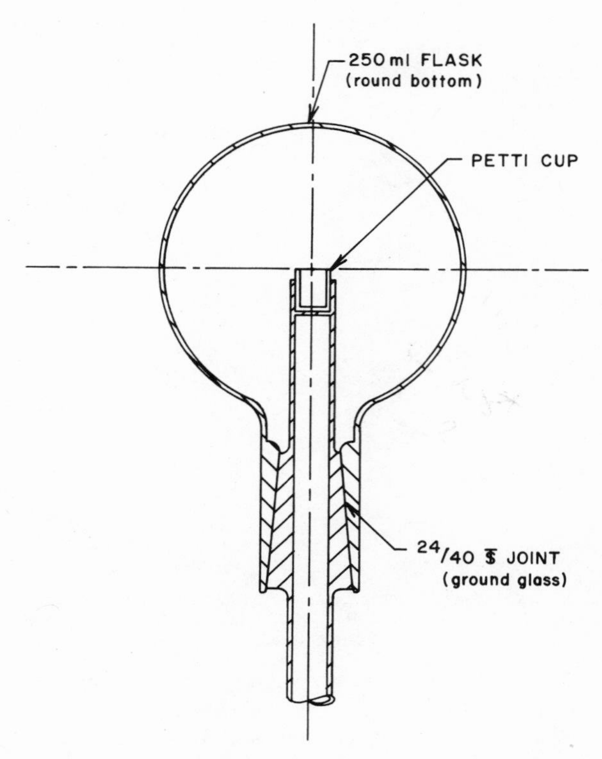

Fig. 2 The apparatus used in the determination of release rates by the static air method, using scintillation counting as the quantification technique.

yielded a mean release rate of 0.34 μg per fiber/day (Table 1) the fibers were subjected to scintillation counting to ascertain the average daily release rate from the residual pheromone. The mean rate obtained was 2.25 μg per fiber/day (Table 1) which is 6.6X greater than the rate obtained from the trapped volatiles.

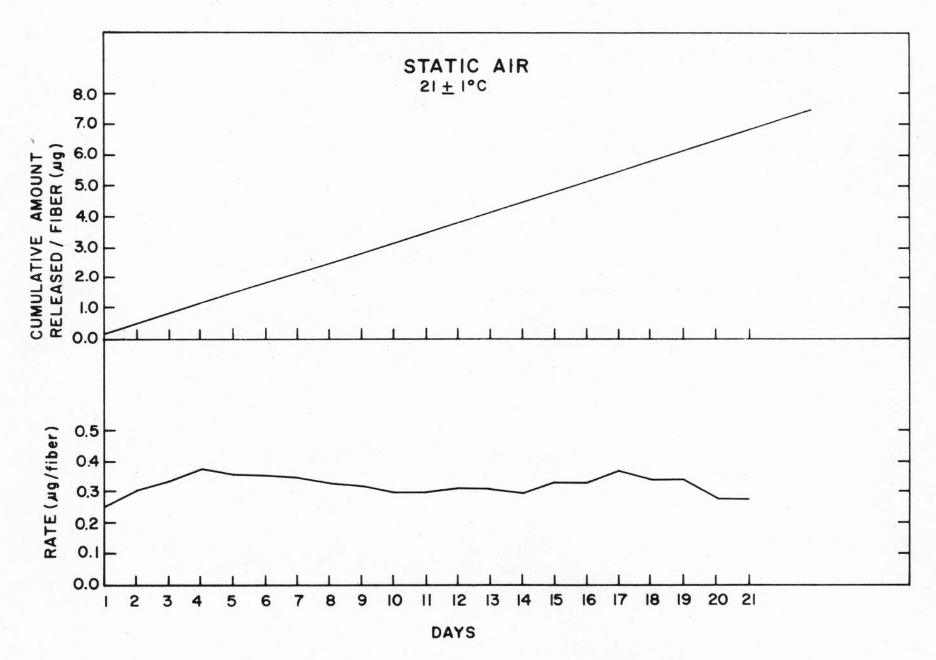

Fig. 3 Release rate data for gossyplure obtained by the static air method, lower trace: daily rate profile; upper trace: cumulative amount released from fiber plotted versus time.

In static air experiments where gas chromatography was the quantifying technique, the aeration and collection of the gossyplure were carried out in 200 ml teardrop-shaped flasks. The petti-cup containing the fibers were either placed in the bottom of the flasks or suspended from modified stoppers. A great deal of preliminary experimentation was carried out with respect to the recovery of gossyplure and other pheromones from glass. These experiments utilized the compounds without fibers, and may be summarized as

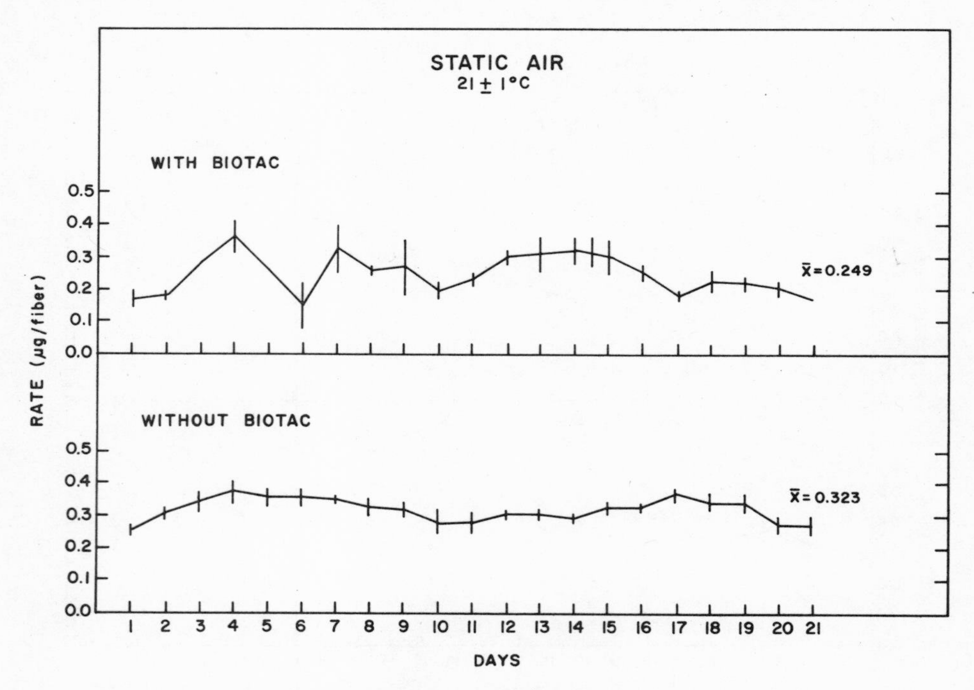

Fig. 4 Comparison of daily rate profiles for the release of gossyplure from fibers coated with Bio-Tac and fibers without Bio-Tac as measured by the static air method.

follows: (a) the position of the petti-cup containing the pheromone within the flask does not affect the release rates; (b) the recovery of gossyplure determined at various time intervals from 6 to 75 h varied between 91-95%; (c) the orientation of the petti-cup; i.e., whether upright or inverted, has a profound effect on the total recovery of gossyplure and on the percentage recovered from the flask walls and the petti-cup. When the petti-cup was upright, 3.4% of the recovered gossyplure came from the walls; this percentage was increased to 16.0 when the petti-cup was inverted; (d) with

(Z)-9-dodecen-1-yl acetate a 99% recovery was recorded with an upright petti-cup but it dropped to 85% when the cup was inverted. The amounts recovered from the walls were respectively, 35.3 and 55.4%, and (e) over 24-h time periods the recovery of (Z)-11-tetradecen-1-yl acetate was 95 ± 3%, and of (E)-11-tetradecenal, 92.5 ± 4%. There are indications that over greater time periods the percentage of recovered aldehydes is somewhat lower.

Nothwithstanding the precautions dictated by the results obtained from the preliminary studies, the average release rate of gossyplure over 25 days was determined at 0.04 μg per fiber/day or 10X less than when the static air method was quantified by scintillation counting.

The apparatus used in the 1st 2 airflow determinations denoted in Table 1 is shown in Figure 5. Air at the rate of 1 liter/min is drawn through the apparatus first being dried and freed of organic impurities by passing through a mixture of resin and molecular sieves. On dispersion from the frit the airstream swirls over and around the petti-cups containing the fibers and passes into the detachable collectors containing the trapping medium. In experiments with radioactive gossyplure 5 fibers were put in each of the 10 petti-cups and 3 collectors in series employed. When the airflow experiments were quantified by gas chromatography, 50 fibers were put in each petti-cup and 2 large collectors used. Although Porapak Q and Tenax have been successfully used to efficiently trap pheromones (e.g. Cross et al., 1980) impurities in the resins and variation in the quality between batches led us to investigate other materials. The use of glass by Baker et al. (1980) led us to experiment with glass beads as an adsorbent for collecting airborne pheromones. The beads used in the experiments reported here were of 1.00 - 1.05 mm in diameter and were held in the collectors with plugs of silanized glass wool; all parts of the apparatus with the exception of the beads were silanized in an attempt to minimize the adsorption of the pheromone on the apparatus walls. Breakthrough of the pheromone from the collectors does not occur as evidenced by insignificant radioactive counts obtained from the 3rd collector. The efficiency of desorption of gossyplure from the beads is shown in Table 2. The detached collectors were eluted with solvent via the adaptor which has a luer-type joint at one end and a standard female $\overline{\text{S}}$ 14/20 at the other. Using the small collectors the mean value of 17 recoveries was 101.78 ± 2.3% whereas with the large collectors the average of 16 recoveries was 97.94 ± 3.6%.

Data from measurements made using this airflow technique with radiolabeled gossyplure are given in Figure 6; over 17 days the average release rate from the fibers both with and without Bio-Tac was 0.736 μg per fiber/day, the uncoated fibers varying from a minimum daily rate of 0.61 μg/fiber to a maximum of 0.79 μg/fiber, while

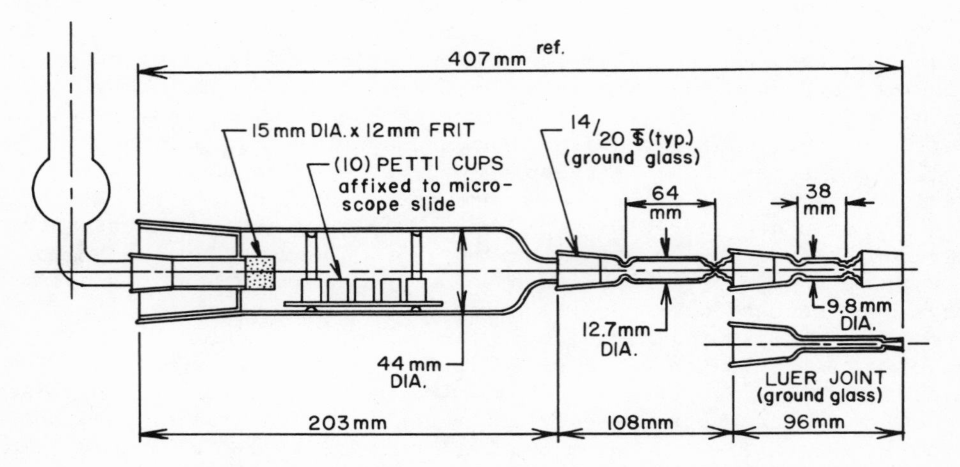

Fig. 5 Apparatus used in the airflow method of determination of pheromone release rates.

Bio-Tac coated fibers had a daily variability of between 0.63 μg/fiber and 0.83 μg/fiber. When fibers were counted for residual activity, the counts translated into an average release rate of 2.02 μg per fiber/day, exhibiting the same trend as did the values from the static air experiments.

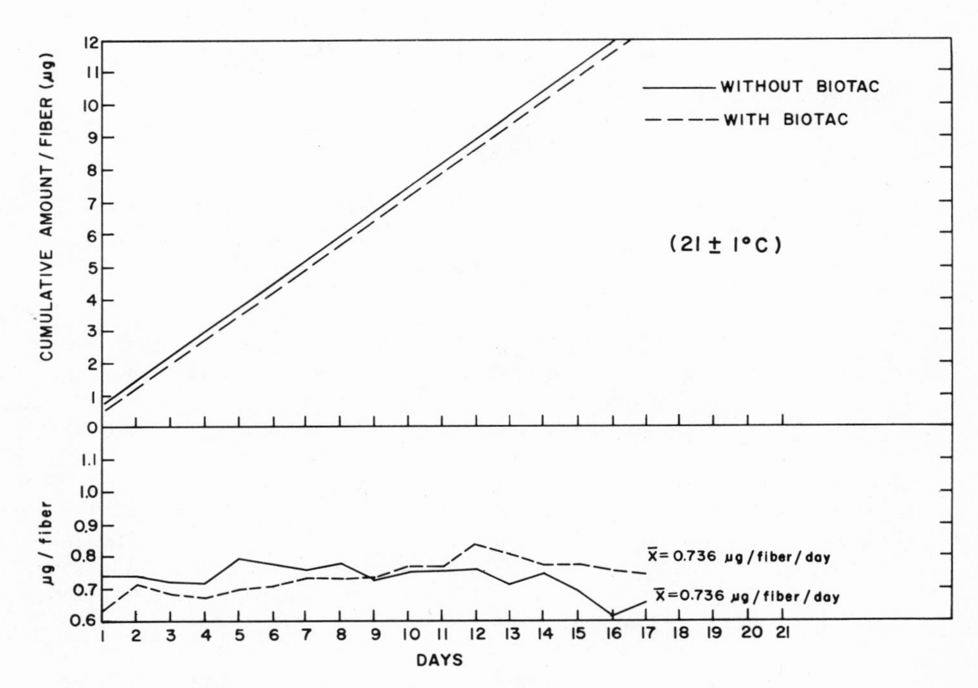

Fig. 6 Release rate data for gossyplure determined by the airflow method, lower trace: daily rate profiles for fibers with and without Bio-Tac; upper trace: cumulative amounts released from fibers coated with, and without Bio-Tac plotted versus time.

Table 2. Efficiency of the Desorption of Gossyplure from Glass Beads.

	Small collectors	Large collectors
Weight of glass beads	2.3 - 3.2 g	14.5 - 18.0 g
Amount of gossyplure	50 mg	50 mg
Volume of iso-octane used for elution	(3 x 5 ml) x 2	(4 x 5 ml) x 2
Percent recovery	$\overline{X}_{17}$ = 101.78 (% SD - 2.3)	$\overline{X}_{16}$ = 97.94 (% SD - 3.6)

Again, when gas chromatography was used in the quantification, the rate obtained, although in agreement with that determined from the static air method, was more than 18 times less than the rates given above.

Summarizing to this point, we have so far obtained what may be considered 3 different release rates for gossyplure evaporating from hollow fibers; these rates are 1.50 - 2.86 µg per fiber/day obtained by methods measuring the pheromone remaining in the fibers; 0.04 µg per fiber/day from both static air and airflow methods as quantified by gas chromatography; and 0.34 - 0.74 µg per fiber/day from both of the effluent collection methods as quantified by scintillation counting. On the positive side the release rate profiles indicate that gossyplure evaporates from the fibers at a constant rate for periods up to 21 days. On the negative side the difference between the smallest and the greatest release rate is more than 70 times, and in instances where it should have been possible to equate the rates obtained from different methods, this was not achieved. For example, in the airflow experiment the rate determined was 0.74 µg per fiber/day, the residue analysis on the same fibers gave counts which translated into a rate of 2.02 µg per fiber/day, which means that 30 µg/fiber of gossyplure is unaccounted for over the 22 days of the experiment. In one airflow run more than half of the pheromone evaporating from the fibers was found on the petti-cups and the walls of the apparatus. The missing 30 µg of gossyplure was probably never washed from the apparatus either because the volume of solvent used was insufficient or not enough washings were employed. Such problems have arisen because airflow apparatus such as the one shown in Figure 5 and the one described by Cross et al. (1980) are too large and have too much area of glass available for adsorption as well as the distance between the pheromone emitter and the collection medium is too great. Apparatus designed to circumvent such serious drawbacks include that reported by Cross (1980a) and the one shown in Figure 7.

The mini-airflow apparatus (Fig. 7) was designed in an attempt to minimize the distance between the substrate and the adsorbent, to offset large areas of glass so that desorption would not require large volumes of solvent, and to generally minimize the loss of volatile material as experienced in the other methods.

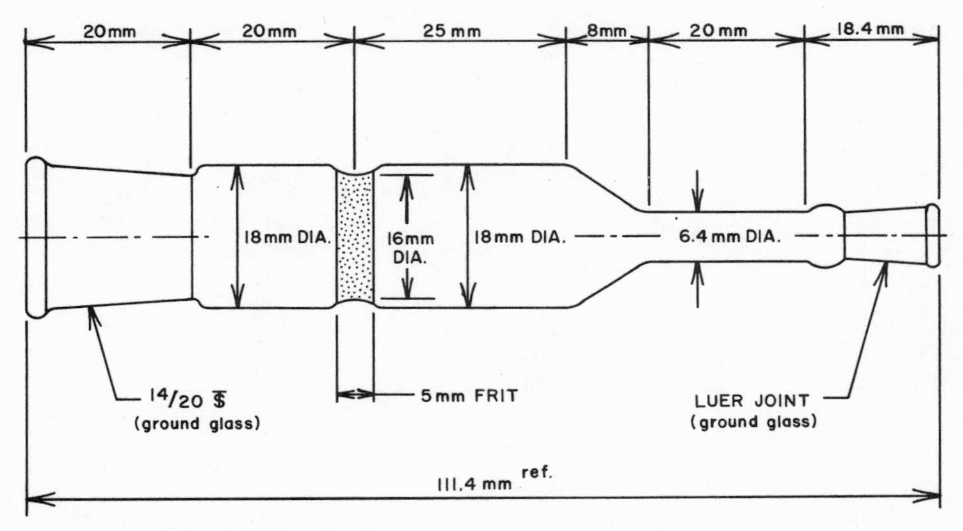

Fig. 7 Mini-airflow apparatus used for the determination of pheromone release rates.

Glass beads (ca. 8 g) are introduced into the larger chamber and held in place by a plug of silanized glass wool. For the experiments reported here 5 fibers were introduced into the smaller chamber, the apparatus attached to a vacuum pump such that dry air freed from organic impurities was pulled through the system at 1 liter/min. Daily, the glassware was detached from the vacuum line, the fibers removed and the beads desorbed by passing solvent through the apparatus from a syringe attached to the luer joint. Each wash eluting from the small chamber end was put directly into a scintillation vial in preparation for counting. Two washes were found to remove the accumulated activity, although 3 washes were routinely used. The glassware was flushed in both directions with nitrogen, the fibers reinserted and the apparatus reattached to the vacuum line.

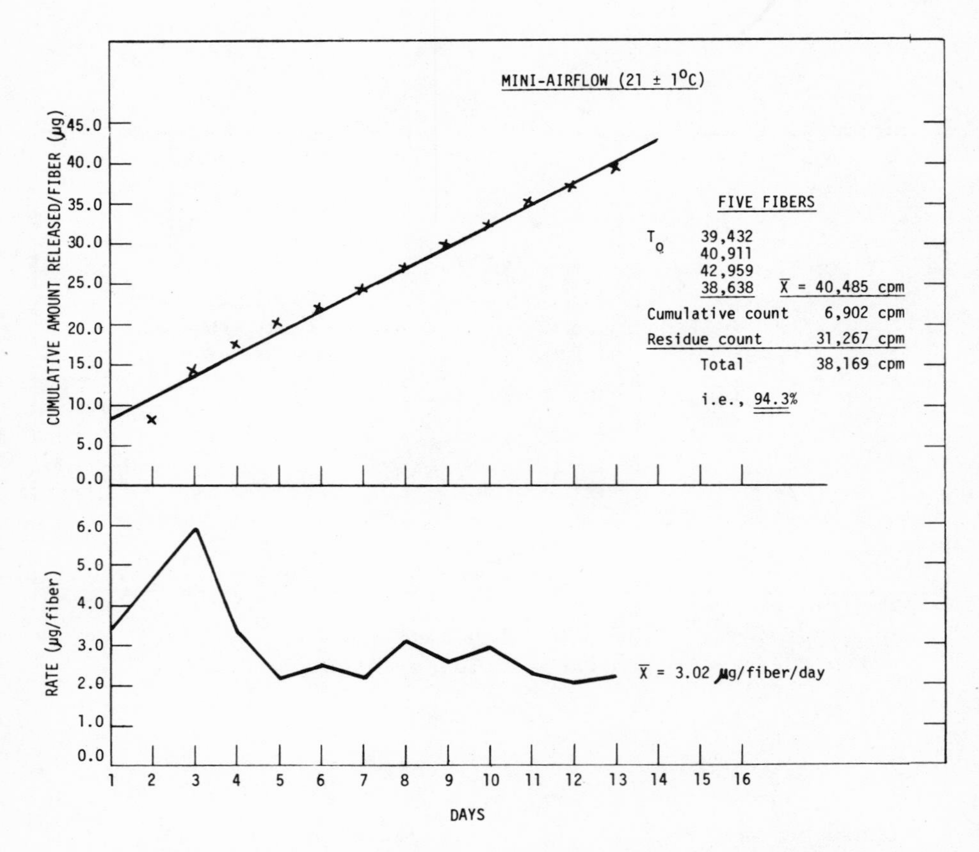

Fig. 8 Data on release rate of gossyplure obtained using the mini-airflow apparatus, <u>lower trace</u>: daily rate profile; <u>upper trace</u>: plot of amount emitted versus time; plus calculation showing percentage recovery of radioactivity.

Data from this system are presented in Figure 8 which exhibits a release rate profile indicative of a mean daily release rate of 3.02 μg/fiber over 13 days. From the profile the calculated average daily release rate from day 4 on is 2.57 μg/fiber, the same as that obtained from the slope of the line in the cumulative amount released versus time plot. Prior to the start of each run 5 groups of 5 fibers are selected, 4 of these groups being used to obtain a time zero value of the activity; the 5th group is used in the experiment. For the run used to generate the data given in Figure 8, the average time zero activity/5 fibers was 40,485 cpm. The activity in the emitted pheromone accounted for 6,901 cpm while the residual count equalled 31,267 cpm giving a total of 38,169 cpm or 94.3% of the average time zero activity. Being able to account for virtually all of the radioactivity indicates that this method reflects the release characteristics of the fibers and can be used in the development of reliable release rate methodology.

CONCLUSIONS

In the frenetic activity which characterizes the field testing of pheromones as viable agents for pest monitoring and control the haste to gather data by the empirical testing of formulations has become almost obsessive. As noted by Roelofs (1979), the greater number of both monitoring and mating disruption tests carried out up until 1977 were completed without knowledge of the release rates of the materials being used. If, however, pheromones are indeed a viable alternative with the promise of commercialization, then those charged with such development must pay more heed to the rather unglamorous studies concerning the release rate of semiochemicals from controlled release systems. Generally, the studies of the release rates have not been well documented, leading to much over-simplification of the problems encountered. The studies summarized in this report indicate some of the many problems to be contended with in the development of reliable methodology.

It is hoped that a method can be developed from the mini-airflow system, utilizing both radiotracer and gas chromatographic quantification, which will be equally suitable for multicomponent blends and simpler pheromones unaffected in efficiency by pheromone functionality and chain length, and applicable to all hollow fiber formulations. Once this has been achieved, studies can then be undertaken to give some insight into the effect of such factors as substrate material, additives, pheromone decomposition rate and climate on the release rate.

REFERENCES

Ashare, E., Brooks, T. W., and Swenson, D. W., 1975, Controlled Release from Hollow Fibers, Proceedings 1975 International Controlled Release Pesticide Symposium (September 1975), p. 42.

Ashare, E., Brooks, T. W., and Swenson, D. W., 1975, Controlled release from hollow fibers, in "Controlled Release Polymeric Formulations," D. R. Paul and F. W. Harris, eds., American Chemical Society Symposium, Series 33, Washington, D.C.

Baker, T. C. Cardé, R. T., and Miller, J. R., 1980, Oriental fruit moth pheromone component release rates measured by glass surface airborne collection, J. Chem. Ecol., 6: (in press).

Beroza, M., Bierl, B. A., James, P., and DeVilbiss, D., 1975, Measuring emission rates of pheromones from their formulations, J. Econ. Entomol., 68(3): 369.

Bierl, B. A., and DeVilbiss, D., 1975, Insect Sex Attractants in Controlled Release Formulations: Measurement and Applications, Proceedings 1975 International Controlled Release Pesticide Symposium (September 1975), p. 230.

Bierl, B. A., DeVilbiss, E. D., and Plimmer, J. R., 1976, Use of pheromones in insect control programs: Slow release formulations, in "Controlled Release Polymeric Formulations," D. R. Paul and F. W. Harris, eds., American Chemical Society Symposium, Series 33, Washington, D.C.

Bierl-Leonhardt, B. A., DeVilbiss, E. D., and Plimmer, J. R., 1979, Rate of release of disparlure from laminated plastic dispensers, J. Econ. Entomol., 72: 319.

Browne, L., Birch, M. C., and Wood, D. L., 1974, Novel trapping and delivery system for airborne insect pheromones, J. Insect Physiol., 20: 183.

Butler, L. I. and McDonough, L. M., 1979, Insect sex pheromones: evaporation rates of acetates from natural rubber septa, J. Chem. Ecol., 5(5): 825.

Byrne, K. J., Gore, W. E., Pearce, G. T., and Silverstein, R. M., 1975, Porapak-Q collection of airborne organic compounds serving as models for insect pheromones, J. Chem. Ecol., 1:1.

Campion, D. G., Lester, R., and Nesbitt, B. F., 1978, Controlled release of pheromones, Pestic. Sci., 9: 434.

Cross, J. H., 1980a, A vapor collection and thermal desorption method to measure semiochemical release rates from controlled release formulations, J. Chem. Ecol., 6: (in press).

Cross, J. H., 1980b, Interpretation of release rate and extraction measurements made on two types of controlled-release-formulation dispensers containing (Z,Z)-3,13-octadecadien-1-ol acetate, J. Chem. Ecol., 6: (in press).

Cross, J. H., Tumlinson, J. H., Heath, R. R., and Burnett, D. E., 1980, Controlled release formulations of semiochemicals: Apparatus and procedure for measuring release rate, J. Chem. Ecol., 6: (in press).

Fitzgerald, T. D., St. Clair, A. D., Daterman, G. E., and Smith, R. G.,1973, Slow release plastic formulation of the cabbage looper pheromone cis-7-dodecenyl acetate: Release rate and biologiactivity, Environ. Entomol., 2: 607.

Flint, H. M., Butler, L., McDonough, L. M., Smith, R. L., and Forey, D. E., 1978, Pink bollworm: Response to various emission rates of gossyplure in the field, Environ. Entomol., 7(1): 57.

Kuhr, R. J., Comeau, A., and Roelofs, W. L., 1972, Measuring release rates of pheromone analogues and synergists from polyethylene caps, Environ. Entomol., 1(5): 625.

Look, M., 1976, Determining release rates of 3-methyl-2-cyclohexane-1-one antiaggregation pheromone of Dendroctonus pseudotsugae (Coleoptera: Scolytidae), J. Chem. Ecol., 2(4): 481.

Maitlen, J. C., McDonough, L. M., Moffitt, H. R., and George, D. A., 1976, Codling moth sex pheromone: Baits for mass trapping and population survey, Environ. Entomol., 6: 199.

McDonough, L. M., 1978, Insect Sex Pheromones: Importance and Determination of Half-Life in Evaluating Formulations, USDA, SEA, ARR-W-1 (May 1978), 20 pp.

Peacock, J. W., Cuthbert, R. A., Gore, W. E., Lanier, G. W., Pearce, G. T., and Silverstein, R. M., 1975, Collection on Porapak-Q of the aggregation pheromone of Scolytus multistriatus (Coleoptera: Scolytidae), J. Chem. Ecol., 1: 149.

Plimmer, J. R., Bierl, B. A., Webb, R. E., and Schwalbe, C. P., 1977, Controlled release of pheromone in the gypsy moth program, in: "Controlled Release Pesticides," H. B. Scher, ed., American Chemical Society Symposium, Series 53, Washington, D.C.

Roelofs, W. L., 1979, "Establishing Efficacy of Sex Attractants and Disruptants for Insect Control," Entomological Society of America, College Park, MD.

Rothschild, G. H. L., 1979, A comparison of methods of dispensing synthetic sex pheromone for the control of oriental fruit moth, Cydia molesta (Busk) (Lepidoptera: Tortricidae), in Australia, Bull. Entomol. Res., 69: 115.

Steck, W. F., Bailey, B. K., Chisholm, M. D., and Underhill, E. W., 1979, 1,2-Dianilinoethane, A constituent of some red rubber septa which reacts with aldehyde components of insect sex attractants and pheromones, Environ. Entomol., 8: 732.

Wiesner, C. and Silk, P., 1979, Testing the release rate characteristics of the Conrel hollow fiber, Canadian Forest Service Research Report, Report of the Spruce Budworm Pheromone Trials Maritimes 1978, pp. 36-42.

THE HERCON DISPENSER FORMULATION AND RECENT TEST RESULTS

Agis F. Kydonieus

Morton Beroza

Hercon Group
Herculite Products, Inc.
1107 Broadway
New York, NY 10010 USA

INTRODUCTION

The use of synthetic replicates of natural sex pheromones to suppress insect pests of agriculture now appears to be rapidly coming to fruition. Reports of many successful demonstrations are scattered throughout recent literature. These trials have repeatedly established the soundness of principles underlying a variety of approaches to the control of a broad spectrum of destructive insect species. Traps monitoring specific insect pests have removed much of the guesswork from the timing of pesticide applications. Mating disruption by permeating the air with pheromones or using analogs has been highly effective in controlling some of our most damaging agricultural insects. Trap cropping and large-scale demonstrations of mass-trapping have been highly successful, and this work continues with vigor.

The recent expansion of pheromone use to the commercial arena has generated a need for large amounts of pheromones, and it has been gratifying to witness the drop in prices of these exotic chemicals as the volume rose. The lower prices are improving the economic outlook for pheromone use to the extent that pheromone use for combatting certain species soon may be less costly than application of pesticides.

Unfortunately, pheromones are not applicable to the control of all insect pests, or in all situations. Nevertheless, there are a number of applications in which pheromones have already shown a strong potential for replacing much of the large volumes of pesticides used in insect pest suppression.

Fortunately, chemists the world over have elucidated the structures and synthesized the sex pheromones of some of our most damaging pest species. Also, exceptional progress has been registered recently in determining identity and amounts of the ingredients of some very important multi-component sex lures.

The advantages of minimizing toxicant use in agriculture through the use of pheromones, generally as part of an overall pest management strategy, are numerous and definitely worthy of our efforts. Through the use of pheromones, we can look forward to greatly reducing exposure of man and wildlife to toxic chemicals, allowing beneficial insects to survive which will help in insect control, and greatly minimizing problems concerning insect resistance to insecticides.

THE HERCON SYSTEM

The key to effectively using pheromones lies in proper formulation and timing of application. Formulation must take into account that each pheromone has different chemical and physical properties. Because pheromones differ in volatility and concentration needed to exert the desired effect, a suitable dispenser must release the chemical at a controlled rate, one optimum for attraction or air permeation, and must continue to emit pheromone for a sufficient time after exposure.

Herculite Products has devised a plastic laminate, now patented and marketed under the name Hercon®, that not only meters out the chemical evenly but also protects it from light, oxidation, and hydrolysis. The chemical is embedded in the center layer (the reservoir) of a plastic laminate sheet, and the lure diffuses out through the outer layers at a controlled rate (Fig. 1). The system is highly versatile and easily regulated because of the dispenser's many parameters that can be varied; for example, we can control the thickness of the layers, the size (area) of the dispenser, the polymer type of both outer and reservoir layers, and the pheromone concentration to give almost any desired rate of pheromone release. At the same time, sensitive compounds like aldehydes are protected because they are inside the dispenser until they surface when they are immediately discharged into the atmosphere. From work with its laminate dispenser the past 7 years, Herculite found that the emission rate of pheromone can be readily predicted from the usual laws governing the chemical diffusion rate. A discussion on how these parameters affect the Hercon system directly has been published (Kydonieus, 1977).

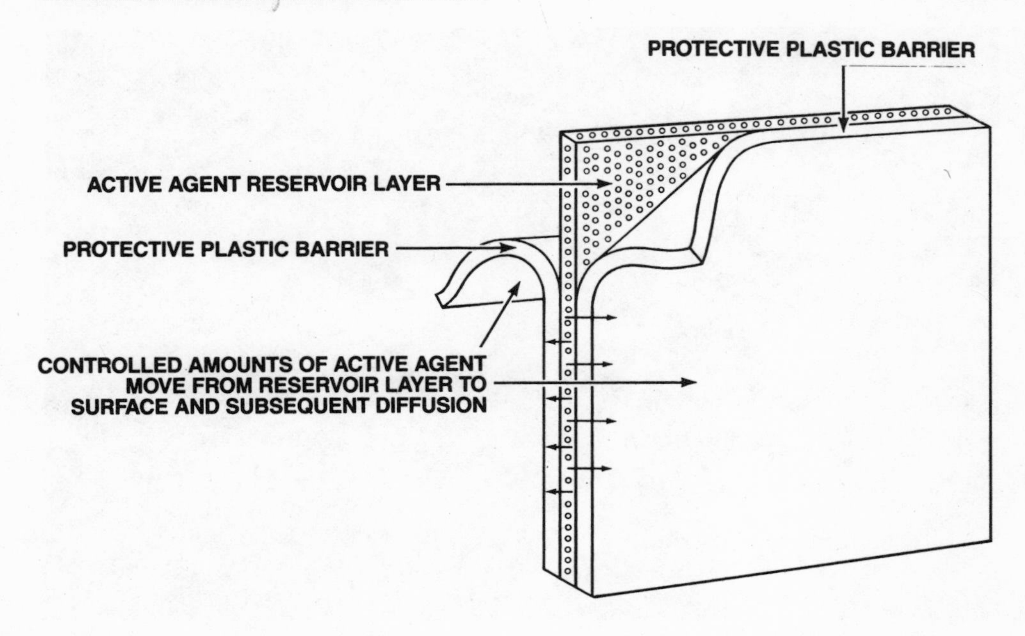

Fig. 1. Hercon controlled-release dispensing system.

Hercon dispensers can be fabricated in a variety of forms: squares, ribbons, tapes, flakes, or confetti (Fig. 2). The squares, generally with 0.5 - 2.5 cm sides, are used in traps for monitoring or mass-trapping insects. In mass-trapping, 5 to 10 traps/acre often can suppress populations, particularly when deployed early in the growing season before the insect population builds up.

The confetti and flakes are used to permeate the air with sex attractant, thereby confusing insects trying to find mating partners. With 10,000 to 20,000 pheromone flakes/acre broadcast by aircraft, insects cannot distinguish the odor of "calling" insects from the synthetic attractant emitted from the many particles in the vicinity. Thus, they cannot find a mate and remain unfulfilled. Mating disruption also has been accomplished by hand placement of several hundred pheromone squares/acre. The confetti or flakes are generally applied by aircraft with a sticker to adhere the particles to the crop foliage being protected. A special apparatus has been designed to dispense flakes efficiently in a broad swath.

Fig. 2. Types of Hercon controlled release dispensers.

Whether squares, flakes or ribbons are used, emission rate of lure from formulations is first determined in the laboratory at 1 or 2 temperatures. This rate is governed by the amount of surface area and the length of edge exposed. Thus, as the size (area) of the individual flake is reduced, more edge becomes exposed and emission rate increases. The next step is to test several candidate formulations in the field to determine their efficacy and duration of effectiveness. The formulation performing best in the field is chosen, or if emission rates of all formulations are too high or too low, a suitable adjustment is made based on the estimated change required. For example, if twice the emission rate is desired based on field performance, a new formulation is prepared and checked in the lab to verify that the desired rate is attained before undertaking the lengthy process of field testing again.

Most of the pheromones identified are long-chain, unsaturated acetates or other esters, aldehydes, alcohols and epoxides. Many of these are highly unstable and readily degraded when exposed to the environment. Evidence shows that aldehydes, which are particularly unstable, are found in the pheromones of some of our most destructive agricultural pests, e.g., the boll weevil (*Anthonomus grandis* Boheman), rice stem borer [*Chilo suppressalis* (Walker)], corn earworm [*Heliothis zea* (Boddie)], spruce budworm [*Choristoneura fumiferana*

(Clemens)], and navel orangeworm [Amyelois transitella (Walker)]. Hercon's laminate has consistently performed well in dispensing these aldehyde pheromones as well as other types of pheromones.

The Environmental Protection Agency (EPA) has granted full registration of Hercon products for suppressing populations of gypsy moth [Porthetria dispar (L.)], boll weevil, and the Japanese beetle (Popillia japonica Newman). The EPA has granted experimental use permits (EUPs) for use of Hercon products against the tobacco budworm [Heliothis virescens (F.)], pink bollworm [Pectinophora gossypiella (Saunders)], western pine shoot borer (Eucosma sonomana Kearfott), and spruce budworm. We anticipate that some EUPs will be elevated to full registration status in 1980. Also, we will be seeking EUPs in 1980 for testing against other insects. Hercon also has fabricated a dispenser for use against the spruce bark beetle (Ips typographus L.), a forest pest of northern Europe.

FIELD PERFORMANCE WITH HERCON PRODUCTS

Pink Bollworm

The pink bollworm is one of the most damaging pests of cotton in Arizona, California and Latin America. It also infests cotton in other parts of the world and has been designated one of the 6 most destructive insects in the world. In 1979, the USDA Western Cotton Research Laboratory at Phoenix, AZ undertook an extensive experiment in which cotton was treated by aircraft using Hercon flakes containing gossyplure to determine the treatment effectiveness in disrupting pink bollworm mating. Twelve 16-ha plots were treated with the pheromone, while 4 plots used as controls were treated with a normal regimen of insecticides. Results were excellent and led Henneberry et al. (1980) to the following conclusions:

> "The results of the present studies demonstrated that the multi-layered, plastic, laminated gossyplure fomulation was an effective vehicle for dissemination of gossyplure under field conditions.
>
> "The promising results of the present study were achieved using economically feasible materials and application methods on a commercial cotton acreage."

Application has been made for full registration of the Hercon flake for pink bollworm control, and prospects for receiving registration for the 1980 growing season are good.

Gypsy Moth

The gypsy moth, *Lymantria dispar* L., accidently imported from Europe, rates as the most serious forest insect pest in northeastern United States. Several field trials for mating disruption using Hercon dispensers before 1979 (by USDA), on which Hercon's EPA registration was based, showed that mating disruption experiments achieved levels of 96% or better for medium to low infestations (Beroza et al., 1975; Webb, 1978).

To check the spread of the gypsy moth from northeastern United States, the USDA Animal and Plant Health Inspection Service sprayed Hercon flakes containing disparlure on about 202 ha of an outlying isolated infestation of the gypsy moth in Wisconsin. As of this date, the treated area showed no evidence of any gypsy moths remaining (C. P. Schwalbe, personal communication). The USDA also carried out additional field trials in Maryland and Massachusetts. Competing with a microcapsular formulation, the standard treatment at the time, Hercon's flake proved much more effective (C. P. Schwalbe and R. E. Webb, personal communications). Since this was the first time the Hercon flake was field-tested against the gypsy moth, the formulation was not optimized. The 1980 flake formulation will be greatly improved; it features a greater emission rate with no increase in pheromone content and a uniform particle size. The flake will be sprayed aerially from an improved apparatus to ensure uniform dispersal of individual particles over a broad swath.

At this juncture, there is little doubt that the USDA, given a suitable directive, could halt spread of the gypsy moth by properly using the synthetic pheromone, disparlure. In fact, infestation borders undoubtedly can be rolled back, like in Wisconsin, and thereby spare the United States from the destruction of its hardwood forests by the moth.

The USDA has also used Hercon dispensers containing (+)-disparlure, the optically active isomer of the gypsy moth pheromone, to map outlying and newly infested areas for combatting the spread of the gypsy moth. With (+)-disparlure, monitoring traps are extremely efficient and reliably signal the presence of gypsy moths, even in sparsely infested areas.

Japanese Beetle

Recent tests demonstrated the effectiveness of the Hercon dispenser for mass-trapping Japanese beetles (M. G. Klein and F. W. Michelotti, personal communication), and EPA registration was obtained. As a result, Herculite sold over a million Hercon dispensers of the pheromone to a company that plans to market these traps through hardware stores and garden outlets in 1980.

Spruce Bark Beetle

In a combined effort, Norway and Sweden conducted what was probably the largest and most ambitious pheromone experiment ever for controlling an insect pest, the spruce bark beetle, a killer of the valuable Norwegian spruce tree. After extensive testing of a number of available dispensers and then conducting field trials in 1978 using 145,000 Hercon dispensers of this pest's multicomponent pheromone, Norwegian and Swedish forest authorities set out over 600,000 traps in 1979 with the help of ca. 1% of Norway's population. The project, which cost over $20 million, was adjudged highly successful, and it is believed that trees in more than 1.2 million ha of forests were saved. Herculite received another order for over 700,000 dispensers for the 1980 season as this exciting and highly rewarding project continues.

Western Pine Shoot Borer

In 1979 in northern California, the U. S. Forest Service aerially treated 404 ha of forest using a Hercon formulation of the western pine shoot borer pheromone. Mating disruption was 99.6% and damage reduction was 88% when compared with a control area (G. E. Daterman, personal communication). These results are probably the best in terms of efficacy reported to date. As a result, a major paper company is planning to test Hercon flakes in a forest infested by the western pine shoot borer during 1980. The current high cost of timber is an added incentive to prevent mating of the western pine shoot borer by pheromone treatment and thus reduce the killing and stunting of trees by this pest.

Boll Weevil

The boll weevil, a very destructive pest of cotton in the United States and Mexico, emits an aggregating pheromone; i.e., it attracts both sexes. This has led investigators to explore the use of the pheromone in mass trapping and trap cropping. Results are most encouraging, particularly when treatment is started early in the season thereby preventing a buildup of the weevil population.

Currently, Herculite offers the only commercial system for dispensing aldehyde-containing pheromones; this includes the boll weevil pheromone. In beltwide tests from North Carolina to Mexico, comparing all available dispensers for the boll weevil pheromone, the Hercon dispenser was the most effective and lasted the longest (T. B. Davich and G. H. McKibben, personal communications). The product was field tested for the past 2 years in the United States and South America; EPA registration (recently received by Hercon) will allow it to be sold in the southeastern United States. Marketing of this Hercon dispenser will also continue in Central and South America.

Other Insect Species

In a small 1979 field test in which available dispensers were tested with the pheromone of the navel orangeworm (an aldehyde), Hercon dispensers were effective for at least 6 weeks. The navel orangeworm is responsible for more than $35 million in damage a year to the California almond crop.

Infesting a wide variety of crops (cotton, corn, tobacco, lettuce and tomatoes), the tobacco budworm is said to be the most damaging insect worldwide. Tests last year in Arizona were inconclusive when damaging levels including those in the control groups were not reached because of local climatic factors. Three EPA experimental use permits were filed for 1980 tests; some will include trials on the closely related corn earworm, another polyphagous and highly destructive agricultural pest. The major ingredient of its pheromone is (*Z*)-11-hexadecenal, the same aldehyde that is present in the tobacco budworm pheromone.

Results from spruce budworm trials using its aldehyde pheromone, and the peachtree borer, *Synanthedon exitiosa*, using its acetate pheromone, also are promising. In a 404-ha test in 1979 done by the U. S. Forest Service in a severely infested area (247,000 insects/ha), the Hercon flake gave 83% disruption of mating for the spruce budworm (Daterman et al., unpublished data); a single pheromone application lasted the entire mating season. Although pheromones should not be applied in other than light infestations, the interruption of mating communication was far greater than expected for such a high population level.

Finally, tests of Hercon materials are being carried out with other pest species, such as the grape moth [*Paralobesia viteana* (Clemens)], oriental fruit moth [*Grapholitha molesta* (Busck)], elm bark beetle [*Scolytus multistriatus* (Marsham)], and the southern pine beetle [*Dendroctonus frontalis* Zimmerman].

CONCLUDING REMARKS

We did not give details of 1979 pheromone trials because individual investigators will wish to report these data.

Pheromones should be used according to principles set forth in the literature; for example, they perform best against low population levels, and they cannot be used in all situations. To avoid problems, insect behavior should be studied while under use conditions, and the situation should be monitored closely by making insect counts (larvae and adults), collecting plant damage (e.g., bolls infested), and, if possible in the initial trials of a dispenser, determining lure content of the dispensers periodically. Full knowledge of the insect-plant-pheromone situation will markedly improve results in using pheromones for pest management. Finally, pheromone use is apt to be most successful when used in conjunction with other pest controls; i.e., as part of an integrated pest management program.

REFERENCES

Beroza, M., Paszek, E. C., DeVilbiss, D., Bierl, B. A., and Tardif, J. G. R., 1975, A 3-layer laminated plastic dispenser of disparlure for use in traps for gypsy moth, Environ. Entomol., 7: 712.

Beroza, M., Hood, C. S., Trefrey, D., Leonard, D., Knipling, E. F., and Klassen, W., 1975, Field trials with disparlure in Massachusetts to suppress mating of the gypsy moth, Environ. Entomol., 4: 705.

Henneberry, T. J., Gillespie, J. M., Bariola, L. A., Flint, H. M., Butler, G. D., Jr., Lingren, P. D., and Kydonieus, A. F., 1980, Recent progress with gossyplure for pink bollworm mating disruption, Proceedings of 8th Desert Cotton Insect Symposium, El Centro, CA.

Kydonieus, Agis F., 1977, The effect of some variables on the controlled release of chemicals from polymeric membranes. Proceedings, American Chemical Society Controlled Release Pesticides Symposium, New Orleans, LA.

FORMULATION, TOXICOLOGY AND REGULATION:

FUTURE THRUSTS FOR DEVELOPMENT OF SEMIOCHEMICALS

IN INSECT PEST CONTROL

The panel agreed on the potential importance of semiochemicals in methods of insect control that provide a safe and environmentally acceptable supplement or alternative to conventional insecticides. Behavior-modifying chemicals such as pheromones, provide insectistastis without insecticidal action. They should be distinguished from insecticides, and suitable terminology should be adopted.

The panel agreed that development of semiochemicals must be continued by the cooperative efforts of scientists in industry, government and universities. To maintain the present momentum of research effort and to ensure that successful and rapid exploitation of this technology can be realized, it may be necessary to provide public funding for the support of large-scale field research programs directed against major pest insects. Part of these costs arise because pheromones are expensive: the panel recommends that improved methods of synthesis and manufacture are an important need that must be met if costs are to be reduced.

Controlled-release formulations are necessary if pheromones are to be used effectively and economically. The development of such formulations must be undertaken by industry in consultation with crop-protection specialists. We recommend that appropriate agencies consider providing support where necessary to accelerate the development of important pest control programs by provision of funding for improved pheromone production technology and development of formulations and by developing scientific personnel for field studies to develop appropriate use patterns and efficacy data needed for registration.

Several international committees, e.g. those established by the European Committees, NATO and the World Health Organization are studying regulatory aspects of pheromones and other alternatives to conventional insecticides. The panel recommends that representatives of these committees meet together to ensure coordination of their activities.

Government agencies can decisively encourage the development of semiochemicals for pest management through regulatory processes, statements of policy and funding for research and development. More favorable attitudes toward the regulation of semiochemicals appear to have been adopted by the U. S. Environmental Protection Agency, although the need to ensure that there are no adverse effects on man and the environment still remains an important, well-recognized goal by all involved in developing the use of semiochemicals for pest management.

On the other hand, legislators and research administrators need to recognize that the costs of developing use patterns and of obtaining the efficacy data needed for registration may be far greater for semiochemicals than for conventional pesticides.

PANEL: J. R. Plimmer, USDA; F. Ritter, TNO (The Netherlands); W. Klassen, USDA; I. Weatherston, Conrel; A. Kydoneius, Hercon; M. Dover, EPA; J. H. Tumlinson, USDA.

OVIPOSITION-DETERRING PHEROMONE OF THE EUROPEAN CHERRY FRUIT FLY: STATUS OF RESEARCH AND POTENTIAL APPLICATIONS

E. F. Boller

Fruit Fly Laboratory
Swiss Federal Research Station for
Fruit Growing, Viticulture & Horticulture
CH-8820 Wädenswil, Switzerland

HISTORY

Uniform egg dispersion has been observed in the European cherry fruit fly, *Rhagoletis cerasi* L., by Wiesmann (1937) and Haefliger (1953) who speculated that a chemical substance might be involved in this phenomenon. Based upon knowledge about research in progress on *R. pomonella* (Walsh) in the United States (Prokopy, 1972), we conducted preliminary laboratory studies in summer of 1971 that showed the male-arresting effect of artificial oviposition devices previously exposed to egg-laying females (Boller, Bush and Derron, unpublished).

Our actual research program was initiated in 1974 that included, at that time, studies on the oviposition behavior of *R. cerasi*, with special emphasis on the role of oviposition-deterring pheromones. The findings concerning its perception by the female, its persistence, its effect on males, its collection from and reapplication to fruit and artificial oviposition devices were described by Katsoyannos (1975). Recently, Staedler, Katsoyannos and Boller (unpublished data) confirmed the observation previously made in *R. pomonella* (Prokopy and Spatcher, 1977; Crnjar et al., 1978) that *R. cerasi* also perceives the oviposition-deterring pheromone principally via the D-hairs on the tarsi.

FIRST APPLICATIONS OF THE OVIPOSITION-DETERRING PHEROMONE IN THE FIELD

Having demonstrated the oviposition-deterring effect of the active principle under laboratory conditions, we conducted small-scale experiments in cherry orchards in summer of 1975 to investigate the effectiveness of the pheromone under field conditions. The experiment was conducted in a cherry orchard of the research station at Wädenswil (Katsoyannos and Boller, 1976). Ten ml of a partly purified pheromone solution, collected from oviposition devices in the laboratory, containing pheromone deposited after ca. 200,000 ovipositions were used as raw material. We prepared a 0.2% aqueous solution adding 0.1% of a wetting agent to obtain a more homogeneous covering of the fruit's waxy surface. Prior to use, we had tested possible effects of the wetting agent on oviposition behavior of R. Cerasi with negative results.

Three branches carrying some 200-500 cherries each and located in the south quadrant of the test trees were selected for treatments with the pheromone in 2 middle and 4 late cultivars. In addition to this partial treatment of trees, the pheromone was applied to 2 entire trees of intermediate size and late cherry varieties. The pheromone solution was sprayed to runoff with a 0.5-liter hand atomizer during the stage at which the cherries were susceptible to oviposition. The timing of the 5 treatments applied was made according to the weather conditions as each spray was applied shortly after periods of heavy rainfall. The results of this 1st experiment are shown in Table 1 and show an effectiveness of 77%.

Based upon the encouraging results achieved in 1975, we decided to expand our pheromone collection and to conduct field experiments on whole trees in summer of 1976. The experiment followed patterns similar to those carried out the previous year. The pheromone solution used had been collected in the laboratory by rinsing large quantities of oviposition devices; it was concentrated by lyophilization to a stock solution that showed biological activity in bioassays (Katsoyannos, 1975) up to a dilution of 10^{-5}. The 1st treatment was made June 14, 1976, when oviposition of the wild population had just started and had reached an average level of 5% of cherries having oviposition punctures. Four trees were sprayed with pheromone solution at a concentration of either 0.2 or 0.02%. Two of the 4 replicates of each treatment were sprayed a 2nd time 1 week later. The 1st spray was followed by 13 mm of rain 2 days later; thereafter, there was no more rain until harvest.

The results of this 2nd field application are given in Table 2, and show effectiveness levels ranging from 71 to 90%. It is evident that the infestation of the pheromone-treated trees remained at a very low level despite the high pest population and the very favorable conditions for oviposition. If we consider that the 5% cherries showing oviposition stings before the 1st application of the pheromone led to an approximate infestation of 2-3% at harvest time, we can conclude that in the best treatments the actual infestation may have been pushed below the economic threshold of 4% if we had started our experiment a few days earlier.

Table 1. Comparative results on infestation rates of cherries treated with oviposition-deterring untreated pheromone and cherries at harvest in 1975. Samples from 8 treated trees and corresponding controls. Total number of cherries examined: 980 pheromone-sprayed; 1310 controls.

	Percent	
Treatment[a]	Infestation[b,c]	Effectiveness
Cherries with 1 or more oviposition punctures		
Sprayed (222)	24.3 ± 8.8	63.2
Control (867)	66.1 ± 8.5	
Total number of oviposition punctures		
Sprayed (254)	28.0 ± 10.0	68.8
Control (1205)	89.8 ± 15.1	
Pupae reared from samples examined at harvest		
Sprayed (95)	9.6 ± 3.0	77.6
Control (577)	42.9 ± 12.2	

[a]Data in ( ) = number cherries examined.
[b]Differences were significant at 0.1% level (Student's *t*-test).
[c]Total percentages of SD computed from the individual percentages of each replicate.

* * *

Table 2. Infestation rates of cherries in 1976 treated with oviposition-deterring pheromone solutions and untreated controls at harvest. Two cherry trees/treatment; 400 cherries examined/sample.

Pheromone concentration	No. sprays	Infestation	Effectiveness
0.2%	2	5.3	90.0
0.02%	2	6.8	87.2
0.2%	1	7.8	85.3
0.02%	1	15.3	71.1
Control	0	53.0	-

At this point, we terminated field experimentation in order to save the limited amount of pheromone for chemical identification.

CHEMICAL ASPECTS OF OVIPOSITION-DETERRING PHEROMONE

In 1975 J. Hurter of the Biochemical Department, Wädenswil Research Station, started to cooperate with our team and agreed to assist us in the development of the necessary purification procedures for the crude pheromone solutions we were collecting in the laboratory. A 1st series of such procedures has been published by Hurter et al., 1976. Although further purification showed only slow progress because of other research priorities we recently were able to speed up this type of work by hiring a chemist for a limited period of time and support his chemical work by a relatively efficient bioassay unit. Referring to Hurter's publication, we can summarize the general physical and chemical attributes of the pheromone as follows. The active principal component is characterized by low volatility, high polarity, and a stability that resists boiling in water, 1N hydrochloric acid or 1N sodium hydroxide. It also appears that we are dealing with a multi-component pheromone. The so far unknown chemical class of the active principle will largely determine whether this oviposition-deterring pheromone or analogues thereof will become available as a potential control agent.

POTENTIAL OF OVIPOSITION-DETERRING PHEROMONES IN THE MANAGEMENT OF FRUIT FLIES

Oviposition-deterring pheromones differ in at least one important aspect from volatile sex pheromones used in mass-trapping or mating disruption. Unlike the volatile sex pheromones that leave the crop unprotected, oviposition-deterrents with their low volatility and apparently considerable persistence provide by their application to the crop itself a similar protective umbrella as do most insecticides. These attributes of oviposition-deterring pheromones have an important influence on the design of novel pest control strategies. Unlike mating disruptants that require either a certain isolation of the treated areas or surfaces the size of which largely depend on the dispersal characteristics of the target species, oviposition-deterring pheromones can be used in a fashion similar to the way we now use most of the conventional pesticides. Hence, they could be used without major limitations at the farm level or even at the lowest level of application for the individual backyard tree. Our recently released final report on alternatives to conventional chemical control of the European cherry fruit fly (Boller et al., 1980) has identified a lack of adequate options at the farm level, whereas biotechnical control by means of visual traps in backyard situations and the sterile insect technique (SIT) in area-wide control programs offer efficient and economically interesting solutions at these 2 different levels of application.

The application of oviposition-deterring pheromones in combination with efficient trapping systems that attract and eliminate deterred females during their increased search for suitable oviposition sites might be developed into powerful integrated control programs to be used in urban backyard situations as well as at the farm level. Like sex pheromones, oviposition-deterring pheromones would lend themselves perfectly for combination with other biological or biotechnical control methods and for integration into existing control programs. They could, for example, be used in conjunction with large-scale SIT eradication or suppression programs for the Mediterranean fruit fly, *Ceratitis capitata* (Wiedemann), to protect delicate soft fruit from sterile stings, which is a severe problem in peaches and apricots. Oviposition-deterring pheromones could serve as a last line of defense in control programs that aim at stages in the life cycle of a fruit fly species from pupal formation through mating of the females. Failures to achieve acceptable levels of control during these stages could be compensated for by oviposition deterrents as the last barrier we can provide before damage occurs. With respect to factors which may limit the long term usefulness of the oviposition-deterring pheromone, I refer to the comprehensive review of Prokopy (1980).

REFERENCES

Boller, E. F., Remund, U., and Katsoyannos, B., 1980, Alternativen zur konventionellen chemischen Bekaempfung der Kirschenfliege. Schlussbericht der Forschungs-periode 1962-1979, *Publ. Eidg.* Forschungsanstalt Wädensil, 107 pp.

Crnjar, R. M., Prokopy, R. J., and Dethier, V. G., 1978, Electrophysiological identification of oviposition-deterring pheromone receptors in *Rhagoletis pomonella* (Diptera: Tephritidae). *J. NY Entomol. Soc.*, 86: 283.

Haefliger, E., 1953, Das Auswahlvermoegen der Kirschenfliege bei der Eiablage (Eine statistische Studie). *Mitt. Schweiz. Entomol. Ges.*, 26: 258.

Hurter, J., Katsoyannos, B., Boller, E., and Wirz, P., 1976, Beitrag zur Anreicherung und teilweisen Reinigung des eiablageverhindernden Pheromons der Kirschenfliege, *Rhagoletis cerasi* L., *Z. Ang. Entomol.*, 80: 50.

Katsoyannos, B. I., 1975, Oviposition-deterring, male-arresting, fruit-marking pheromone in *Rhagoletis cerasi* L., *Environ. Entomol.*, 4: 801.

Katsoyannos, B. I., and Boller, E. F., 1976, First field application of oviposition-deterring marking pheromone of European cherry fruit fly, *Environ. Entomol.*, 5: 151.

Prokopy, R. J., 1972, Evidence for a marking pheromone-deterring repeated oviposition in apple maggot flies, *Environ. Entomol.*, 1: 326.

Prokopy, R. J., 1980, Epideictic pheromones influencing spacing patterns of phytophagous insects, _in_: "Semiochemicals: Their Role in Pest Control," D. A. Nordlund, R. L. Jones and W. J. Lewis, eds., John Wiley & Sons, New York.

Prokopy, R. J., and Spatcher, P. J., 1977, Location of receptors for oviposition-deterring pheromone in _Rhagoletis pomonella_, _Ann. Entomol. Soc. Am._, 70: 960.

Wiesmann, R., 1937, Die Orientierung der Kirschenfliege, _Rhagoletis cerasi_ L., bei der Eiablage (Eine sinnesphysiologische Utersuchung). _Landw. Jahrb. Schweiz._, 51: 1080.

ELUCIDATION AND EMPLOYMENT OF SEMIOCHEMICALS IN THE MANIPULATION OF ENTOMOPHAGOUS INSECTS

Donald A. Nordlund

W. J. Lewis

Harry R. Gross, Jr.

Southern Grain Insects Research
Laboratory, AR-SEA, USDA
Tifton, GA 31793 USA

INTRODUCTION

Most of this colloquium has been dedicated to discussions of pheromones--semiochemicals that mediate intraspecific interactions and various approaches for their use in monitoring or controlling insect pests. There are numerous intraspecific interactions, but there are considerably more interspecific interactions that are also mediated by semiochemicals. Semiochemicals that mediate interspecific interactions are known as allelochemics.

Price (1980) has done a superb job of describing the complex web of semiochemical mediated interactions that exist in any particular ecosystem. Because there are so many chemically mediated interactions in an ecosystem and so many semiochemicals floating around in any particular ecosystem, it is difficult to understand how they can be effective. It is obvious that they must generally function with some degree of specificity. Because of their specificity, theoretically at least, semiochemicals are particularly attractive as agents of pest control. If we can just learn enough about these semiochemicals and the interactions they mediate, we should be able to use them to great advantage. To quote Shorey (1977):

> "An insect does not think; it reacts. The reactions are usually triggered by external stimuli and modified by

a host of environmental variables, internal physiologic variables, and some rudimentary learning. The reactions are often highly stereotyped and cause the insect to perform appropriate behaviors that enhance species survival when appropriate stimuli are encountered. Much of the sensory world of the insect involved in stimulation or inhibition of such behaviors as mating, feeding, and egg-laying is chemical. The reactions of the insects to these chemicals are so predictable that if man could learn enough about the attendant behaviors, he could literally make the insects jump through a hoop."

In recent years, with the advances in chemical technology, hundreds of semiochemicals have been identified; and many more are in various stages of the identification process. Identification of a behaviorally active chemical, however, is not an end in itself. To quote Shorey (1977) again:

"It cannot be stressed too strongly that the key to devising efficient systems for the management of insect pests by chemically modifying their behavior is the acquisition of an intimate knowledge of the insects' own normal use of chemicals. This important factor is too often overlooked. Once a pheromone or other behaviorally active chemical is identified, there is a tendency to feel that the research is all over and that the chemical can be used as a bait in traps, or distributed through fields, causing insect control. Rather, the identification of the chemical should open the door to more necessary research to determine whether the normal behavior of the insects can be interfered with and manipulated to our advantage."

Generally, we would all agree that our knowledge of behavior has lagged behind our knowledge of chemistry. This unfortunate situation has developed, we think, because we have had an overly simplistic view of the behavioral responses that occur in most interactions. For example, we originally thought that sex attraction in any particular species resulted from a response to a single chemical. On further examination, this seldom proved to be the case.

The heightened awareness, in the last decade or so, of the involvement of allelochemics in parasitoid-host and predator-prey interactions has taken us from a view that parasitoids and predators must depend on random encounter with hosts or prey to the view that their behavior is regulated by numerous stimuli that enhance the probability of encounter. This realization has stimulated interest in the possibility that entomophagous insects can be manipulated to improve their performance in biological control programs.

Biological control efforts have traditionally relied on 3 basic approaches--importation, conservation, and augmentation. Conservation, of course, is an essential component of the other 2 approaches as well as an approach unto itself. Importation has given us some rather spectacular successes in orchard-type situations, where annual disturbances of the agroecosystems are minimal (van den Bosch and Messenger, 1973). In the annual row-crop situations, however, the story is different; and success with any of the 3 approaches has been less than desired. There are several reasons for this difference, such as the very limited diversity associated with monoculture and the annual disturbances that occur in row-crop agroecosystems. We believe that one of the most important contributing factors may be the rapidly expanding surface area of available host plant material. This phenomenon occurs in all annual row-crop agroecosystems and greatly exaggerates the typical lag phase in the parasitoid-host and predator-prey relationships (Lewis and Nordlund, 1980). This brings us to the need, particularly in annual row-crop agroecosystems, for the augmentation and manipulation of beneficial insects and for further investigations to unravel the involvement of allelochemics in these interactions.

HOST OR PREY SELECTION BEHAVIOR

The host-selection behavior of parasitoids and the prey-selection behavior of predators, with the exception of ambush predators, are generally very similar. For parasitoids, successful parasitization has been divided into host habitat location, host location, host acceptance, host suitability, and host regulation (Salt, 1935; Flanders, 1953; Doutt, 1959; Vinson, 1975). We will concentrate our discussion on habitat location and host or prey location, keeping in mind that semiochemicals play important roles in the other steps as well. Each of these general steps may involve several behaviors and stimuli. We visualize the basic sequences of host-finding activities to be something like the process shown in Figure 1. The stimuli involved may be either chemical or physical, but it is safe to say that, generally, chemical stimuli predominate (Lewis et al., 1976; Vinson, 1976).

Habitat Location

When an adult entomophagous insect emerges, it may or may not find itself in a habitat populated by its host or prey. If it is to survive and reproduce, it must have the capacity to locate a habitat in which its host or prey might be present. As stated by Laing (1937), the movement of an entomophage that is outside the habitat of its host or prey must be either entirely random with respect to that habitat, or be influenced by some environmental character or characters typical of that habitat. Habitat location is a very important first step in the selection of hosts or prey. The host-specificity of parasitoids with many potential hosts is, apparently, largely an effect of host-habitat location which places greater limitation on the number of hosts actually attacked than does the suitability of hosts (Picard and Rabaud, 1914; Flanders, 1962). The same probably also applies to predators. In spite of its importance and the fact that entomophagous insects will not generally search in habitats that do not have the appropriate structural or chemical characteristics even if hosts or prey are present, little effort has been expended in this area.

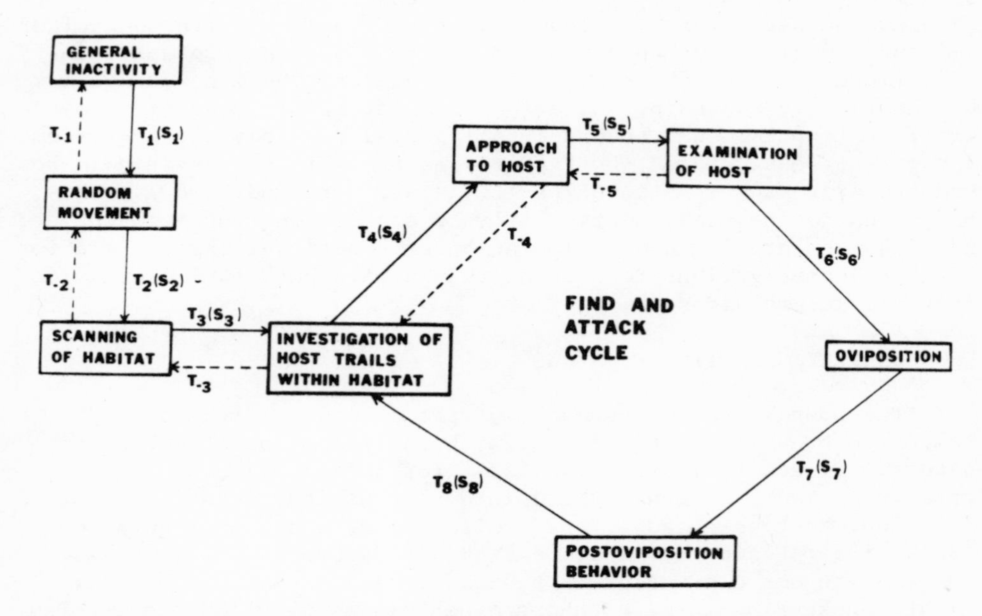

Fig. 1. Basic sequence of host-finding activities of females of parasitic insects: T_1 to T_8 and T_1 to T_4 = transition between the indicated behavioral acts. S_1 to S_6 = stimuli releasing the indicated behavioral patterns. S_2 = olfactory, visual, and physical cues associated with host plants or other characteristics of the habitat. S_3 = primarily chemical cues from frass, moth scale, or other products associated with the presence of host insects. S_4 = olfactory, visual, and other chemical or visual cues from host insect. S_5 and S_6 = olfactory, tactile, auditory, or a combination of these cues from host individual. (Lewis et al., 1976)

Many types of stimuli may influence these behaviors. It is obvious that other factors, such as light intensity and humidity, play a role. Semiochemicals from the food of the host or prey also play a significant role. Picard and Rabaud (1914) were probably the first to suggest that the food plant of the host influenced host habitat selection. When the source of these semiochemicals is a food plant of the host or prey, the chemicals are called synomones.

Camors and Payne (1972) demonstrated that α-pinene alone attracted Heydenia unica Cook and Davis, a parasitoid of Dendroctonus frontalis Zimmerman. Both male and female Diaeretiella rapae M'Intosh are attracted to fresh collard leaves and to very low concentrations of mustard oil (allyl isothiocyanate), which occurs in collard leaves (Read et al., 1970). The coccinellid, Anatis ocellata (L.), a predator of aphids that are found on pine needles, is attracted to chemicals found in pine needles (Kesten, 1969).

Greany et al. (1977) reported that female Biosteres longicaudatus Ashmead, a parasitoid of tephritid fruit fly larvae, are attracted to rotting fruit, a probable location of host larvae. They found that the female parasitoid is attracted by acetaldehyde, ethanol, and acetic acid, all of which are fermentation products produced by the fungus Monilinia fructicola (Wint.) honey. Because the interaction between the fungus and the parasitoid is not understood, we are unable to classify these chemicals more fully than to say that they are allelochemics.

Walker (1940) found that Collyria coxator (Villers) (=Calcitrator) was much more effective in controlling the population of its host, Cephus pygmaeus L., in wheat fields than in barley fields. Microplitis croceipes (Cresson) will search for its host, Heliothis zea (Boddie), in cotton but not generally in corn.

These few examples represent little more than a scratch on the surface of this important area. For a more thorough review, see Vinson (1980). Some very interesting possibilities for using this type of semiochemical in biological control exist. It might be possible, through the use of appropriate plant sprays, the magic of plant breeding, or possibly intercropping, to attract parasitoids or predators into crops in which they would otherwise not search but which may harbor suitable hosts or prey. Plant breeders should be aware of the importance of plant characteristics to the activities of organisms in the 3rd trophic level, otherwise varieties that have a slight resistance to a particular pest may be developed, but increased damage may still occur if the plants do not provide a suitable habitat for parasitoids or predators of pest insects.

HOST OR PREY LOCATION

After locating an appropriate habitat, the entomophage must be able to find suitable hosts or prey. This process is also mediated, generally, by various allelochemics. Those chemicals released by host or prey which assist the entomophage are known as kairomones.

Kairomones are known to play many important roles in the host or prey location behavior of many entomophagous insects. Considerably more (though not nearly enough) is known about this process than is known about the habitat-location process.

Kairomones may actually attract the entomophage. Ullyett (1953), for instance, reported that whenever he broke open a pupae of *Euproctis terminalis* Walker in a forest, a swarm of *Pimpla bicolor* Bouche females would appear. Much work has been done with *Venturis canescens* (Gravenhorst) (Thorpe and Jones, 1937; Williams, 1951) and with *Drino bohemica* Mesnil (Monteith, 1958). These parasitoids preferentially walk up the olfactometer arm having host odor.

In some field-trap studies, parasitoids were attracted to host sex pheromones (Mitchell and Mau, 1971; Sternlight, 1973) or host aggregation pheromones (Rice, 1968, 1969; Lanier et al., 1972; Bedard, 1965; Kennedy, 1979). Numerous authors have reported that various bark beetle predators are attracted to bark beetle aggregation pheromones (Greany and Hagen, 1980 and references therein). Attraction, however, is not the only or necessarily the most important way in which kairomones function.

In many other cases the parasitoids and predators are stimulated into an intensive search behavior by contact with kairomones. McLain (1979), for example, found that 3 species of predatory pentatomids follow trails made by a variety of caterpillars. It would not be surprising to find that these trail substances are also epideictic pheromones for the caterpillars. Kairomones which elicit this type of response have been found in host traces near eggs, host feces, host mandibular and labial gland secretions, and other host products. Rather than list numerous examples, we will discuss a few in detail and refer you to Greany and Hagen (1980) and Weseloh (1980).

For many years now *Trichogramma* spp. have been recognized as having great potential for the biological control of many important lepidopterous pests. Their use in the field, however, in inoculative or inundative releases, has failed to give consistent results.

Laing (1937) demonstrated that *Trichogramma evanescens* Westwood perceived chemicals left by adult moths. Lewis et al. (1971) affirmed that chemicals left by *H. zea* moths increased the rates at which *T. evanescens* found and parasitized *H. zea* eggs. Lewis et al. (1972) reported that moth scales are the source of the kairomones that stimulate this host-seeking response, and Jones et al. (1973) identified tricosane as the most active of several chemicals for *T. evanescens*. Tricosane is not active for all species of *Trichogramma*, however, so in our continuing behavioral studies, the crude moth-scale extract was generally used. Most of this research has been directed at developing techniques for using these kairomones to improve the performance of *Trichogramma* spp. in biological control programs.

Our initial studies were aimed at determining exactly how these kairomones increased the rate of parasitization. First, we were able to show that the primary process by which they function is by releasing and maintaining the appropriate host-seeking response rather than by attracting and serving as a steering mechanism (Lewis et al., 1975a,b). We also found that parasitoids in this host-seeking state distributed their eggs more efficiently among the hosts that were contacted (less superparasitization).

Because the kairomones elicit and maintain host-seeking behavior in the parasitoids, we assumed that a solid spray of the field would be the best approach for using them. Initially, this approach worked well when the host egg density was very high (due to the artificial application of host eggs). But when we began working with lower host egg densities, such as occur naturally in the early season, we found that a solid application could actually reduce the rate of parasitization below that which was obtained in the absence of the kairomone treatment.

Upon closer examination (Lewis et al., 1979; Beevers et al., 1980) of the behavior of *Trichogramma* in response to continuous contact with these kairomones, we found that the response involved increased turning and an intensive examination of a fairly small area. This intensive search response impeded movement from one oviposition site to another, unless the sites were quite close to each other. We then changed the application strategy and attempted to develop some simulated moth scales. These simulated moth scales consisted of diatomaceous earth particles (40/50 mesh) impregnated with the moth-scale extract. Using these simulated moth scales, we found that we could provide the parasitoids with sufficient contact with the kairomones to insure that they remain in a host-seeking state. They are thus retained in the habitat but do not concentrate their search in a way that might impede their movement from one oviposition site to the next. We also demonstrated that the optimum distribution of the kairomone is dependent on host density, that solid treatment is not effective at low host densities, and that simulated moth scales can be used to increase rates of parasitization at intermediate or low densities.

Release of flying beneficial insects for biological control has always been a problem because of their initial escape response. Because of this response, many of the insects released into a target area may never actually function in the target area. Gross et al. (1975) reported that kairomones can be used to stimulate a host-seeking response prior to release of the insect into a target area. This prerelease stimulation eliminates the initial escape response and insures that the parasitoids begin searching for suitable hosts immediately. Prerelease stimulation increased the rates of parasitization by *Trichogramma achaeae* Nagaraja and Nagarkatti, *Trichogramma pretiosum* Riley, and the larval parasitoid *M. croceipes* (which is stimulated into an intensive host-seeking behavior by kairomones found in the frass of H. zea larvae). This approach certainly deserves consideration in any beneficial insect release program.

Nordlund et al. (1976) reported some effects of kairomones on the productivity and longevity of T. pretiosum. They found that, in the laboratory, female T. pretiosum that were constantly exposed to the kairomones and host eggs lives significantly longer and produced more progeny than females exposed only to host eggs. These findings suggest that kairomones may have important uses in rearing programs.

Interactions involving allelochemics can be surprisingly complex. For example, we are now beginning to realize that some kairomones or kairomone precursors can be sequestered in some pest insects when they feed on certain food plants but not on others. Roth et al. (1978) reported that female Lixophaga diatreae (Townsend) are stimulated by frass from its host Diatraea saccharalis (F.) if the host fed on sugarcane but not if it fed on a soybean flour-wheat germ diet. Sauls et al. (1979) and Nordlund and Sauls (unpublished data) found that frass from H. zea larvae that fed on plant material was significantly more stimulatory to female M. croceipes than frass from larvae reared on a laboratory diet, and that frass from larvae reared on different host plant materials differed in stimulatory activity. Frass from larvae reared on corn was not stimulatory at all. Interestingly enough, M. croceipes females are not found to search in corn.

Predaceous Chrysopa carnea Stephens nymphs also have been found to be stimulated into a prey-seeking behavior by kairomones found in the moth scales of H. zea (Lewis et al., 1977). There is an obvious hierarchy of behavioral responses to the chemicals in the scales, but acceptance of H. zea eggs is stimulated by an additional kairomone, probably in the accessory gland secretion, which is associated with egg deposition (Nordlund et al., 1977).

CONCLUSION

We have just scratched the surface of the involvement of semiochemicals in the behavior of entomophagous insects. Little has been said about the complex hierarchy of behaviors that are usually involved in the host or prey selection behavior of most entomophagous insects. Nothing was mentioned about the involvement of epideictic pheromones, which can play a very important role (van Lenteren, 1980). Neither was much said about the fact that many of these semiochemicals may function in several different ways; for example, as sex pheromones, epideictic pheromones, and kairomones. Nevertheless, there are several ways in which we can begin using these chemicals in pest control efforts.

Semiochemicals can be useful in testing beneficial insects that are being considered for importation against important pest species. We should make sure that the beneficial insect will search effectively in the crop system in which it is to be released and that it will attack the target pest in that crop before an extensive investment is made in the project.

Plant breeders and host plant resistance researchers should be made aware of the importance of beneficial insects and the importance of synomones, from the host plant, to their effectiveness. Attempts to produce plant varieties on which the important beneficial insects in that crop will actively search should be encouraged, and we must avoid producing plant varieties on which they will not search.

To manipulate the host-searching behavior of entomophagous insects, we feel that using kairomones which stimulate host seeking generally offer the most promising approach. Properly employed, they can be used to maintain active populations of entomophagous insects in the target area. This can be particularly important early in the season, when naturally occurring host populations are low, and for release programs. By supplying factitious hosts, nonviable hosts, or even small populations of the host species (Parker et al., 1971) in conjunction with the kairomones, we could build up the population of beneficial insects before the pest insect population becomes a problem. Kairomones can also be used to improve the efficiency of insect release programs and of entomophagous insect rearing programs.

We believe that, because of the artificial and unstable nature of the annual row-crop agroecosystem, augmentation and manipulation are necessary for the successful use of entomophagous insects as biological control agents in these systems. This is true even if imported entomophages are used (Lewis and Nordlund, 1980). Thus, a thorough understanding of the involvement of semiochemicals and the behaviors they mediate is essential.

Many more examples could be cited to show how synomones and kairomones, in particular, mediate the host or prey selection behavior of entomophages. However, few can be cited that provide much more than a superficial examination of the interaction. A perusal of the available literature points out one obvious fact--we have just begun. Before we can hope to make entomophagous insects jump through hoops, we have a tremendous amount of research to do. The probability of success is great if we are willing to make the necessary commitment.

REFERENCES

Bedard, W. D., 1965, The biology of *Tomicobia tibialis* (Hymenoptera: Pteromalidae) parasitizing *Ips confusus* (Coleoptera: Scolytidae) in California. Contrib. Boyce Thompson Institute Plant Research, 23: 77.

Beevers, M., Lewis, W. J., Gross, H. R., Jr., and Nordlund, D. A., 1980, Kairomones and their use for management of entomophagous insects: X. Laboratory studies of *Trichogramma pretiosum* Riley with a kairomone extracted from *Heliothis zea* (Boddie) moth scales. *J. Chem. Ecol.*, (In press).

Camors, F. B., and Payne, T. L., 1972, Resistance of *Heydenia unica* (Hymenoptera: Pteromalidae) to *Dendroctonus frontalis* (Coleoptera: Scolytidae) pheromones and a host-tree terpene. *Ann. Entomol. Soc. Am.*, 65: 31.

Doutt, R. L., 1959, The biology of parasitic Hymenoptera, *Ann. Rev. Entomol.*, 4: 161.

Flanders, S. E., 1953, Variation in susceptibility of citrus-infesting coccids to parasitization, *J. Econ. Entomol.*, 46: 266.

Flanders, S. E., 1962, The parasitic Hymenoptera: Specialists in population regulation, *Can. Entomol.*, 94: 1133.

Greany, P. D., Tumlinson, J. H., Chambers, D. L., and Boush, G. M., 1977, Chemically mediated host finding *Biosteres* (*Opius*) *longicaudatus*, a parasitoid of tephritid fruit fly larvae, *J. Chem. Ecol.*, 3: 189.

Greany, P. D., and Hagen, K. S., 1980, Prey selection, *in*: "Semiochemicals: Their Role in Pest Control," D. A. Nordlund, R. L. Jones, and W. J. Lewis, eds., Wiley, New York.

Gross, H. R., Jr., Lewis, W. J., Jones, R. L., and Nordlund, D. A., 1975, Kairomones and their use for management of entomophagous insects: III. Stimulation of *Trichogramma achaeae*, *T. pretiosum*, and *Microplitis croceipes* with host-seeking stimuli at time of release to improve their efficiency, *J. Chem. Ecol.*, 1: 431.

Jones, R. L., Lewis, W. J., Beroza, M., Bierl, B. A., and Sparks, A. N., 1973, Host-seeking stimulants (kairomones) for the egg parasite, *Trichogramma evanescens*, *Environ. Entomol.*, 2: 593.

Kennedy, B. H., 1979, The effect of multilure on parasites of the European elm bark beetle, *Scolytus multistriatus*, *Bull. Entomol. Soc. Am.*, 25: 116.

Kesten, U., 1969, Morphologie und biologie van *Anatis ocellata* (L.), *Z. Angew. Entomol.*, 63: 412.

Laing, J., 1937, Host-finding by insect parasites. I. Observations on the finding of hosts by *Alysia manducator*, *Mormoniella vitripennis*, and *Trichogramma evanescens*, *J. Anim. Ecol.* 6: 298.

Lanier, G. N., Birch, M. C., Schmitz, R. F., and Furniss, M. M., 1972, Pheromones of *Ips pini* (Coleoptera: Scolytidae): Variation in response among three populations. Can. Entomol., 104: 1917.

Lewis, W. J., Sparks, A. N., and Redlinger, L. M., 1971, Moth odor: A method of host-finding by Trichogramma evanescens, J. Econ. Entomol., 64: 557.

Lewis, W. J., Jones, R. L., and Sparks, A. N., 1972, A host-seeking stimulant for the egg parasite, Trichogramma evanescens. Its source and demonstration of its laboratory and field activity, Ann. Entomol. Soc. Am., 65: 1087.

Lewis, W. J., Jones, R. L., Nordlund, D. A., and Sparks, A. N., 1975a, Kairomones and their use for management of entomophagous insects: I. Evaluation for increasing rates of parasitization by Trichogramma spp. in the field, J. Chem. Ecol., 1: 343.

Lewis, W. J., Jones, R. L., Nordlund, D. A., and Gross, H. R., Jr., 1975b, Kairomones and their use for management of entomophagous insects: II. Mechanisms causing increase in rate of parasitization by Trichogramma spp., J. Chem. Ecol., 1: 349.

Lewis, W. J., Jones, R. L., Gross, H. R., Jr., Nordlund, D. A., 1976, The role of kairomones and other behavioral chemicals in host finding by parasitic insects, Behav. Biol., 16: 267.

Lewis, W. J., Nordlund, D. A., Gross, H. R., Jr., Jones, R. L., and Jones, S. L., 1977, Kairomones and their use for management of entomophagous insects: V. Moth scales as a stimulus for predation of Heliothis zea (Boddie) eggs by Chrysopa carnea Stephens larvae, J. Chem. Ecol. 3: 483.

Lewis, W. J., Beevers, M., Nordlund, D. A., Gross, H. R., Jr., and Hagen, K. S., 1979, Kairomones and their use for management of entomophagous insects: IX. Investigations of various kairomone-treatment patterns for Trichogramma spp., J. Chem. Ecol. 5: 673.

Lewis, W. J., and Nordlund, D. A., 1980, The employment of parasitoids and predators for fall armyworm control, Fla. Entomol., (In press).

McLain, K., 1979, Terrestrial trail following by three species of predatory stink bugs, Fla. Entomol., 62: 152.

Mitchell, W. C., and Mau, R. F. L., 1971, Response of the female southern green stink bug and its parasite Trichopoda pennipes, to male stink bug pheromones, J. Econ. Entomol., 64: 856.

Monteith, L. G., 1958, Influence of host and its food plant on host-finding by Drino bohemica Mesn. (Diptera:Tachinidae) and interaction of other factors, Proceedings 10th International Congress of Entomology, 2: 603.

Nordlund, D. A., Lewis, W. J., Jones, R. L., and Gross, H. R., Jr., 1976, Kairomones and their use for management of entomophagous insects: IV. Effect of kairomones on productivity and longevity of Trichogramma pretiosum Riley (Hymenoptera: Trichogrammatidae), J. Chem. Ecol., 2: 67.

Nordlund, D. A., Lewis, W. J., Jones, R. L., Gross, H. R., Jr., and Hagen, K. S., 1977, Kairomones and their use for management of entomophagous insects: VI. An examination of the kairomones for the predator Chrysopa carnea Stephens at the oviposition

sites of Heliothis zea (Boddie), J. Chem. Ecol., 3: 507.
Parker, F. D., Lawson, F. R., and Pinnell, R. E., 1971, Suppression of Pieris rapae utilizing a new control system: Mass releases of both the pest and its parasites, J. Econ. Entomol., 64: 721.
Picard, F., and Rabaub, E., 1914, Sur le parasitisme externe des braconides (Hym.), Bull. Entomol. Soc. France, 1914: 266.
Price, P. W., 1980, Semiochemicals in evolutionary time, in: "Semiochemicals: Their Role In Pest Control," D. A. Nordlund, R. L. Jones, and W. J. Lewis, eds., Wiley, New York.
Read, D. P., Feeny, P. P., and Root, R. B., 1970, Habitat selection by the aphid parasite Diaeretiella rapae (Hymenoptera: Braconidae) and hyperparasite Charips brassicae (Hymenoptera: Cynipidae), Can. Entomol., 102: 1567.
Rice, R. E., 1968, Observations on host selection by Tomicobia tibialis Ashmead (Hymenoptera:Pteromalidae), Contrib. Boyce Thompson Institute, 24: 53.
Rice, R. E., 1969, Response of some predators and parasites of Ips confusues (Lec.) (Coleoptera:Scolytidae) to olfactory attractants, Contrib. Boyce Thompson Institute, 24: 189.
Roth, J. P., King, E. G., and Thompson, A. C., 1978, Host location behavior by the tachinid, Lixophaga diatraeae, Environ. Entomol., 7: 794.
Salt, G., 1935, Experimental studies in insect parasitism. III. Host selection, Proc. Roy. Soc. Ser. B. Biol. Sel. 117: 413.
Sauls, C. E., Nordlund, D. A., Lewis, W. J., 1979, Kairomones and their use for management of entomophagous insects: VIII. Effect of diet on the kairomonal activity of frass from Heliothis zea (Boddie) larvae for Microplitis croceipes (Cresson), J. Chem. Ecol., 5: 363.
Shorey, H. H., 1977, Interactions of insects with their chemical environment, in: "Chemical Control of Insect Behavior," H. H. Shorey and J. J. McKelvey, Jr., eds., Wiley, New York.
Sternlight, M., 1973, Parasitic wasps attracted by the sex pheromone of their coccid hosts, Entomophaga 18: 339.
Thorpe, W. H., and Jones, F. G. W., 1937, Olfactory conditioning and its relation to the problem of host selection, Proc. Roy. Soc. Ser. B., 124: 56.
Ullyett, G. C., 1953, Biomathematics and insect population problems, Entomol. Soc. S. Africa Mem. 2: 1.
van den Bosch, R., and Messenger, P. S., 1973, Biological Control, Intertex Educational Publication, New York.
van Lenteren, J. C., 1980, Host discrimination, in: "Semiochemicals: Their Role in Pest Control," D. A. Nordlund, R. L. Jones and W. J. Lewis, eds., Wiley, New York.
Vinson, S. B., 1975, Biochemical coevolution between parasitoids and their hosts, in: "Evolutionary Strategies of Parasitic Insects and Mites," P. W. Price, ed., Plenum, New York.
Vinson, S. B., 1976, Host selection by insect parasitoids, Ann. Rev. Entomol., 21: 109.

Vinson, S. B., 1980, Habitat location, in: "Semiochemicals: Their Role In Pest Control," D. A. Nordlund, R. L. Jones, and W. J. Lewis, eds., Wiley, New York.

Walker, M. G., 1940, Notes on the distribution of *Cephus pyamaeus* Linn. and its parasite, *Collyria calcitrator* Grav., *Bull. Entomol. Res.*, 30: 551.

Weseloh, R. M., 1980, Host location by parasitoids, in: "Semiochemicals: Their Role In Pest Control," D. A. Nordlund, R. L. Jones, and W. J. Lewis, eds., Wiley, New York.

Williams, J. R., 1951, The factors which promote and influence the oviposition of *Nemeritis canescens* Grav. (Ichneumonidae, Ophioniae), Proc. Roy. Entomol. Soc. London, Ser. A., 26: 49.

OVIPOSITION-DETERRING PHEROMONE SYSTEM OF APPLE MAGGOT FLIES

Ronald J. Prokopy

Department of Entomology
University of Massachusetts
Amherst, Mass. 01003 USA

INTRODUCTION

In a recent review (Prokopy, 1981), I discussed examples from the literature which suggest that in many phytophagous insects, there is a range within which the density of individuals on an exhaustible unit of food resource is optimal, with selection favoring a high degree of individual fitness and full exploitation of the benefits of the resource, but not overcrowding. I proceeded to survey literature indicating that a considerable number of phytophagous insects--spanning at least 6 orders, 16 families, and 33 pest species--release epideictic pheromones which elicit dispersal of arriving conspecifics away from food resources already occupied at densities near or above the upper end of the optimal range. In some species, such pheromones deter conspecifics from landing or feeding, while in others, they deter oviposition. I suggested that use of synthetic epideictic pheromones to disrupt landing, feeding, or oviposition behavior might eventually occupy a place alongside of pheromones which disrupt mating behavior as a selective, effective technique in the management of certain phytophagous insect pests. Such a technique could be especially valuable as a complement to techniques aimed at pest management through mass trapping or disruption of mating behavior.

In this paper, I will restrict my discussion to the realm of oviposition deterring pheromones (ODP's). The first evidence for the existence of ODP's is apparently that of Yoshida (1961), who demonstrated their release by two species of *Callosobruchus* bean weevils. Since then, ODP's have been discovered in at least 24 additional species. These are: apple maggot fly, *Rhagoletis pomonella* (Prokopy, 1972); walnut husk fly, *R. completa* (Cirio,

1972); black cherry fruit fly, R. fausta (Prokopy, 1975b); European cherry fruit fly, R, cerasi (Katsoyannos, 1975); eastern and western cherry fruit flies, R. cingulata and R. indifferens (Prokopy et al., 1976); blueberry maggot fly, R. mendax (Prokopy et al., 1976); two species of dogwood berry flies, R. cornivora and R. tabellaria (Prokopy et al., 1976); rose hip fly, R. basiola (Averill and Prokopy, 1980b); Caribbean fruit fly, Anastrepha suspensa (Prokopy et al., 1977); Mediterranean fruit fly, Ceratitis capitata (Prokopy et al., 1978); an anthomyiid fly, Hylemya spp. (Zimmerman, 1979); Mediterranean flour moth, Ephestia kuehniella (Corbet, 1973); dried currant moth, E. cautella, cocoa moth, E. elutella, and Indian meal moth, Plodia interpunctella (Mudd and Corbet, 1973); the noctuid Hadena bicruris (Brantjes, 1976); corn earworm, Heliothis zea (Gross and Jones, unpub. data); cabbage looper, Trichoplusia ni (Renwick and Radke, 1980); the cabbage butterflies Pieris brassicae and P. rapae (Rothschild and Schoonhoven, 1977); Monarch butterfly, Danaus plexippus (Dixon et al., 1978); and dry bean weevil, Acanthoscelides obtectus (Szentesi, unpub. data).

My particular focus will be on the ODP system of R. pomonella. Female R. pomonella deposit ODP in a trail on the surface of a host fruit during dragging of the extended ovipositor immediately following boring and oviposition into the fruit flesh (one egg is laid per ovipositional bout). The approach taken in our studies and some of the findings outlined here could serve as a useful guide to those initiating ODP studies in other taxa. Brief reference will be made to corresponding knowledge of ODP's of other species.

My overview will consist of current knowledge of the following aspects of the R. pomonella ODP system: chemical identification of ODP; reception of ODP; site of ODP production; dynamics of ODP release, residual activity, and protection of larval food resources from overcrowding; female foraging activities in relation to ODP-marked and unmarked oviposition sites; inter-population and inter-species recognition of ODP; and finally, the influence of trail substance containing ODP on the behavior of R. pomonella males and Opius lectus parasitoid females. No field studies have yet been conducted on application of ODP for management of R. pomonella. The fine work of Boller and colleagues on ODP application for control of R. cerasi in Switzerland will be covered in the next paper.

CHEMICAL IDENTIFICATION OF PHEROMONE

Behavioral Bioassay Method

A key element in accurate identification of pheromonal components is a reliable technique of evaluating the activity of candidate fractions. Our technique is as follows. We collect apples and fruit of the native host, hawthorne (Crataegus spp.), infested by R. pomonella larvae in nature, store the resulting pupae 5-15 months at 3 C until completion of diapause, and maintain the emerging adults at ca. 25°C and 60% RH in laboratory cages provided with water and a 4:1 mixture of sucrose and enzymatic yeast hydrolysate as food (we use this procedure in all laboratory phases of our ODP studies). When mature (ca. 14 days after eclosion), mated females intended for use in bioassays are supplied with uninfested hawthorne fruit (storable at 3°C up to one year) for oviposition ca. 24 hr prior to the start of the assay. On the day of assay, a candidate female is presented with (i.e., allowed to walk onto) first a clean fruit, followed by an ODP-marked fruit, and then a second clean fruit. If the female attempts boring into both of the clean fruits and rejects the ODP-marked fruit, the female is then employed in an assay.

For assay, 0.2 ml of the candidate ODP fraction or control solution (each in 5% methanol) is swabbed onto a 7-mm-diam uninfested hawthorne fruit. Each assay female is offered a succession of six fruits, four of which represent different candidate fractions and two of which represent controls (a standard solution of crude ODP extract concentrated to 125 ovipositor dragging bout equivalents per ml, and water). The fruits are offered in random order and blind fashion, with no fruit other than the single assay fruit, and no flies other than the assay fly, present in the cage. Data on acceptance (a boring attempt) or rejection (leaving fruit without attempting to bore within the standard 10-min assay period) are recorded for each fruit. Following the sixth fruit, the female is offered a clean fruit. If she rejects it, then all data from that female are likewise rejected, on grounds that her physiological state may not have been conducive to oviposition throughout the entire assay. If she accepts the clean fruit, then all data are accepted. A major advantage of this method is that in no case is an assay female allowed to contaminate a test fruit with deposition of ODP following egglaying. Instead, a small piece of paper is placed under the fore-tarsi, and immediately following egg deposition, the female is transferred to another fruit, where she commences and completes ovipositor dragging.

Over the past 12 months (March 1979-March 1980), 102 tests of this sort have been conducted in our laboratory (32 females assayed per test). In 84 of these, there was good consistency among tests in the percentage of females attempting to bore into the controls: the mean acceptance of crude extract ODP controls was 21.2% (s.d. = 7.5), and that of water controls was 73.8% (s.d. = 11.9). However, in the 18 tests conducted during July and August of 1979, the mean acceptance of crude extract ODP controls rose to 39.0% (s.d. = 15.0), while that of water controls fell to 63.0% (s.d. = 12.3). During these two months, there were frequent afternoon storms, accompanied by noticeably "altered" activity of females used in bioassays conducted in periods preceding or during storms, and by reduced discrimination between ODP-marked and unmarked fruits.

This suggests that results of laboratory behavioral bioassays of ODP conducted under conditions of low or fluctuating barometric pressure may not be entirely valid. Lanier and Burns (1978) reported reduced responsiveness of bark beetle adults to aggregation pheromones under changing atmospheric pressure conditions compared with stable pressure conditions.

Pheromonal Components

Initial assays revealed the ODP of *R. pomonella* to be highly soluble in methanol, less though still substantially soluble in water and acetone, and not soluble in methylene chloride, ether, or hexane. A methanol extract of ODP deposited on fruit following ca. 200,000 *R. pomonella* ovipositions has been fractionated by the cooperating chemists (Drs. J. H. Tumlinson and E. Dundulis of Gainesville, Florida) using distillation, chromatographic, and other separation techniques. Behavioral bioassay results to date suggest that the ODP consists of a mixture of a residual principal component(s) and a relatively volatile minor component(s). As yet, we have no indication of the chemical class(es) to which these components belong.

There has been some progress toward the chemical identity of ODP's of other species (Mudd and Corbet, 1973; Oshima et al., 1973) including *R. cerasi* (Hurter et al., 1976), but in no case has chemical identity yet been established.

RECEPTION OF PHEROMONE

Results of initial behavioral tests (Prokopy, 1972) suggested that *R. pomonella* females are unable to perceive ODP prior to direct contact. This suggestion was confirmed in an electrophysiological assay (Roelofs and Prokopy, unpub. data), wherein female antennal sensilla exhibited a strong response to the odor of ammonia (a food-type stimulus) and male sex pheromone (Prokopy, 1975a) but no response whatsoever to the "odor" of ODP, even at high concentration (ca. 300 ovipositor dragging bout equivalents per ml).

We closely examined under magnification the behavior of 100 females after arrival on a host fruit (Prokopy, Cooley, and Weeks, unpub. data). We found that in more than 60% of the observed cases, the head is lowered to where the mouthparts touch the fruit surface during the latter portion of the ca. 3-190 sec (Prokopy, 1972) pre-oviposition searching behavior period. Occasionally (ca. 18% of cases), the head is lowered to such an angle that the antennae also are in contact with the fruit, and occasionally (ca. 20% of cases), the ventral surface of the posterior part of the abdomen (but almost never the ovipositor) contacts the fruit. These values were the same irrespective of whether the fruit was marked with ODP or not. These observations suggested that the tarsi and/or mouthparts may bear the ODP receptors. Through structural ablation (Prokopy and Spatcher, 1977), we obtained strong evidence that ODP receptors are indeed present on the fore-tarsi. Fore-tarsal ablation reduced the level of response to ODP by ca. 50%, while ablation of mid-tarsi, hind-tarsi, mouthparts, antennae, or ovipositor had no detectable effect on level of response to ODP.

Studies on the neurophysiology of ODP reception were initiated in cooperation with Drs. V.G. Dethier, R.M. Crnjar and L.S. Bowdan. The location of chemosensory sensilla on the mouthparts and tarsi was mapped using light and scanning electron microscope techniques (Crnjar et al., 1981). Using a class capillary micro-electrode, we then employed a hair-tip recording technique (Hodgson et al., 1955) for measuring the electrophysiological response of individual sensilla to aqueous and methanol washes of ODP collected from infested fruit. Aqueous and methanol washes from pin-pricked clean fruit served as controls. In each case, the methanol was evaporated off before testing. Recording from every type of tarsal and labellar chemoreceptor, and from more than 400 females, we found that the D hairs on the ventral surface of the 2nd, 3rd, and 4th tarsomeres of the fore-, mid-, and hind-tarsi, as well as the short marginal hairs of the labellum, are highly responsive to the principal component of ODP (Crnjar et al., 1978, 1981). Various other types of chemoreceptors on the tarsi and mouthparts have proven unresponsive, or at most only slightly responsive to this component. The fact that ablation of fore-tarsi repeatably reduces by ca. 50% the degree of behavioral responsiveness to ODP (while ablation of mid- or hind-tarsi results in no such reduction), combined with the facts that D hairs are equally populous on the fore-, mid-, and hind-tarsi and are equally responsive to the principal component of ODP, suggests that discrimination of an ODP message from a non-ODP one involves central neural processes. Recently, we have begun to use this hair-tip recording technique in conjunction with our previously established behavioral technique for assaying the activity of candidate fractions of ODP.

Irrespective of concentration above threshold level, solutions of salt (NaCl) applied to tarsal D hairs elicit electrophysiological spike patterns consistently distinguishable from the principal component of ODP, while solutions of sucrose sometimes elicit patterns distinguishable from this component (Bowdan et al., unpublished data). This knowledge has proven important in interpretation of certain behavioral bioassay data. For example, we have found that 0.2 ml aqueous solutions of 0.1-3.0 molar NaCl and 0.2 ml aqueous solutions of 0.05-0.5 molar sucrose, when swabbed onto entire surfaces of hawthorne fruit, elicit levels of oviposition deterrence (ca. 40 to 80%) comparable to 0.2 ml solutions of crude extract ODP at concentrations of 75-150 ovipositor dragging bout equivalents per ml (Prokopy et al., 1981). The deterrence elicited by salt apparently results from stimulation of salt receptor neurons to fire, which in many insects elicits deterrence to feeding (Dethier, 1976). The deterrence elicited by sucrose apparently results from frequent firing of sugar receptor neurons and the observed propensity of females to engage in lengthy bouts of feeding behavior rather than oviposition type behavior. Clearly, then, electrophysiological assays can be a valuable aid in discriminating between major ODP components and other substances which may deter egglaying in behavioral bioassays of *R. pomonella*.

The pathway of ODP reception has been partially characterized in two other species: *P. brassicae*, wherein ODP is perceived olfactorily from a distance via at least two types of sensilla on each segment of the antennae, and is perceived also upon direct contact by B-hair receptors on the five tarsomeres of the fore-tarsi (Behan and Schoonhoven, 1978); and *R. cerasi*, where as in *R. pomonella*, tarsal D hairs are the principal receptors (Städler et al., 1980).

SITE OF PHEROMONE PRODUCTION

To ascertain the site(s) of production of ODP, component structures of the reproductive and digestive tracts of more than 600 mature females were extirpated, rinsed, sonicated in water, and centrifuged. In multiple series of tests, the supernatant of each extract type was behaviorally and electrophysiologically assayed for ODP activity (Prokopy et al., 1981).

The behavioral assays revealed that extracts of the crop, midgut, malphighian tubules, rectal glands, hindgut, and ovaries elicited significant oviposition deterrence when applied at a rate of 14 female equivalents per fruit, while extracts of the salivary glands, foregut, bursa and vagina, spermathecae, accessory glands, ovipositor, haemolymph, and whole body minus the digestive and reproductive tracts did not elicit significant deterrence. The electrophysiological assays of tarsal D-hair responses confirmed the midgut, malphighian tubules, and hindgut as sites of accumulation of the principal ODP component but did not reveal any of this component in the crop, rectal glands, or ovaries, nor in any of the structures that were inactive in the behavioral bioassays. Females receiving food containing amaranth dye, which is confined to the digestive tract, deposited amaranth-dyed pheromone trails, substantiating that the trail substance emanates from the digestive tract. Additional electrophysiological assays suggested that midgut and hindgut tissues are sites of production of the principal ODP component, that the feces of mature females contain just as large an amount of ODP as the trail substance itself, and that the hindgut and feces of males contain none of the principal ODP component. Together, these findings suggest that the pathway of ODP production and release involves secretion of the principal ODP component from midgut and hindgut tissue into the gut contents, which are released through ovipositor dragging after egglaying as well as through defecation.

We cannot yet explain the anomaly between the behavioral and electrophysiological assay results with respect to activity of rectal gland and ovary extracts. Analysis revealed the concentration of combined NaCl and KCl salts in these structures, and in the trail substance, to be less than 0.01 molar, well below the threshold level of ODP receptor detection (0.1 molar). This suggests that salts accumulated or released by the females are not of sufficient amount, in and of themselves, to elicit oviposition deterrence. Possibly minor components of ODP are produced in the ovaries or rectal glands. The frequency of receptor firing in response to crop extract suggests a concentration of ca. 0.1 molar sucrose therein. Such an amount of sucrose is great enough to have elicited considerable feeding and thus may have been the principal factor accounting for the significantly less oviposition into fruit treated with crop extract.

Because the great bulk of fly food (insect honeydew) and female fecal deposition occurs on host foliage rather than on the fruit, the question now arises as to what influence ODP in feces on foliage might have on female dispersal from populated host trees. Conceivably, the influence may be no more than slight, inasmuch as: (a) the volume of substance released during a bout of fecal deposition is only about one eighth the volume released during a bout of ovipositor dragging, and the total volume of feces excreted per day by an ovipositing female (ten ovipositions) is only ca. one-tenth that of the total volume of oviposition-deterring trail substance; and (b) typically, the surface area of foliage is much greater than the surface area of fruit on a tree.

The presence of ODP in the gut contents and feces raises the possibility that metabolites of the flies' food may be precursors of some components of ODP. The gut-inhabiting symbiotic bacterium *Pseudomonas melophthora* is known to be involved in nutrient metabolism in *R. pomonella* (Allen et al., 1934, Miyazaki et al., 1968). Perhaps such symbiotes may in some way be involved in the process of production of certain ODP components. One example of this sort is production of the pheromone verbenol by bacteria in the gut of the bark beetle *Ips paraconfusus* (Brand et al., 1975).

Knowledge of the physiology of ODP production in other species includes evidence that ODP in *E. kuehniella* is produced in larval mandibular glands (Corbet, 1971), ODP in *P. brassicae* is produced in female accessory glands (Behan and Schoonhoven, 1978), and ODP in *R. cerasi* is contained in the hindgut (Boller, pers. communic.).

DYNAMICS OF PHEROMONE RELEASE, RESIDUAL ACTIVITY, AND PROTECTION OF LARVAL FOOD RESOURCES FROM OVERCROWDING

To study factors affecting release of ODP trail substance, it is first necessary to be able to measure the amount of trail substance deposited after a single ovipositional bout. To accomplish this, we have developed a technique wherein a recently ODP-marked fruit is dusted with dry magnetic toner powder of the sort used in copying machines (Averill and Prokopy, 1980a). This renders the trail substance readily apparent and measurable. Preliminary behavioral bioassays suggest a good positive correlation between the area of trail substance and oviposition-deterring activity.

The average trail is 53 mm long x 0.05 mm wide (area = 2.7 mm^2), and the average duration of a dragging bout is 46 sec. Although the degree of variability in trail area is high even among successive dragging bouts by the same female, we have found through sufficient replication that starved or older (28 day) females deposit significantly less trail substance than fed or younger (14 day) females, and that females deposit just as much trail substance on small (12 mm) or ODP-marked hawthorne fruit as on large (19 mm) or clean fruit (Averill and Prokopy, 1980a). Inasmuch as the gut is the principal or exclusive site of accumulation of the principal ODP component, and inasmuch as food consumption by 28-day females is substantially less than that by 14-day females (Webster et al., 1979), it is not surprising that starved or older flies should release less ODP trail substance.

Under dry conditions at ambient summer temperatures, the ODP has a half life of about seven days (Averill and Prokopy, unpub. data). The fact that ODP is quite soluble in water raises a question (yet to be answered) as to degree of residual activity under varying rainfall or dew conditions. The partial solubility of ODP in acetone would suggest that some ODP component(s) may bind to fruit surface components and therefore be less susceptible to rain washing.

The ODP may need to deter egglaying only for the length of the egg incubation period (3-4 days) to be of selective advantage in giving the first larva a competitive edge over subsequent larvae. If so, then one would expect ODP detection to be accompanied by detection of a developing larva within the fruit. Our laboratory bioassay data suggest that females are unable to detect first instar larvae within fruit, but may be able to detect second and third instar larvae, or their effects (Averill and Prokopy, unpub. data).

Assuming that deposition and recognition of ODP were originally of selective advantage to a female in deterring oviposition during revisitation of a fruit already containing one of her own eggs, then one wonders about the advantage to the female under conditions of prolonged rainfall. Possibly the apparent ecological disadvantage of partial solubility of ODP in water is somewhat compensated by production of water-soluble ODP at low energy cost. If such were true, it would be a distinct advantage, in that the average female deposits ODP after each of the ca. ten eggs per day which she lays over the 20-30 day period of her reproductive lifetime. Of the three studies conducted to date on distribution of R. pomonella eggs among fruits in nature, two (LeRoux and Mukerji, 1963; Cameron and Morrison, 1974) revealed a uniform egg distribution, as might have been mediated by ODP, and one (Reissig and Smith, 1978) a random egg distribution. Greater rainfall during the latter study may have diminished the effectiveness of ODP in mediating uniform egg distribution among available fruits. Sampling procedures (van Lenteren et al., 1978) may also have been a factor.

Rarely is more than one R. pomonella larva able to complete development in a typical-size (12 mm diam) Crataegus spp. fruit, compared with ca. 15 larvae completing development in a 80-mm-diam apple (Averill and Prokopy, unpub. data). Earlier, I hypothesized that the amount of ODP deposited after a single oviposition may be related to the amount of food or space required by the larva to develop into a fit adult (Prokopy, 1972). At present, we are attempting to evaluate this hypothesis quantitatively.

Among other tephritids, it is known that the ODP's of R. fausta, A. suspensa, and C. capitata retain good activity under dry ambient conditions for at least six days (Prokopy, 1975b; Prokopy et al., 1977, 1978), while under similar conditions, the ODP of R. cerasi remains substantially active for at least 12 days (Katsoyannos, 1975). The female-released ODP indicating occupancy of a Melandrium flower by a conspecific egg of the moth H. bicruris lasts only one day; but this suffices, inasmuch as ovipositing females show a clear preference for one day old flowers over those two or more days old (Brantjes, 1976). The ODP deposited on the egg surface by P. brassicae females has a remarkably long residual activity of at least seven weeks, well beyond the ca. two week egg incubation period (Schoonhoven et al., 1980). As yet, no one has related in any species the amount of ODP released after egg-laying to the amount of food or space required for optimal development of the larval progeny.

FEMALE FORAGING ACTIVITIES IN RELATION TO PHEROMONE-MARKED AND UNMARKED OVIPOSITION SITES

The outcome of a female's reaction to an ODP-marked fruit is undoubtedly determined by a variety of mutually interacting factors, including: quantity and quality of fruit stimuli indicative of suitability of oviposition site; quantity and quality of ODP in relation to fruit size; physiological state of the female with respect to number of mature eggs in ovaries and time period elapsed since previous oviposition; and availability and distribution of unmarked and ODP-marked fruit within the habitat.

Fruit shape, size, color, surface and internal structure and condition, and chemistry may all constitute stimuli eliciting *R. pomonella* oviposition (Prokopy, 1977). However, no one has yet characterized the relationship, if any, between quantity or quality of these stimuli and favorability of the fruit for larval development.

The quantity of ODP trail substance deposited may vary with age, food intake, and population origin (wild vs. laboratory) of the females, as pointed out in the previous and following sections. With respect to fruit size, we have preliminary evidence suggesting that degree of deterrence to boring is a function of the frequency with which a female crosses a trail of ODP vs. the frequency with which she encounters unmarked areas of the fruit surface. In some cases, crossing an ODP trail six times during pre-oviposition searching behavior on an isolated 11-mm-diam hawthorne fruit is sufficient to elicit deterrence (Prokopy, unpub. data). Under laboratory conditions, we found that one ODP trail on a 9-, 15-, and 21-mm-diam cherry elicits moderate, slight, and no deterrence, respectively (Prokopy, 1972).

We recently found that *R. pomonella* females require prior experience with ODP before they are able to discriminate between ODP-marked and unmarked host fruit (Roitberg and Prokopy, 1981). The adaptive advantage of this prior experience requirement is unknown. This is apparently the first case in a phytophagous insect wherein pheromone recognition may depend upon prior contact with the pheromone.

The physiological state of experienced females has a strong bearing on degree of responsiveness to ODP, as females deprived of oviposition sites for two days have a much greater tendency to bore into ODP-marked fruit than non-deprived females (Prokopy, 1972).

We have encountered very large differences in reaction to a given amount of ODP by females under natural or semi-natural conditions compared with laboratory conditions. For example, early studies showed that one fresh ODP trail on a 15-mm-diam cherry was sufficient to elicit complete deterrence under natural field conditions but only slight deterrence under laboratory conditions (Prokopy, 1972). More recent studies have confirmed this pattern, showing that a much greater amount of ODP is necessary to elicit the same level of deterrence in the laboratory compared with the field. Even where laboratory-caged females are continuously provided with ample unmarked fruit prior to testing, they still bore into fruit marked with one ODP trail much more often than do females on host trees. We have evidence to suggest that the comparative abundance of neighboring fruits and the comparative frequency of visits to neighboring fruits prior to arrival on the test fruit are important factors contributing to this difference in female reaction. These findings clearly indicate the inadvisability of relating levels of female response to varying concentrations of ODP in the laboratory to the field situation.

To study the fruit-foraging behavior of females, tests were conducted in large outdoor field cages placed over individual host trees ca. 2.5-m-tall, devoid of naturally growing fruit but augmented with varying numbers of pre-picked hawthorne fruit (7 mm diam) hung from the tree branches. We found that individually released (on leaves) *R. pomonella* females exhibited the following sorts of responses (Roitberg et al., 1981). When the tree was devoid of fruit, females departed within ca. nine min (all females assayed in this and other tests had oviposited once within 20 min of testing). When there were 32 unmarked fruits per tree, females remained ca. 87 min, making ca. 89 visits to the fruits and ovipositing in ca. 31% of them. When there were 32 ODP-marked fruits per tree (two ODP trails per fruit 24 hr earlier), females remained on the tree only ca. 14 min, making ca. 31 visits to the fruits, and ovipositing in only ca. 0.4% of them. These data show that females react to a tree having ODP-marked fruit similarly to a tree without fruit, leaving the tree after a comparatively few minutes. We plan to extend this approach beyond intra-tree oviposition site foraging behavior into the realm of inter-tree foraging behavior. To my knowledge, there is no published information on factors influencing female foraging activities in other phytophagous species which release ODP.

INTER-POPULATION AND INTER-SPECIES RECOGNITION OF PHEROMONE

Among those wild populations behaviorally assayed to date, *R. pomonella* from different hosts and different regions have shown complete cross-recognition of one another's ODP's (Prokopy et al., 1976, unpub. data). *R. pomonella* also has shown excellent recognition of the ODP's of its sibling species, *R. mendax* and *R. cornivora*. In behavioral as well as electrophysiological assays, *R. pomonella* has shown partial recognition of *R. basiola* ODP, and essentially no recognition of the ODP's of *R. fausta* and *R. cingulata* (the latter three species do not recognize *R. pomonella* ODP) (Prokopy et al., 1976; Averill and Prokopy, 1980b; Bowdan et al., unpub. data). The latter findings are of considerable ecological interest in view of the fact that within the past 100 years or so, populations of *R. pomonella* and *R. basiola* in Massachusetts have shifted from native hosts onto the introduced host *Rosa rugosa* (Prokopy and Berlocher, 1981). Similarly, populations of *R. pomonella*, *R. fausta*, and *R. cingulata* in Wisconsin have shifted from native hosts onto the introduced host Montmorency sour cherry. On trees of each of these introduced host types, we have observed cases of simultaneous fruit colonization by all respective *Rhagoletis* species. In such cases, any advantage of intra-specific regulation of larval spacing through ODP deposition may be reduced by effects of unregulated inter-specific competition.

The only laboratory population of *R. pomonella* assayed to date was cultured as larvae on apples for ca. 15 generations in Geneva, New York. Females were found to deposit a type or amount of ODP much less active than ODP deposited by lab-caged wild flies (Prokopy et al., 1976). A lab population of *C. capitata* flies cultured on artificial media for more than 200 generations in Honolulu, Hawaii, was almost completely unresponsive to an amount of ODP that effectively deterred lab-caged wild flies (Prokopy et al., 1978). It is uncertain if these differences between lab and wild flies are due to dietary factors, to selection having operated in different directions under different circumstances, or to genetic drift.

In other families, Mudd and Corbet (1973) have evidence suggesting excellent cross-recognition of ODP's among *Ephestia* and *Plodia* flour moths, while Rothschild and Schoonhoven (1977) have data hinting at partial recognition of *P. brassicae* ODP by *P. rapae* females.

INFLUENCE OF PHEROMONAL SUBSTANCE ON BEHAVIOR OF MALES AND PARASITOIDS

Not only does *R. pomonella* trail substance containing ODP function as an oviposition deterrent, but also it functions to arrest the males, apparently providing a signal to males that a female has recently been in the vicinity (Prokopy and Bush, 1972). This enhances the probability of courtship encounters between males and females (Prokopy and Bush, 1973; Smith and Prokopy, 1980). The component(s) of the trail substance which arrests the males affects male behavior for no more than a few hours (Prokopy and Bush, 1972), and hence may not be the principal component(s) which deters female oviposition. The short duration of effect on males renders it unlikely that a male would remain for very long in an area already deserted by females. The ODP-containing trail substances of *R. cerasi* and *C. capitata* likewise have male-arresting effects (Katsoyannos, 1975; Prokopy and Hendrichs, 1979).

R. pomonella trail substance containing ODP acts to retain and elicit antennal tapping of female *Opius lectus* parasitoids on ODP-marked fruit (Prokopy and Webster, 1978). *Opius lectus* females oviposit in the eggs and young larvae of *R. pomonella* in the fruit flesh. Whether or not the trail substance component(s) having this kairomonal effect on *O. lectus* is one of the ODP components is unknown, but the residual activity of the kairomonal effect is similar to that of the oviposition-deterring effect (at least seven days), extending through the initial stages of larval development. The trail substance has no detectable effect on female *O. alloeus* which attack older larvae of *R. pomonella*, some two to three weeks after egg deposition (Prokopy and Webster, 1978).

With respect to other species, Corbet (1971) found that the ODP-containing mandibular gland secretion of *E. kuehniella* elicits ovipositional thrusts by female *Venturia canescens* parasitoids. Components of ODP-containing larval frass of *H. zea* elicit host-searching activity of female *Microplitis croceipes* parasitoids (Lewis et al., 1976), while *H. zea* female scales deposited on and near the egg surface and containing ODP have components which stimulate host searching activity of female *Trichogramma evanescens* and *T. pretiosum* parasitoids, and larvae of the predator *Chrysopa carnea* (Lewis et al., 1972, 1977; Nordlund et al., 1977). The probability is high that other insect-released substances whose only described biological significance to date is that of having kairomonal effects on parasitoids or predators are in fact substances which contain ODP released by the host.

CONCLUSIONS

In this review, I have attempted to outline our current understanding of the oviposition-deterring pheromone system of *R. pomonella*. It is hoped that the approach taken in studying this system, and some of the experimental results will be of assistance to those investigating ODP systems of other species.

Certain findings in *R. pomonella* suggest potential impediments in future studies on this and other taxa. First, the negative influence which uncontrollable atmospheric conditions during periods of stormy weather appears to have on *R. pomonella* female ability to discriminate behaviorally between ODP-marked and unmarked fruits in laboratory bioassays presents a problem of repeatability of behavioral bioassay results. Second, the fact that salt and sucrose can elicit real or apparent deterrence of *R. pomonella* oviposition when swabbed onto fruit necessitates use of electrophysiological assays in conjunction with behavioral bioassays to ensure against misinterpretation of the nature of an oviposition-deterring substance. Third, the much greater behavioral sensitivity of wild *R. pomonella* females in nature compared with wild females in lab cages to the same concentration of ODP indicates that levels of behavioral response obtained in lab cages may not be applicable to nature. Fourth, the failure of lab-cultured *R. pomonella* females to release a quantity or quality of ODP equal to that of wild females raises a question as to the wisdom of employing lab-cultured individuals in ODP studies.

Ultimately, the success with which synthetic ODP's can be employed in future integrated management programs for control of R. pomonella and other species will depend on a thorough knowledge of the oviposition-site foraging activities of the females. Studies of female foraging activity in relation to ODP-marked and unmarked oviposition sites offer an excellent opportunity to evaluate existing models of optimum foraging theory of animals (Pyke et al.,1977; Hassel and Southwood, 1978; Hubbard and Cook, 1978), and to develop new models. There exist very few published studies among phytophagous insects of empirically derived optimum foraging models dealing with reproductive success.

It is doubtful that synthetic ODP's alone would lead to truly effective long-term management of R. pomonella or any other pest in a large agro-ecosystem. In the absence of acceptable untreated egglaying sites, the threshold level for oviposition at treated sites might become so low that deterred females would, over time, eventually oviposit in spite of ODP. Problems of physiological habituation or adaptation to large-scale applications of ODP, and possible selection for reduced responsiveness to ODP within the population, might also arise. For long-term success, therefore, application of ODP will in most cases probably require integration with use of attractive olfactory-visual stimuli to capture out deterred females. In the case of R. pomonella, use of red spheres (visual mimics of oviposition sites, Prokopy, 1968, 1975c) baited with synthetic fruit volatiles (Prokopy et al., 1973; Fein et al., 1978) and synthetic male sex pheromone (Prokopy, 1975a; Roelofs et al., unpub. data), all of which are attractive to the females, holds much promise as an effective trap. The design of optimum patterns of ODP and attractant trap station distribution for achieving management will ultimately depend on the success with which the foraging behavior of females can be accurately modeled.

ACKNOWLEDGMENTS

Portions of the research on R. pomonella reported here were supported by the Rockefeller Foundation, the Science and Education Administration of the U.S. Department of Agriculture under Grant No. 7800168 from the Competitive Grants Research Office and Cooperative Agreement No. 12-14-1001-1205, and Massachusetts Agricultural Experiment Station Project 380. Appreciation is expressed to the following persons for their constructive criticisms of an earlier draft of this manuscript; Ms. A. L. Averill, Mr. B. D. Roitberg, and Drs. E. S. Bowdan, J. S. Elkinton, and P. D. Greany.

REFERENCES

Allen, T. C., Pinckard, J. A., and Riker, A. J., 1934, Frequent association of Phytomonas melophthora with various stages in the life cycle of the apple maggot, Rhagoletis pomonella, Phytopathology 24:228-238.

Averill, A. L., and Prokopy, R. J., 1980a, Release of oviposition deterring pheromone by apple maggot flies, J. N. Y. Entomol. Soc. (In Press).

Averill, A. L., and Prokopy, R. J., 1980b, Oviposition deterring pheromone in Rhagoletis basiola, J. Fla. Entomol. Soc. (In Press)

Behan, M., and Schoonhoven, L. M., 1978, Chemoreception of an oviposition deterrent associated with eggs in Pieris brassicae, Entomol. Exp. App. 24:163-179.

Brand, J. M., Bracke, J. W., Markovitz, A. J., Wood, D. L. and Brown, L. E., 1975, Production of verbenol pheromone by a bacterium isolated from bark beetles, Nature 254:136-137.

Brantjes, N. B. M., 1976, Prevention of superparasitation of Melandrium flowers by Hadena, Oecologia 24:1-6.

Cameron, P. J., and Morrison, F. O., 1974, Sampling methods for estimating the abundance and distribution of all life stages of the apple maggot, Rhagoletis pomonella, Can. Entomol. 106: 1025-1034.

Cirio, U., 1972, Osservazioni sul comportamento di ovideposizione della Rhagoletis completa in laboratario, Proc. 9th Congr. Italian Entomol. Soc. (Siena):99-117.

Corbet, S. A., 1971, Mandibular gland secretion of larvae of the flour moth, Anagasta kuehniella, contains an epideictic pheromone and elicits oviposition movements in a Hymenopteran parasite, Nature 232:481-484.

Corbet, S. A., 1973, Oviposition pheromone in larval mandibular glands of Ephestia kuehniella, Nature 243:537-538.

Crnjar, R. M., Prokopy, R. J., and Dethier, V. G., 1978, Electrophysiological identification of oviposition-deterring pheromone receptors in Rhagoletis pomonella, J. N. Y. Entomol. Soc. 86: 283-284.

Crnjar, R. M., Dethier, V. G., and Prokopy, R. J., 1981, Morphology and electrophysiological mapping of oviposition-deterring pheromone receptors in Rhagoletis pomonella, J. Insect Physiol. (Submitted).

Dethier, V. G., 1976, The Hungry Fly, Harvard Univ. Press, Cambridge, Mass.

Dixon, C. A., Erickson, J. M., Kellert, D. N., and Rothschild, M., 1978, Some adaptations between Danaus plexippus and its food plant, with notes on Danaus chrysippus and Euploea core, J. Zool. Lond. 185:437-467.

Fein, B. L., Reissig, W. H., and Roelofs, W. L., 1978, Attraction of apple maggot Rhagoletis pomonella females to apple volatiles in wind tunnel bioassays, J. N. Y. Entomol. Soc. 86:286.

Hassell, M. P., and Southwood, T. R., 1978, Foraging strategies of insects, Ann. Rev. Ecol. Syst. 9:75-98.

Hodgson, H. S., Lettvin, J. Y., and Roeder, K. D., 1955, Physiology of a primary chemoreceptor unit, Science 122:417-418.

Hubbard, S. F., and Cook, R. M., 1978, Optimal foraging by parasitoid wasps, J. Anim. Ecol. 47:593-604.

Hurter, J., Katsoyannos, B., Boller, E. F., and Wirz, P., 1976, Beitrag zur Anreicherung und teilweisen Reinigung des eiablageverhindernden Pheromons der Kirschenfliege, Rhagoletis cerasi L. (Dipt., Trypetidae), Z. Angew. Entomol. 80:50-56.
Katsoyannos, B. I., 1975, Oviposition-deterring male-arresting fruit marking pheromone in Rhagoletis cerasi, Environ, Entomol. 4:801-807.
Lanier, G. N., and Burns, B. W., 1978, Barometric flux: Effects on the responsiveness of bark beetles to aggregation attractants, J. Chem. Ecol. 4:139-147.
LeRoux, E. J., and Mukerji, M. K., 1963, Notes on the distribution of immature stages of the apple maggot, Rhagoletis pomonella, on apples in Quebec. Ann. Entomol. Soc. Quebec 8:60-70.
Lewis, W. J., Jones, R. L., and Sparks, A. N., 1972, A host-seeking stimulant for the egg parasite Trichogramma evanescens: Its source and a demonstration of its laboratory and field activity. Ann. Entomol. Soc. Amer. 65:1087-1089.
Lewis, W. J., Jones, R. L., Gross, H. R., and Nordlund, D. A., 1976, The role of kairomones and other behavioral chemicals in host finding by parasitic insects, Behav. Biol. 16:267-289.
Lewis, W. J., Nordlund, D. A., Gross, H. R., Jones, R. L., and Jones, S. L., 1977, Kairomones and their use for management of entomophagous insects. V. Moth scales as a stimulus for predation of Heliothis zea eggs by Chrysopa carnea larvae. J. Chem. Ecol. 3:483-487.
Miyazaki, S., Boush, G. M., and Baerwald, R. J., 1968, Amino acid synthesis by Pseudomonas melophthora, bacterial symbiote of Rhagoletis pomonella, J. Insect Physiol. 14:513-518.
Mudd, A., and Corbet, S. A., 1973, Mandibular gland secretion of larvae of the stored products pests Anagasta kuehniella, Ephestia cautella Plodia interpunctella, and Ephestia elutella, Entomol. Exp. Appl. 16:291-293.
Nordlund, D. A., Lewis, W. J., Todd, J. W., and Chalfant, R. B., 1977, Kairomones and their use for management of entomophagous insects. VII. The involvement of various stimuli in the differential response of Trichogramma pretiosum to two suitable hosts, J. Chem. Ecol. 3:513-518.
Oshima, K., Honda, H., and Yamamoto, I., 1973, Isolation of an oviposition marker from Azuki bean weevil, Callosobruchus chinensis (L.) Agric. Biol. Chem. 37:2680.
Prokopy, R. J., 1968, Visual responses of apple maggot flies, Rhagoletis pomonella: Orchard studies, Entomol. Exp. Appl. 11:403-422.
Prokopy, R. J., 1972, Evidence for a marking pheromone deterring repeated oviposition in apple maggot flies, Environ. Entomol. 1:326-332.
Prokopy, R. J., 1975a, Mating behavior in Rhagoletis pomonella. V. Virgin female attraction to male odor, Can. Entomol. 107:905-908.
Prokopy, R. J., 1975b, Oviposition-deterring fruit marking pheromone in Rhagoletis fausta, Environ. Entomol. 4:298-300.

Prokopy, R. J., 1975c, Apple maggot control by sticky spheres, J. Econ. Entomol. 68:197-198.

Prokopy, R.J., 1977, Stimuli influencing trophic relations in Tephritidae, Colloq. Intern. Cent. Nat. Res. Sci. (France)265:305-336.

Prokopy, R. J., 1981, Epideictic pheromones influencing spacing patterns of phytophagous insects, in"Semiochemicals: Their Role in Pest Control," (D. A. Nordlund, R. L. Jones, and W. J. Lewis, ed.) Wiley Press, New York (In Press).

Prokopy, R. J., and Berlocher, S. H., 1981, Establishment of Rhagoletis pomonella on rose hips in New England. Can. Entomol. (In press).

Prokopy, R. J., and Bush, G. L. 1972, Mating behavior in Rhagoletis pomonella. III. Male aggregation in response to an arrestant. Can. Entomol. 104:275-283.

Prokopy, R. J., and Bush, G. L., 1973, Mating behavior in Rhagoletis pomonella, IV. Courtship, Can. Entomol. 105:873-891.

Prokopy, R. J., and Hendrichs, J., 1979, Mating behavior of Ceratitis capitata on a field-caged host tree, Ann. Entomol. Soc. Am. 72:642-648.

Prokopy, R. J., and Spatcher, P. J., 1977, Location of receptors for oviposition deterring pheromone in Rhagoletis pomonella, Ann. Entomol. Soc. Am. 70:960-962.

Prokopy, R. J., and Webster, R. P. 1978, Oviposition-deterring pheromone in Rhagoletis pomonella: A kairomone for its parasitoid Opius lectus, J. Chem. Ecol. 4:481-494.

Prokopy, R. J., Moericke, V., and Bush, G. L., 1973, Attraction of apple maggot flies to odor of apples, Environ. Entomol. 2:743-749.

Prokopy, R. J., Reissig, W. H., and Moericke, V., 1976, Marking pheromones deterring repeated oviposition in Rhagoletis flies, Entomol. Exp. Appl. 20:170-178.

Prokopy, R. J., Greany, P. D., and Chambers, D. L., 1977, Oviposition-deterring pheromone in Anastrepha suspensa, Environ. Entomol. 6:463-465.

Prokopy, R. J., Ziegler, J. R., and Wong, T. T. Y., 1978, Deterrence of repeated oviposition by fruit marking pheromone in Ceratitis capitata, J. Chem. Ecol. 4:55-63.

Prokopy, R. J., Averill, A. L., Bardinelli, C. M., Bowdan, E. S., Cooley, S. S., Crnjar, R. M., Spatcher, P. J., and Weeks, B. L., 1981, Site of oviposition deterring pheromone production in Rhagoletis pomonella, J. Insect Physiol. (Submitted).

Pyke, G. J., Pulliam, H. R., and Charnov, E. L., 1977, Optimal foraging theory: A selective review of theory and tests, Quart. Rev. Biol. 52:137-154.

Reissig, W. H., and Smith, D. C., 1978, Bionomics of Rhagoletis pomonella in Crataegus, Ann. Entomol. Soc. Am. 71:155-159.

Renwick, J. A. A. and Radke, C. D. 1980. An oviposition deterrent associated with frass from feeding larvae of the cabbage looper, Trichoplusia ni. Environ. Entomol. 9:318-320.

Roitberg, B. D., and Prokopy, R. J. 1981, Experience required for pheromone recognition by the apple maggot fly, Nature (In Press).

Roitberg, B. D., van Lenteren, J. C., van Alphen, J. J., Jallis, F., and Prokopy, R. J., 1981, Foraging behavior of apple maggot flies: A parasite of hawthorne fruit. J. Anim. Ecol. (Submitted).

Rothschild, M., and Schoonhoven, L. M., 1977, Assessment of egg load by Pieris brassicae (Lepidoptera: Pieridae), Nature 226:352-355.

Schoonhoven, L. M., Sparnay, T., van Wissen, W., and Meerman, J., 1980, Seven weeks persistence of an oviposition deterring pheromone, J. Chem. Ecol. (In Press).

Smith, D. C., and Prokopy, R. J., 1980, Mating behavior in Rhagoletis pomonella, VI. Site of early season encounters, Can. Entomol. 112:585-590.

Städler, E., Katsoyannos, B. I., and Boller, E. F., 1980, Das Makierungspheromon der Kirschenfliege: Erste Electrophysiologische Untersuchung der Tarsalen Sinnesorgan der Weibchen. Mitt. Schweiz. Entomol. Ges. (In Press).

van Lenteren, J. C., Bakker, K., and van Alphen, J. J. M., 1978, How to analyze host discrimination, Ecol. Entomol. 3:71-75.

Webster, R. P., Stoffolano, J. G., and Prokopy, R. J., 1979, Long-term intake of protein and sucrose in relation to reproductive behavior of wild and laboratory cultured Rhagoletis pomonella. Ann. Entomol. Soc. Am. 72:41-46.

Yoshida, T., 1961, Oviposition behaviors of two species of bean weevils and interspecific competition between them, Mem. Fac. Lib. Arts Educ., Miyazaki Univ. 11:41-65.

Zimmerman, M., 1979, Oviposition behavior and the existence of an oviposition deterring pheromone in Hylemya, Environ. Entomol. 8:277-279.

A GLOSSARY OF TERMS USED TO DESCRIBE CHEMICALS THAT MEDIATE INTRA- AND INTERSPECIFIC INTERACTIONS

Donald A. Nordlund

Southern Grain Insects Research
Laboratory, AR-SEA, USDA
Tifton, GA 31793 USA

Allelochemic - A substance significant to organisms of a species different from its source, for reasons other than food as such (Whittaker, 1970).

Allomone - A substance produced or acquired by an organism which, when it contacts an individual of another species in the natural context, evokes in the receiver a behavioral or physiological reaction adaptively favorable to the emitter but not to the receiver (Brown, 1968).

Apneumone - A substance emitted by a nonliving material that evokes a behavioral or physiological reaction adaptively favorable to a receiving organism, but detrimental to an organism of another species, which may be found in or on the nonliving material (Nordlund and Lewis, 1976).

Arrestant - A chemical that causes organisms to aggregate in contact with it, the mechanism of aggregation being kinetic or having a kinetic component. An arrestant may slow the linear progression of the organism by reducing actual speed of locomotion, or by increasing the turning rate (Dethier et al., 1960).

Attractant - A chemical that causes an organism to make oriented movements toward its source (Dethier et al., 1960).

Deterrent - A chemical which inhibits feeding, mating, or oviposition when in a place where an organism would, in its absence, feed, mate, or oviposit (Dethier et al., 1960).

Epideictic Pheromone - A pheromone which elicits dispersal from presently or potentially overcrowded resources (e.g., food space, and refugia), and thereby acts to partition intra-specific foraging activities (Prokopy, 1980).

Hormone - A chemical that is produced by tissue or endocrine gland and that controls various physiological processes within an organism (Starling, 1905).

Kairomone - A substance produced or acquired by an organism which, when it contacts an individual of another species in the natural context, evokes in the receiver a behavioral or physiological reaction adaptively favorable to the receiver but not to the emitter (Brown et al., 1970).

Pheromone - A substance that is secreted by an organism to the outside that causes a specific reaction in a receiving organism of the same species (Karlson and Butenandt, 1959; Karlson and Luscher, 1959).

Repellent - A chemical which causes an organism to make oriented movements away from its source (Dethier et al., 1960).

Semiochemical - A chemical involved in the chemical interaction between organisms (Law and Regnier, 1971).

Stimulant - (a) Locomotor Stimulant - a chemical that causes, by a kinetic mechanism, organisms to disperse from a region more rapidly than if the area did not contain the chemical. (b) Feeding, Mating, or Ovipositional Stimulant - a chemical that elicits feeding, mating, or oviposition in an organism (feeding stimulant is synonymous with the term "Phagostimulant," coined by Thorsteinson [1953]) (Dethier et al., 1960).

Synomone - A substance produced or acquired by an organism which, when it contacts an individual of another species in the natural context, evokes in the receiver a behavioral or physiological reaction adaptively favorable to both emitter and receiver (Nordlund and Lewis, 1976).

REFERENCES

Brown, W. L., Jr., 1968, An hypothesis concerning the function of the metapleural glands in ants, *Am. Nat.*, 102: 188.

Brown, W. L., Jr., Eisner, T., and Whittaker, R. H., 1970, Allomones and kairomones: Transpecific chemical messengers, *BioScience*, 20: 21.

Dethier, V. G., Browne, L. Barton, and Smith, C. W., 1960, The designation of chemicals in terms of the responses they elicit from insects, *J. Econ. Entomol.*, 53: 134.

Karlson, P., and Butenandt, A., 1959, Pheromones (ectohormones) in insects, Ann. Rev. Entomol., 4: 39.
Karlson, P., and Luscher, M., 1959, "Pheromones" a new term for a class of biologically active substances, Nature, 183: 155.
Law, J. H., and Regnier, F. E., 1971, Pheromones, Ann. Rev. Biochem., 40: 533.
Nordlund, D. A., and Lewis, W. J., 1976, Terminology of chemical releasing stimuli in intraspecific and interspecific interactions, J. Chem. Ecol., 2: 211.
Prokopy, R. J., Epideictic pheromones influencing spacing patterns of phytophagous insects, in: "Semiochemicals: Their Role in Pest Control," D. A. Nordlund, R. L. Jones, and W. J. Lewis, eds., Wylie, New York (1980).
Starling, E. H., 1905, The chemical correlation on the functions of the body, Lancet, 2: 339.
Thorsteinson, A. J., 1953, The chemotactic responses that determine host specificity in an oligophagous insect (Plutella maculipennis (Curt.) Lepidoptera), Can. J. Zool., 31: 52.
Whittaker, R. H., The biochemical ecology of higher plants, in: "Chemical Ecology," E. Sondheimer and J. B. Simeone, eds., Academic Press, New York (1970).

List of Contributors

M. R. Alder - ICI Ltd, Plant Protection Division, Jealotts Hill, Research Station, Bracknell, Berkshire RG12 6EY, England

G. S. Arida - International Rice Research Institute, P. O. Box 933, Manila, Philippines

Heinrich Arn - Swiss Federal Research Station for Horticulture and Viticulture, CH-8820 Wädenswil, Switzerland

Alf Bakke - Norwegian Forest Research Institute, N-1432 AS-NLH, Norway

Louis A. Bariola - Western Cotton Research Laboratory, AR-SEA, USDA, Phoenix, Arizona, U.S.A.

William D. Bedard - Pacific Southwest Forest and Range Experiment Station, Forest Service, USDA, Berkeley, California 94701, U.S.A.

P. S. Beevor, - Tropical Products Institute, 56-62 Gray's Inn Road, London WCIX 8LU, England

M. H. Benn - Department of Chemistry, The University of Calgary, Calgary, Alberta T2N 1N4, Canada

Morton Beroza - Hercon Group, Herculite Products, Inc., 1107 Broadway, New York, New York 10010, U.S.A.

Barbara A. Bierl-Leonhardt - Agricultural Environmental Quality Institute, AR-SEA, USDA, Beltsville, Maryland 20705, U.S.A.

E. F. Boller - Fruit Fly Laboratory, Swiss Federal Research Station for Fruit Growing, Viticulture and Horticulture, CH-8820 Wädenswil, Switzerland

John H. Borden - Pestology Centre, Department of Biological Sciences, Simon Fraser University, Burnaby, B. C., Canada V5A 1S6

Thomas W. Brooks - Albany International Controlled Release Division, 110 A Street, Needham Heights, Massachusetts 02194, U.S.A.

W. E. Burkholder - Stored Product and Household Insects Laboratory, AR-SEA, USDA, Department of Entomology, University of Wisconsin, Madison, Wisconsin 53706, U.S.A.

D. G. Campion - Centre for Overseas Pest Research, College House, Wrights Lane, London W8 5SJ, England

Ring T. Cardé - Department of Entomology and Pesticide Research Center, Michigan State University, East Lansing, Michigan 48824, U.S.A.

J. A. Coffelt - Insect Attractants, Behavior, and Basic Biology Research Laboratory, AR-SEA, USDA, P. O. Box 14565, Gainesville, Florida 32604, U.S.A.

John H. Cross - Insect Attractants, Behavior, and Basic Biology Research Laboratory, AR-SEA, USDA, P. O. Box 14565, Gainesville, Florida 32604, U.S.A.

Roy T. Cunningham - Tropical Fruit and Vegetable Research Laboratory, AR-SEA, USDA, Honolulu, Hawaii 96804, U.S.A.

Roy A. Cuthbert - USDA Forest Service, Delaware, Ohio 43015, U.S.A.

G. E. Daterman - Pacific Northwest Forest and Range Experiment Station, USDA, Forest Service, Corvallis, Oregon 97331, U.S.A.

Charles C. Doane - Albany International Controlled Release Division, P. O. Box 537, Buckeye, Arizona 85326, U.S.A.

Michael J. Dover - Benefits and Field Studies Division, Office of Pesticide Programs, U. S. Environmental Protection Agency, Washington, D. C. 20460, U.S.A.

V. A. Dyck - International Rice Research Institute, P. O. Box 933, Manila, Philippines

Joseph S. Elkinton - Department of Entomology, University of Massachusetts, Amherst, Massachusetts 01003, U.S.A.

Hollis M. Flint - Western Cotton Research Laboratory, AR-SEA, USDA, Phoenix, Arizona 85040, U.S.A.

Janice M. Gillespie - Herculite Products, Inc., 2321 North 27th St., Phoenix, Arizona 85006, U.S.A.

M. A. Golub - Albany International Controlled Release Division. 110 A Street, Needham Heights, Massachusetts 02194, U.S.A.

Harry R. Gross, Jr. - Southern Grain Insects Research Laboratory, AR-SEA, USDA, Tifton, Georgia 31793, U.S.A.

D. R. Hall - Tropical Products Institute, 56-62 Gray's Inn Road, London WCIX 8LU, England

D. D. Hardee - Pest Management Specialists, Inc., P. O. Box 364, Starkville, Mississippi 39759, U.S.A.

A. W. Hartstack - Cotton Pest Control Equipment Research, AR-SEA, USDA, College Station, Texas 77843, U.S.A.

Thomas J. Henneberry - Western Cotton Research Laboratory, AR-SEA, USDA, Phoenix, Arizona 85040, U.S.A.

Y. Y. Huang - Department of Chemistry, The University of Calgary, Calgary, Alberta T2N 1N4, Canada

P. Hunter-Jones - Centre for Overseas Pest Research, College House, Wrights Lane, London W8 5SJ, England

Claude Jaccard - Station Fédérale de Recherches, Agronomiques de Changins, CH-1260 Nyon, Switzerland

Hiroo Kanno - Laboratory of Applied Entomology, Hokuriku National Agricultural Experiment Station, Joetsu, Niigata 943-01, Japan

John W. Kennedy - John W. Kennedy Consultants, Inc., 608 Washington Boulevard, Laurel, Maryland 20810, U.S.A.

Michael G. Klein - Japanese Beetle and Horticultural Insect Pests Research Laboratory, AR-SEA, USDA, Wooster, Ohio 44691, U.S.A.

E. F. Knipling - Pest Management, AR-SEA, USDA, Beltsville, Maryland 20705, U.S.A.

Agis F. Kydonieus - Hercon Group, Herculite Products, Inc., 1107 Broadway, New York, New York 10010, U.S.A.

Gerald N. Lanier - State University College of Environmental Science and Forestry, Syracuse, New York 13210, U.S.A.

J. E. Leggett - Cotton Insects Research Laboratory, AR-SEA, USDA Florence, South Carolina 29502, U.S.A.

R. Lester - Tropical Products Institute, 56-62 Gray's Inn Road, London WCIX 8LU, England

W. J. Lewis - Southern Grain Insects Research Laboratory, AR-SEA, USDA, Tifton, Georgia 31793, U.S.A.

Reidar Lie - Borregaard Industries, Ltd., N-1701 Sarpsborg, Norway

Pete D. Lingren - Western Cotton Research Laboratory, AR-SEA, USDA, Phoenix, Arizona 85040, U.S.A.

E. P. Lloyd - Boll Weevil Eradication Research Laboratory, AR-SEA, USDA, Raleigh, North Carolina 27607, U.S.A.

D. F. Lockwood - North Carolina Agricultural Research Service, North Carolina State University, Raleigh, North Carolina 27650, U.S.A.

Harold F. Madsen - Agriculture Canada Research Station, Summerland, British Columbia, VOH 1Z0, Canada

R. W. Mankin - Insect Attractants, Behavior, and Basic Biology Research Laboratory, AR-SEA, USDA, P. O. Box 14565, Gainesville, Florida 32604, U.S.A.

G. J. Marrs - ICI Ltd., Plant Protection Division, Jealotts Hill, Research Station, Bracknell, Berkshire RG12 6EY, England

G. H. McKibben - Boll Weevil Eradication Research Laboratory, AR-SEA, USDA, Raleigh, North Carolina 27607, U.S.A.

John R. McLaughlin - Insect Attractants, Behavior, and Basic Biology Research Laboratory, AR-SEA, USDA, P. O. Box 14565, Gainesville, Florida 32604, U.S.A.

John A. McLean - Faculty of Forestry, University of British Columbia, Vancouver, B. C., Canada V6T 1W5

L. J. McVeigh - Centre for Overseas Pest Research, College House, Wrights Lane, London W8 5SJ, England

Everett R. Mitchell - Insect Attractants, Behavior, and Basic Biology Research Laboratory, AR-SEA, USDA, P. O. Box 14565, Gainesville, Florida 32604, U.S.A.

B. F. Nesbitt - Tropical Products Institute, 56-62 Gray's Inn Road, London WC1X 8LU, England

Donald A. Nordlund - Southern Grain Insects Research Laboratory, AR-SEA, USDA, Tifton, Georgia 31793, U.S.A.

M. A. Novak - Department of Entomology, New York State Agricultural Experiment Station, Geneva, New York 14456, U.S.A.

Thomas L. Payne - Department of Entomology, Texas Agricultural Experiment Station, Texas A & M University, College Station, Texas 77843, U.S.A.

John W. Peacock - USDA Forest Service, Delaware, Ohio 43015, U.S.A.

Jack R. Plimmer - Organic Chemical Synthesis Laboratory, AR-SEA, USDA, Beltsville, Maryland 20705, U.S.A.

Ronald J. Prokopy - Department of Entomology, University of Massachusetts, Amherst, Massachusetts 01003, U.S.A.

Stefan Rauscher - Swiss Federal Research Station for Horticulture and Viticulture, CH-8820 Wädenswil, Switzerland

W. L. Roelofs - Department of Entomology, New York State Agricultural Experiment Station, Geneva, New York 14456, U.S.A.

G. H. L. Rothschild - Division of Entomology, C.S.I.R.O., P. O. Box 1700, Canberra, A.C.T. 2601, Australia

C. J. Sanders - Canadian Forestry Service, Great Lakes Forest Research Centre, P. O. Box 490, Sault Ste. Marie, Ontario

C. Sartwell - Pacific Northwest Forest and Range Experiment Station, USDA Forest Service, Corvallis, Oregon 97331, U.S.A.

Augustin Schmid - Station Fédérale de Recherches, Agronomiques de Changins, CH-1260 Nyon, Switzerland

E. L. Soderstrom - Stored-Product Insects Laboratory, AR-SEA, USDA, 5578 Air Terminal Drive, Fresno, California 93727, U.S.A.

L. L. Sower - Pacific Northwest Forest and Range Experiment Station, USDA Forest Service, Corvallis, Oregon 97331, U.S.A.

J. Stockel - INRA Station de Zoologie, Centre de Recherches de Bordeaux, 33140 Pont de la Maye, France

F. Sureau - INRA Station de Zoologie, Centre de Recherches de Bordeaux, 33140 Pont de la Maye, France

Sadahiro Tatsuki - Laboratory of Insect Physiology and Toxicology, The Institute of Physical and Chemical Research, Wako, Saitama 351, Japan

K. W. Vick - Insect Attractants, Behavior, and Basic Biology Research Laboratory, AR-SEA, USDA, P. O. Box 14565, Gainesville, Florida 32604, U.S.A.

J. Weatherston - Albany International Controlled Release Division, 110 A Street, Needham Heights, Massachusetts 02194, U.S.A.

J. A. Witz - Cotton Pest Control Equipment Research, AR-SEA, USDA, College Station, Texas 77843, U.S.A.

David L. Wood - Department of Entomological Sciences, University of California, Berkeley, California 94720, U.S.A.

INDEX (Scientific names)

INDEX (Common names)

INDEX (Chemicals)